Indoor Allergens

Assessing and Controlling Adverse Health Effects

Andrew M. Pope, Roy Patterson, and Harriet Burge, *Editors*

Committee on the Health Effects of Indoor Allergens

Division of Health Promotion and Disease Prevention

INSTITUTE OF MEDICINE

NATIONAL ACADEMY PRESS
Washington, D.C.   1993

NATIONAL ACADEMY PRESS • 2101 Constitution Avenue, N.W. • Washington, D.C. 20418

NOTICE: The project that is the subject of this report was approved by the Governing Board of the National Research Council, whose members are drawn from the councils of the National Academy of Sciences, the National Academy of Engineering, and the Institute of Medicine. The members of the committee responsible for the report were chosen for their special competencies and with regard for appropriate balance.

This report has been reviewed by a group other than the authors according to procedures approved by a Report Review Committee consisting of members of the National Academy of Sciences, the National Academy of Engineering, and the Institute of Medicine.

The Institute of Medicine was chartered in 1970 by the National Academy of Sciences to enlist distinguished members of the appropriate professions in the examination of policy matters pertaining to the health of the public. In this, the Institute acts under both the Academy's 1863 congressional charter responsibility to be an adviser to the federal government and its own initiative in identifying issues of medical care, research, and education.

This project was supported by funds from the Environmental Protection Agency, National Institute of Environmental Health Sciences, National Institute of Allergy and Infectious Diseases, National Heart, Lung, and Blood Institute, and Agency for Toxic Substances and Disease Registry.

Library of Congress Cataloging-in-Publication Data

Indoor allergens : assessing and controlling adverse health effects /
   Andrew M. Pope, Roy Patterson, and Harriet Burge, editors ;
   (Committee on the Health Effects of Indoor Allergens, Division of
   Health Promotion and Disease Prevention, Institute of Medicine).
      p. cm.
   Includes bibliographical references and index.
   ISBN 0-309-04831-1
   1. Respiratory allergy. 2. Indoor air pollution. 3. Allergens.
I. Pope, Andrew MacPherson, 1950- . II. Patterson, Roy, 1926-
III. Burge, Harriet. IV. Institute of Medicine (U.S.). Committee
on the Health Efects of Indoor Allergens.
   [DNLM: 1. Hypersensitivity—complications. 2. Allergens—adverse
effects. 3. Air Pollution, Indoor—prevention & control. 4. Air
Pollution, Indoor—adverse effects. WA 754 I4115 1993]
RC589.I53   1993
616.2'02—dc20
DNLM/DLC
for Library of Congress   93-744
                                            CIP

Additional copies of this book are available from the National Academy Press, 2101 Constitution Avenue, NW, Box 285, Washington, DC 20055. Call 800-624-6242 or 202-334-3313 (in the Washington Metropolitan Area).

Printed in the United States of America

The serpent has been a symbol of long life, healing, and knowledge among almost all cultures and religions since the beginning of recorded history. The image adopted as a logotype by the Institute of Medicine is based on a relief carving from ancient Greece, now held by the Staatlichemuseen in Berlin.

COMMITTEE ON THE HEALTH EFFECTS
OF INDOOR ALLERGENS

Preface

This report deals with the growing concern of many in this country about the indoor environment and its relationship to human health and comfort. Among the various health issues facing the citizens of the United States, the problem of indoor airborne allergens is one of the more serious. A high percentage of the population becomes sensitive to indoor allergens and suffers chronic or intermittent allergic disease. Most of these disease conditions can be classified as mild or moderate, but many are severe, and some are fatal. A perspective on this issue is provided by the estimated $6.2 billion annual cost of asthma-related illness in the United States. Although costs related to allergic asthma are by no means the largest component of this total, they are extensive enough to warrant attention to the role of indoor allergens in this condition.

Allergic reactions to indoor allergens can produce inflammatory diseases of the eyes, nose, throat, and bronchi, which are medical problems that come under the headings of allergic conjunctivitis, allergic rhinitis, allergic asthma, and hypersensitivity pneumonitis (extrinsic allergic alveolitis) respectively. The overall objective of our committee was to provide a comprehensive evaluation of current knowledge of these allergic diseases as they relate to airborne allergens in the indoor environments, and methods for their control.

The range of allergens to be found indoors is broad; it is set against the backdrop of an equally broad range of indoor environments. Our committee focused on *airborne* allergens, or aeroallergens, in residential, hospital, school, and office environments. We examined what is known about the adverse effects on human health caused by indoor allergens that elicit allergic reactions, characterizing the magnitude of these problems nationally and the populations that are commonly affected. We also identified specific causative agents and their sources and reviewed testing methodologies for indoor allergens, including their applicability and interpretation. Industrial environments were addressed only to the extent that a few specific industrial chemicals have been shown to be immunogenic and provide models for potential exposures in other environments. The report does not cover indoor allergens that are not airborne, such as contact allergens that elicit a lymphocyte-mediated contact dermatitis, for example, from dyes or cosmetics. Also not discussed are allergic reactions to therapeutic drugs.

A specific exclusion from the committee's scope of work was the phenomenon known as multiple chemical sensitivity (MCS). MCS was not addressed in any detail because the committee's focus was on the relationship between allergic agents and conditions mediated by the immune system. So far, scientific evidence that would support such a relationship with respect to MCS is lacking. As stated in a recent National Research Council report (NRC, 1992a, p. 138), "there is insufficient evidence to ascribe an immune etiology to this disorder." In addition, the sponsors of this project specifically excluded MCS from the charge to the committee.

As this report on indoor allergens evolved, the importance of the committee's multidisciplinary structure became apparent. Engineers, aerobiologists, epidemiologists, psychologists, and physicians subspecializing in allergy, pulmonary medicine, and immunology and epidemiology interacted and communicated with enlightening candor. The result was an educational process for all involved. The resulting coordinated effort attempts to provide a comprehensive, balanced report on a complex of health issues that appear to be assuming increasing importance. The recommendations generated by the committee stand as recommendations adopted by the committee as a whole. They are thus consensus opinions developed in the best interests of the public, the health care professions, the engineers and architects concerned with the structures in which we live, and the government agencies charged with carrying out the mandates of the people. The committee's recommendations are directed broadly, rather than toward any specific agencies, institutions, or offices, because the committee was not asked by the sponsors to address this aspect of the problem and did not believe that the additional focus was necessary or appropriate.

Finally, it should be noted that the conclusions and recommendations of this study are based on the knowledge and expertise available at this time. There remains a great need for research and further understanding of allergens and indoor environments. Notwithstanding such a need, it must also be emphasized that some aspects of the diseases discussed in this report cannot be altered even by major emphasis on control of indoor allergens.

For example, the rising death rate from asthma, a problem of international scope, extends beyond the issue of indoor allergens, because the severe inflammatory disease of the airways in many asthmatics often is not the result of allergen exposure. Control of severe asthma in patients with potentially fatal disease may involve addressing other problems, including psychiatric disease and socioeconomic factors such as poverty, the geographic distribution of medical care, and illicit drug use. To place too much emphasis on indoor allergens when other such major problems are present is not the intent of this report.

It is our hope and intention that this report will prove a spur to action. Many of the adverse health effects resulting from occupying indoor environments can be prevented. What is needed is a plan that embraces education of health professionals, engineers, building designers, and the general public—a plan that will not only improve the quality of life for millions of allergy sufferers but that will also result in substantial savings in many areas of the nation's health care costs.

> Roy Patterson, M.D., *Chair*
> Harriet Burge, Ph.D., *Vice-Chair*
> Committee on the Health Effects
> of Indoor Allergens

Acknowledgments

The committee wishes to acknowledge and express their gratitude for the support and assistance of a few very helpful and generous individuals. H. Jenny Su, Harvard School of Public Health; Thomas L. Creer, Ohio University; and Mary Kay O'Rourke, University of Arizona, prepared provocative, informative background documents for the committee. These documents were extremely valuable in generating vigorous discussion and productive thought in relevant areas. The committee also thanks Jay Slater of Children's Hospital, District of Columbia, for his assistance in sorting out the issues related to latex allergy. Jim Frazier is acknowledged for his persevering scientific curiosity, which ultimately led to the conduct of this study.

The sponsors of this project are acknowledged for their assistance in providing information and guidance as well as for identifying this as an important area and requesting that the IOM conduct the study. In particular, the following sponsoring agencies and individuals are acknowledged: from the Environmental Protection Agency, Robert Axelrad and Pauline Johnston; from the National Institute on Allergy and Infectious Disease, Robert Goldstein, Eugene Zimmerman, and Marshall Plaut; from the National Heart, Lung, and Blood Institute, Suzanne Hurd; from the National Institute of Environmental Health Sciences, Kenneth Olden and C. W. Jameson; and from the Agency for Toxic Substances and Disease Registry, Barry Johnson and Stephen Von Allmen.

Lastly, a debt of gratitude is owed to the IOM staff. No effort of this kind can be accomplished without the hard work and dedication of a talented staff. The committee thanks the following IOM staff members: Laura Baird, director of the IOM Library; Leah Mazade, technical editor; Polly Buechel, study assistant; Gary Ellis, former director of the Division of Health Promotion and Disease Prevention (through 12/24/92), and Andrew Pope, study director.

Contents

Indoor Allergens

Executive Summary

ABSTRACT *Americans spend almost all of their time indoors in environments that are increasingly airtight and that often contain sources of allergens—such as dust mites, fungi, house pets, rodents, cockroaches, and certain chemicals. An increased incidence, prevalence, and severity of asthma and other allergic diseases are associated with exposure to these agents and constitute major components of the public health problems related to indoor air quality in the United States.*

Methods for detecting and characterizing specific allergic diseases include taking a thorough medical history and performing skin tests, in vitro antibody tests, and pulmonary function tests. Environmental testing and exposure assessment can identify contaminants and their sources, environmental exposure media, routes of entry into the body, intensity and frequency of contact, and spatial and temporal concentration patterns. Using published data, a first-ever risk assessment in this field shows a positive relationship between cumulative exposure to mite allergen and the risk of sensitization.

Proper education of allergy sufferers and medical personnel, coupled with appropriate building design, construction, management, and engineering, can prevent many of the problems associated with indoor allergens and alleviate those that remain. Continued vigorous investigation of the problems related to indoor allergens is essential to improved public health. Recommendations focus on the need for improved education and expanded research.

INTRODUCTION

People have always considered the indoor environment a refuge from the dangers of the outdoors, offering protection from weather, wild animals, and, in the late nineteenth and twentieth centuries, air pollution—long viewed as primarily an outdoor phenomenon. In recent years, however, questions have been raised about potential health problems arising from indoor air pollution. The resultant concerns are related to modern lifestyles and the fact that most Americans spend more than 90 percent of their time inside a building of one kind or another. Indeed, if hours in transit between locations are considered, most Americans are indoors nearly 24 hours a day (Figure 1). In addition, there is a trend in newer construction toward energy-efficient, relatively airtight, structures. These factors taken together have raised increasing concern about the potential adverse health effects of indoor air quality.

Motivated by this concern, several agencies of the federal government asked the Institute of Medicine (IOM) to undertake an assessment of the public health significance of indoor allergens. The IOM responded by assembling a committee of experts in such fields as allergy and immunology, epidemiology, mycology, engineering, industrial health, pulmonology, education, and public policy. The study undertaken by the committee had three primary objectives: (1) to identify airborne biological and chemical agents found indoors that can be directly linked to allergic diseases; (2) to assess

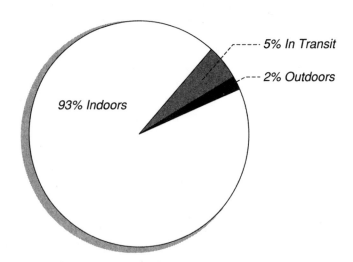

FIGURE 1 Amount of time spent indoors, outdoors, and in transit by most Americans. Sources: NRC, 1981; Spengler and Sexton, 1983.

the health impacts of these allergens; and (3) to determine the adequacy of the knowledge base that is currently available on this topic.

This Executive Summary presents an overview of the committee's findings, followed by a section that presents all of the committee's recommendations and research agenda items from throughout the report.

The Allergic Reaction

An allergen is a biological or chemical substance that causes an allergic reaction. An allergy, in turn, is the immune-mediated state of hypersensitivity that results from exposure to an allergen. The symptoms commonly associated with allergy and allergic reactions are conjunctivitis (red, irritated eyes), rhinitis (a stuffy, runny nose, or "hay fever"), and bronchitis (cough and congestion). Allergic reactions are classified and described in many ways, including: on the basis of the allergenic substance that causes them (such as cat dander, or pollen); according to the specific immunologic reactants (such as the types of antibodies or immune system cells involved in the reaction); or on the basis of the resulting disease (such as allergic asthma or hypersensitivity pneumonitis).

The most common allergic reaction is mediated by the immunoglobulin E (IgE) antibodies of the immune system. Mast cells, which contain histamine and other chemicals involved in the inflammatory response, also play a key role. After an initial exposure to a particular allergen, an individual who develops an allergy becomes "sensitized" to that allergen. In biological terms, sensitization occurs when IgE antibody specific for that allergen attaches to the surface of the individual's mast cells—making the individual "sensitive" (or "hypersensitive") to additional exposures. Subsequent exposure of the sensitized individual to the same allergen causes the mast cells to respond by releasing histamine and other inflammatory-response agents. These agents, in turn, interact with the surrounding tissues, causing an allergic reaction and the symptoms that we commonly associate with allergies (see Figure 2).

Everyone produces some IgE, but people with genetic predispositions to allergy (a condition called *atopy*) produce significantly greater quantities of it. In addition, the production of IgE antibody can continue for years after an encounter with an allergen. Thus, for example, someone who had an allergic reaction to penicillin as a child could still be allergic to the drug as an adult.

The reason why some people develop allergies and others do not is unclear. We do know, however, that genetic factors have a major influence on whether an individual develops allergy-related medical problems. For example, if one parent has allergies and the other does not, the chances are one in three that each of their children will have allergies. If both parents have allergies, it is very likely (seven in ten) that their children will exhibit some allergic symptoms.

Another illustration of the process of developing allergic disease is presented in Figure 3. This figure shows the series of steps, beginning with the sensitization of a genetically predisposed or susceptible individual, and the potential interaction with other, nonallergenic agents (e.g., environmental tobacco smoke). Additional exposure to the sensitizing allergen then leads to a mild, moderate, or severe allergic reaction, depending on the amount of exposure. Exposure to nonallergens that irritate the respiratory tract (e.g., environmental tobacco smoke) can promote the development of allergic reactions and disease.

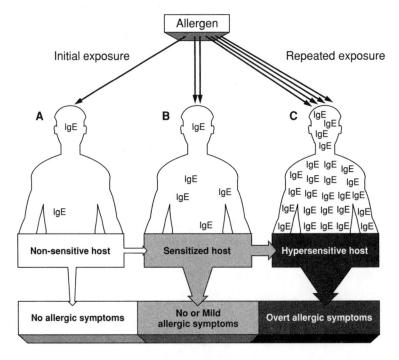

FIGURE 2   Hypersensitivity and IgE-mediated allergy. (A) Upon initial exposure to an allergen, there will be no overt manifestations of allergic disease because the patient is nonsensitized. However, this allergen will initiate an immune response that results in the synthesis of IgE and sensitization of the susceptible (atopic) host. (B) Upon subsequent, repeated exposures, this sensitized individual will synthesize increased amounts of IgE, thus becoming hypersensitive, although mild allergic symptoms may or may not be present. (C) From this point onward, reexposure to this specific allergen will provoke the overt manifestations of allergic disease. Source: Fireman and Slavin, 1991.

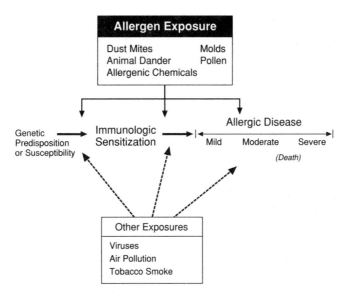

FIGURE 3 Development of allergic disease, illustrated schematically. A genetically susceptible individual is exposed to an allergen and becomes immunologically sensitized. At this stage the person is asymptomatic, but the sensitization may be detected by skin tests or laboratory tests. Over time, a proportion of sensitized individuals will develop one of a group of allergic diseases. Exposure to allergen is understood to be a major factor at each stage of the pathogenesis of these diseases.

MAGNITUDE OF THE PROBLEM

The most common allergic diseases caused by indoor allergens are allergic rhinitis, sinusitis, asthma, and allergic skin diseases (dermatitis). Figure 4 shows the generally accepted ranges of prevalence for these conditions in the United States. Based on these and other data, we can predict that one out of five Americans will experience allergy-related illness at some point during their lives and that indoor allergens will be responsible for a significant share of these cases. As such, indoor allergens are a major public health problem in the United States.

Some interesting facts about allergies and the indoor environment follow:

• More than 50 million Americans—one out of five—suffer from allergic rhinitis (hay fever) and other allergic diseases, many of which are related to exposure to allergens in indoor environments.

• Allergic rhinitis is the single most common chronic disease experienced by humans.

• The level of allergen exposure in the home relates to the risk of

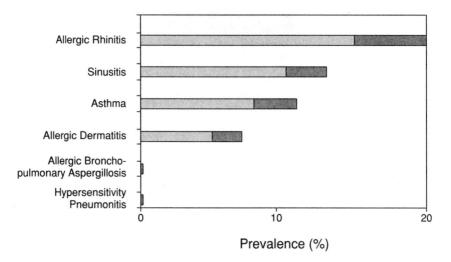

FIGURE 4   Estimated range of prevalence of diseases in total U.S. population that are commonly, although not invariably, related to allergy. Darker screen indicates the range in published data. Source: NHLBI, 1991.

becoming sensitized to indoor allergens, the risk of developing an allergic disease, and the severity of the allergic disease.

• Population-based and case-control studies suggest that indoor allergens are a major reason for trips to hospital emergency rooms.

• Effective environmental control reduces disease severity.

Allergy also plays a key but sometimes unrecognized role in triggering asthma, a disease that deserves special public health attention because of its prevalence, documented cost, and potential severity. The following information provides perspective on the magnitude and cost of asthma:

• An estimated 20 million to 30 million Americans—8 to 12 percent of the American population—have asthma.

• Among chronic diseases, asthma is the leading cause of school absenteeism.

• Asthma kills approximately 4,000 people per year (an increase of 33 percent in the past decade); in 1988, 4,580 people died from asthma in the United States.

• Asthma mortality rates among African Americans are 2–3 times greater than among Caucasions; 5 times greater among children.

• In 1985, an estimated 1.8 million people required emergency room services for asthma, 48% of these were children under 18 years of age.

- In 1990, the estimated cost of asthma-related illness was $6.2 billion.
- Some cases of asthma that require emergency care can be traced to poor control of home allergens.
- With proper medical care and education, asthma can be controlled in the great majority of cases.

Figures 5, 6, and 7 show recent trends in asthma prevalence, mortality, and hospitalization rates, respectively.

Other allergic diseases related to indoor exposures, such as allergic bronchopulmonary aspergillosis and hypersensitivity pneumonitis, occur less frequently than allergic rhinitis and asthma, but are often severe and difficult to control. A role for indoor allergens has also been suggested in chronic sinusitis and bronchitis, sick building syndrome, and other nonspecific syndromes, as well as in acute respiratory illness.

The economic and social impacts of allergic disease in the United States are significant. The persistence of allergy-related symptoms and the occurrence of complications, such as otitis media and sinusitis, result in an overall increase in the number of physician visits and the use of medications.

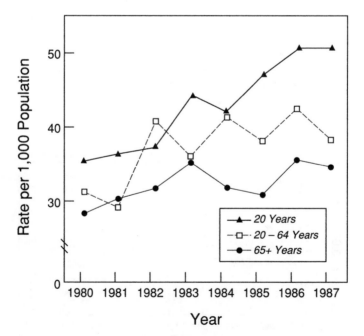

FIGURE 5 Trends in Asthma Prevalence. Source: NHLBI, 1991.

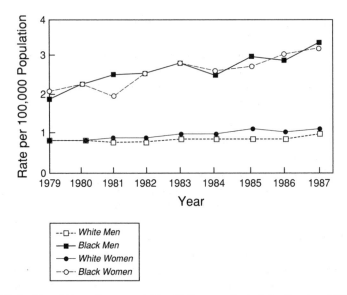

FIGURE 6 Trends in asthma mortality, U.S. age-adjusted death rates, 1979–1987. Source: NHLBI, 1991.

Work and school absenteeism as a result of allergic reactions of the upper airway contributes to the economic burden posed by indoor allergens.

Although complete data are lacking, sufficient information exists to conclude that indoor allergens constitute a substantial public health problem. Given the amount of time most Americans spend indoors, expanded efforts to assess, prevent, and control the health threats posed by indoor allergens are warranted.

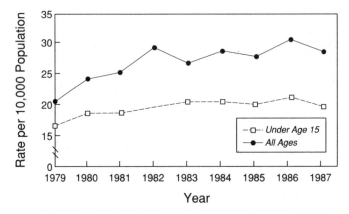

FIGURE 7 Trends in asthma hospitalization rates. Source: NHLBI, 1991.

AGENTS, SOURCES, AND SOURCE CONTROLS

The major sources of indoor allergens in the United States are house dust mites, fungi and other microorganisms, domestic pets (cats and dogs), and cockroaches. The allergens produced by these organisms become airborne and can cause the allergic diseases mentioned previously, such as hay fever, asthma, and hypersensitivity pneumonitis.

The dust mite, a microscopic organism that lives primarily in carpeting and on mattresses and upholstery, produces several of the most common residential allergens. Although there are regional and local differences in dust mite concentrations, they are thought to be one of the most important allergenic causes of childhood asthma. Dust mite allergens, which have been purified and characterized, are contained in fecal balls that accumulate in bedding and on other surfaces.

Fungi release allergen-containing spores and other products indoors and outdoors. Although there are thousands of different fungi that can contaminate indoor air, purified allergens have been recovered from only a few; none has been completely characterized. Fungus spore allergens are known to cause hay fever, asthma, and hypersensitivity pneumonitis; other fungal products can be irritating or toxic, possibly exacerbating allergic conditions.

Allergens released by house pets have been purified and characterized and are an important cause of allergic rhinitis and asthma. Research has shown that cat allergens become airborne on very small particles and remain in buildings long after the cat has departed.

Cockroach allergen (which has also been purified and characterized) is derived from the insect's body parts and feces. In many inner-city areas, cockroach allergens probably play a significant role in the development of allergic rhinitis and asthma.

In addition to allergens derived from biological sources, some reactive allergenic chemicals can cause allergic disease. In general, these chemicals are more prevalent in the industrial workplace than the home. However, a limited number of household products may contain immunogenic agents (e.g., isocyanates in bathtub refinishing products).

The best way to prevent and/or control allergen-caused disease is to prevent exposure to the allergenic agent. Avoidance of specific allergens can lessen the probability of initial sensitization and, for individuals with known sensitivity, it can improve their condition dramatically by reducing the cascade of symptoms.

The bedroom is one area where steps to reduce exposure to allergens can have a beneficial effect on health. In the case of dust mite allergen, for example, covering mattresses and pillows with impermeable materials is an effective way to limit exposure (Box 1). Such avoidance measures are important tools for managing allergic disease and asthma caused by dust mite allergen.

BOX 1   Avoidance Measures for Mite Allergen

A. Bedrooms
- Cover mattresses and pillows with impermeable covers
- Wash bedding regularly at 130° F
- Remove carpets, stuffed animals, and clutter from bedrooms
- Vacuum clean weekly (wearing a mask)*

B. Rest of the House
- Minimize carpet and upholstered furniture; do not use either in basements
- Reduce humidity below 45 percent relative humidity or 6 g/kg
- Treat carpets with benzyl benzoate or tannic acid

*There is a temporary increase in potential exposure to allergens associated with the vacuuming process. The net potential for exposure should be reduced by vacuuming, however, and is considerably less than the cumulative effects of not vacuuming. Wearing a mask while vacuuming should help reduce exposure while vacuuming.

Drugs such as antihistamines, cromolyn, and topical steroids can be used to suppress allergic symptoms. Immunotherapy ("allergy shots") stimulates the production of protective antibodies and is especially effective for certain well-defined allergens.

MECHANISMS OF IMMUNE FUNCTION

The immunologic nature of allergic reactions was recognized early in this century, but many of the specific types of antibodies and cellular mechanisms involved in such reactions have only recently been defined. Research on the immune mechanisms of allergic disease has led to the hope that the immune responses that occur in allergic diseases can be interrupted at the molecular level. In addition, modern research techniques have allowed scientists to define the protein sequences of many allergens, providing a focus for studies into new methods for detecting and quantifying specific allergenic agents in the environment.

MEDICAL TESTING METHODS

There are several types of medical tests available for the detection and diagnosis of allergic disease. These include skin tests, in vitro antibody tests, and pulmonary function tests. In allergy practice, however, the medical history is the primary diagnostic tool. Laboratory studies, including skin and in vitro tests for specific antibodies, have relevance only when correlated with the patient's medical history. Thus, although confirmation of a diagnosis of allergy generally requires an immunologic (laboratory) test to help identify a specific allergenic agent, an immunologic response is not sufficient to diagnose allergic disease; it means only that a prior sensitizing exposure has occurred. Treatment should always be directed toward current symptomatology, not merely toward the results of specific immunologic laboratory tests.

Medical History

Allergists use a variety of methods to obtain a history. These include open-ended, nondirected question-and-answer sessions, a series of questions ordered according to a formal protocol to ensure completeness, a structured questionnaire history completed by the physician, and a structured questionnaire history completed by the patient. Many allergists use a combination of these methods.

Despite universal agreement about the primary importance of a patient's allergy history, medical textbooks devote little or no space to this topic, and research on the subject seems to be nonexistent.

Skin Tests

The most common in vivo allergy test is the skin test. It is used both diagnostically for individual patients and by epidemiologists to develop estimates of disease incidence and prevalence in a population. Skin testing should be conducted with appropriate positive and negative controls. And for safety reasons, appropriately trained personnel and adequate equipment need to be available to treat any adverse systemic reactions resulting from the test.

One of the major problems with performing allergen skin tests is the lack of well-characterized, standardized reagents. Certain indoor allergenic extracts (e.g., dust mite and cat) now contain standardized amounts of major allergens, but others, such as fungi, are crude preparations of arbitrarily chosen fungi.

In Vitro Tests

There are several in vitro allergy tests used for diagnosis and in clinical research. The radioallergosorbent test (RAST) and the enzyme-linked immunosorbent assay (ELISA) can be used to evaluate levels of IgE or IgG directed against an allergen. Precipitin assays such as the double-immunodiffusion method of Ouchterlony, are often used to detect high levels of specific IgG, as can occur in patients with hypersensitivity pneumonitis. Specific cell-mediated immunity can be assessed by several methods, including the lymphocyte transformation assay and assays involving the production of mediators or cytokines. These tests are generally regarded strictly as research assays, however, and are not usually used to evaluate patients or populations.

As with skin tests, the accuracy of any immunodiagnostic test depends on the characteristics of the test reagents, and in particular on the allergen reagent. Standardization and characterization of allergen reagents used for immunodiagnostic tests are imperative. Similarly, the existence and characterization of control antibody, whether polyclonal or monoclonal, would be valuable for standardization and quality control of immunodiagnostic tests. Ideally, minimal standards for quality control should be devised for labs reporting results of tests to detect specific immunologic responses to indoor allergens.

Pulmonary Function Tests

Diseases such as asthma and hypersensitivity pneumonitis are characterized by decrements in lung function that vary over time. Pulmonary function tests, which measure these decrements and any improvements that may occur, have many applications in clinical medicine and research related to indoor allergens. These tests include spirometry (the most reliable pulmonary function test), peak-flow measurement, lung-volume measurement using gas dilution or body plethysmography, diffusing-capacity testing, exercise studies, and rhinomanometry. The choice of the appropriate pulmonary function test depends on the requirements of the specific application.

One drawback of many pulmonary function tests is that they must be administered by medical personnel or trained technicians. Peak-flow measuring devices are the exception. These devices are less reliable than some other technologies, but they are easily portable and can be self-administered—making them useful in the diagnosis and management of asthma.

ASSESSING EXPOSURE AND RISK

The process of assessing exposure to indoor allergens is complex. The relevant contaminants, contaminant sources, and environmental exposure media must be identified. The manner in which allergens are transported through each medium and how the contaminants are chemically and physically transformed must be determined. In addition, the allergen's routes of entry into the body, the intensity and frequency of contact between the

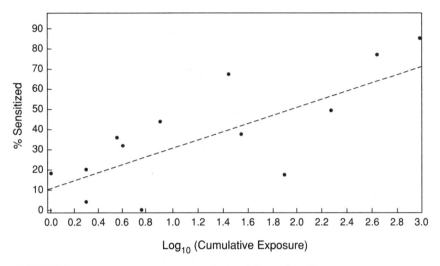

FIGURE 8   An exposure-response analysis of dust mite allergens.

allergen and the affected individuals, and the allergen's spatial and temporal concentration must be assessed.

A determination of the relationship between exposure and the development of allergic disease can be used to estimate the level of risk for developing allergic disease within a population. This relationship could be expressed as the number of new allergy cases expected at a given exposure level or the distribution of expected change in severity of symptoms with changing exposure levels. In an example using published data on dust mite allergy, this committee reports a positive relationship between cumulative exposure to dust mite allergen and the risk of allergic sensitization (see Figure 8). This is a finding that has long been suspected to exist, but has not been previously demonstrated. The residual sensitization predicted by these calculations would occur irrespective of exposure to dust mite allergen, and is consistent with the knowledge that other factors may also result in sensitization.

ENGINEERING CONTROL STRATEGIES

An in-depth understanding of how building systems and structures operate and perform is essential for assessing and controlling indoor air quality problems. The reduction and/or elimination of human exposure is probably best achieved by simultaneously controlling allergen sources and improving building ventilation, i.e., the design, operation, and maintenance of heating, ventilation, and air-conditioning (HVAC) systems.

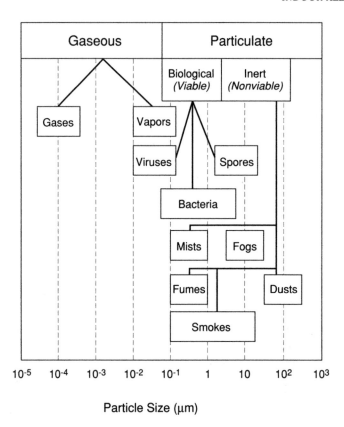

Particle Size (μm)

FIGURE 9 Dimensional continuum of potential contaminants. Source: Adapted from Woods, 1982.

Most HVAC systems have dry, panel-type barrier filters, which are installed primarily to protect the system's mechanical air-handling equipment. Such filters remove few if any particles smaller than 1 to 2 μm (see Figure 9). Most pollens, many fungus spores, and dust mite fecal particles are larger than this. However, some allergens, such as cat dander and fragments of fungi, travel on particles less than 1 μm in size. Therefore, for optimal allergen control, it may be reasonable to upgrade these central filters and to use high-efficiency portable air cleaners in closed, confined, clean areas (e.g., the bedrooms of allergic patients).

The improved design, maintenance, and operation of residences and other buildings should reduce the incidence, prevalence, and severity of allergic disease. A recent report by the Environmental Protection Agency (*Building Air Quality: A Guide for Building Owners and Facility Managers*, 1991a) is an excellent source of practical information.

THE ROLE OF EDUCATION

Improved education is essential to the prevention and control of indoor allergic disease. This is true both for common allergic conditions of low-to-moderate severity and for less common diseases that are more severe and/or potentially fatal. Medical education at all levels (medical schools, postgraduate training programs, and continuing medical education programs for practicing physicians) should emphasize the importance of recognizing and properly managing these diseases. Nurses, physician assistants, and other nonphysician health care providers require similar education and training. In addition, patients and the general public should be made aware of when and where to seek medical help for allergy-related health problems.

One recent publication, the National Asthma Education Program's *Guidelines for the Diagnosis and Management of Asthma* (NHLBI, 1991), presents detailed recommendations for diagnosing and managing asthma. It also emphasizes the importance of education and of identifying causative agents, such as indoor allergens, that may initiate and exacerbate asthma symptoms. An evaluation of the impact of these guidelines awaits their wide dissemination and acceptance. Meanwhile, efforts to improve what is already known about preventing and controlling allergic diseases should continue. This knowledge should be disseminated to health care providers, allergy patients, and the public at large.

RECOMMENDATIONS AND RESEARCH AGENDA

This report describes what is known about the adverse effects on human health caused by indoor allergens, the magnitude of the problem nationally, the specific causative agents and their sources, the testing methods currently used for identifying allergens and diagnosing related diseases, and finally the need for increased knowledge and awareness of indoor allergens among primary care physicians, patients, and others. In describing these issues, the committee identified and developed a list of research agenda items and 15 priority recommendations. The recommendations focus primarily on the need to improve awareness and education, and as such are also relatively inexpensive and easy to implement compared to the research agenda items. The research agenda focuses on the longer term, more expensive, and more technical aspects of fundamental research and data collection.

Although cost-benefit analyses were beyond the scope of the committee's charge in conducting this study the committee's findings have been separated into recommendations and research agenda items, as described above, based on relative ease of implementation and estimations of associated costs. Policy development and further priority setting, however, should be based on investigations of the probable costs and benefits of the various strategies for prevention and control. The cost analyses should include both direct and indirect costs (i.e., medical costs as well as lost work), and benefits should include a full assessment of the effects on improving the quality of life, as well as on reducing morbidity and mortality.

All of the committee's recommendations and research agenda items are presented below with brief introductions. The recommendations are presented first, followed by the research agenda.

Recommendations

AGENTS, SOURCES, SOURCE CONTROLS, AND DISEASES

Dust Mite, Cockroach, and Other Arthropoda

With most Americans spending the great majority of their time indoors—and much of that in their own homes—it is not surprising that the bulk of inhaled foreign protein is associated with indoor air. The evidence shows that a large proportion of asthmatics are allergic to indoor allergens and that several changes in the way we live indoors may have affected the levels of these allergens. These changes include increased mean temperatures, reduced ventilation (with consequent increased humidity), fitted carpets, and cool-wash detergents which have led to water temperatures for washing bedding that do not kill dust mites.

Once identified, reducing exposure to allergenic "trigger factors" has been a standard part of the treatment of allergic disease for many years. Since approximately half of existing cases of asthma have been attributed to allergenic factors, it is reasonable to expect that asthmatics who require more than occasional treatment might also have allergies that induce their asthma.

> **Recommendation: Conduct appropriate allergy evaluations of asthmatics who require more than occasional treatment. Where allergic factors are present, ensure that these patients are given specific, practical information about how to reduce their exposure to arthropod and other allergens.**

MEDICAL TESTING METHODS

Skin Tests

Despite some relatively minor shortcomings, the value of skin testing has been well established over the past century. When correlated with an appropriate clinical history, skin prick tests often are a useful way of screening for the presence of allergic disease. Using appropriate positive and negative controls, intradermal testing can be used to demonstrate low levels of sensitization when allergy is clinically suspected. However, studies are necessary to determine the optimal concentrations and methods for skin testing and to address the relationship between defined skin test reactions and disease.

With respect to the specificity and sensitivity of skin tests within a population, more research is needed to determine the predictive values of both prick and intradermal tests. Moreover, the doses and criteria for positive skin tests used for such studies need further definition, together with criteria to define the relationship of a positive skin prick test to the wheal area or erythema that appears.

Recommendation: Encourage the development and use of improved standardized methods for performing and interpreting skin tests.

Pulmonary Function Tests

There are simple, reliable measures of lung function that may be used for studying diseases caused by indoor allergens. Indeed, objective measures of respiratory function should be a part of protocols to determine the efficacy of therapeutic strategies for these diseases. Predicted values for pulmonary function fall along a normal distribution curve with 95 percent confidence intervals for FEV_1 and FVC of approximately 80–120 percent. The lower limit of variation in population studies for the midlevel expiratory flow rate (FEF_{25-75}) is approximately 60 percent.

Spirometry is limited in its ability to detect impairment of ventilatory function in asymptomatic individuals because of the wide range of normal values, even with predicted levels that control for age, sex, and height. Significant inaccuracy can result from errors in spirometry performance, almost all of which lead to underestimation of the true respiratory function. These tests can, however, help to evaluate the effects on an individual of sensitization to specific allergens. They can also help to diagnose respiratory diseases that may be caused or worsened by indoor allergens and to assess disease severity, which is often critically important in clinical decisionmaking. Serial pulmonary function testing in the home or workplace can demonstrate causal relationships between the indoor environment and respiratory illness. Serial pulmonary function testing coupled with bronchoprovocation can demonstrate the causal relationship between specific allergens and respiratory responses.

In epidemiological studies, measures of environmental factors and of pulmonary function can be evaluated for associations that suggest causal relationships. Pulmonary function tests may also be used to assess the efficacy of therapy, determine response to treatment, or determine the effect of environmental modification. Such tests are required when physicians are asked to determine impairment resulting from a respiratory disease for insurance or benefit systems such as workers' compensation and social security disability. Finally, estimates of disease incidence or prevalence often result from epidemiological studies in which pulmonary function tests are used to ascertain disease.

Recommendation: Include pulmonary function tests in epidemiological studies to help improve estimates of disease incidence and prevalence. Because they are portable and can be self-administered, tests that utilize peak-flow measurements are most desirable for this purpose.

One drawback of many pulmonary function tests is that they must be administered by technicians. Peak flow measurements are less reliable but are highly portable, can be self-administered, and are therefore often more sensitive in the diagnosis of asthma.

Recommendation: Include objective measures of respiratory function in experimental protocols designed to determine the efficacy of therapeutic strategies (e.g., pharmacotherapy, environmental modification, avoidance) used to treat respiratory diseases caused by indoor allergens.

ASSESSING EXPOSURE AND RISK

Methods for determining the effects of indoor allergens can be divided into two general categories: patient testing and environmental testing. Data from both kinds of testing can be useful to the physician in directing the treatment, control, and prevention of allergic disease. There are, however, no effective means currently available to physicians or other medical professionals for obtaining quantitative information on environmental exposures.

Recommendation: Establish effective mechanisms for medical professionals to acquire assessments of potential exposure to indoor allergens in residential environments.

ENGINEERING CONTROL STRATEGIES

The fundamental objectives of environmental control are to prevent or minimize occupant exposures that can be deleterious and to provide for the comfort and well-being of the occupants. Well-designed and maintained HVAC systems will exclude most aeroallergens (e.g., pollen, fungal spores) from interior spaces. Poorly designed or maintained systems, however, can provide for amplification and/or infiltration and dissemination of allergens.

Inappropriate control strategies have been associated with nearly all problem buildings.

Recommendation: Improve the design, installation, use, and maintenance of residential and commercial HVAC equipment, for both new and existing construction, in order to minimize allergen reservoirs and amplifiers. These improvements should be based on recommendations developed by the American Society of Heating, Refrigerating and Air-Conditioning Engineers (ASHRAE).

Carpeting can provide niches for both the accumulation and production of allergens, and has been characterized by some as a "cultivation medium" for microorganisms when wetted. Carpeting can also serve as a reservoir for pollen and pollen fragments. The magnitude of the potential significance of carpeting as a source and reservoir of indoor allergens indicates that it should be given consideration as a serious problem.

Recommendation: Expand the scope of the Carpet Policy Dialogue Group of the Environmental Protection Agency to consider the serious problem of carpets as a source and reservoir of indoor allergens.

Standards have been established by the American Society of Heating, Refrigerating and Air-Conditioning Engineers (ASHRAE) for acceptable temperature, humidity, and ventilation as they relate to human comfort. However, little attention is given in these standards to the protection of buildings, furnishings, and construction materials from water damage, and the potential for subsequent adverse health effects.

Recommendation: Develop consensus standard recommendations for controlling moisture in naturally and mechanically ventilated buildings. These recommendations, designed to help control microbial and arthropod aeroallergens and allergen reservoirs, should be developed by ASHRAE and be included in their Standard Series 55 (thermal environmental conditions for human occupancy) and Standard Series 62 (ventilation for acceptable indoor air quality).

THE ROLE OF EDUCATION

Education is an important component in the prevention and control of allergen-induced diseases. Considering that a large percentage of hospital admissions for asthma can be prevented by educating physicians and patients in the proper control measures, the need for emphasis on education becomes obvious. By disseminating information to patients, to health care providers, and to building design, construction and operations professionals, prevention of diseases associated with indoor allergens becomes not only realistic, but may offer a cost-effective means of reducing morbidity.

Patients

In developing and implementing educational interventions for patients, consideration should be given to identifying populations such as those with severe asthma who are more motivated and more likely to benefit from intervention.

Recommendation: Identify population groups most likely to benefit from educational and allergen-avoidance interventions. This effort should be based on an understanding of what allergens serve as risk factors for different individuals.

Socioeconomic, educational, and ethnic characteristics are important variables that should be considered in developing effective educational intervention programs. Programs that focus on these factors in tailoring self-management programs should greatly enhance both the acquisition and the performance of self-management competencies.

Recommendation: Develop focused educational programs for allergic populations with different socioeconomic and educational characteristics. Such programs should help patients:

- **understand allergic-disease risk factors;**
- **predict the occurrence of such risk factors;**
- **adopt behaviors required to avoid or control these factors; and**
- **develop self-management skills to translate and use the knowledge they acquire to control allergic risk factors in different contexts.**

A relapse prevention component should be included in these programs as well as follow-up studies to assess patient acquisition of allergy-related knowledge and the need for additional educational efforts.

Health Care Providers

Curricula vary in medical schools, often with little focus given to the topic of allergy diagnosis, prevention, and control—an unfortunate situation that should be corrected, especially considering the relationship of allergy to asthma. In addition, improved medical education is important because the majority of health care of the allergic patient is delivered by primary care providers, and the primary care provider is often the patient's main source of information about allergy control.

Recommendation: Incorporate the diagnosis and management of allergic diseases in the curricula and training materials for medical school students, residents in primary care practice, and subspecialists

**who will subsequently care for patients with allergen-based allergic
disease. Nurses, physician assistants, and other non-physician health
care providers should receive similar education and training.**

Allergic disease should receive additional emphasis at all levels of medical
education, across specialties, and in clinical practice. One mechanism to
help promote this concept would be to enlist the support and interest of
scientific and medical societies.

**Recommendation: Encourage scientific and medical societies with
expertise in allergy, pulmonary medicine, public health, and occupational and environmental medicine to continue to assess and promote the development of prevention strategies for allergic disease.**

Engineers, Architects, and Building Maintenance Personnel

Concerns about the design and operation of heating, ventilation, and air
conditioning systems have focussed traditionally on the comfort of the building
occupants and the efficiency of the operation of the equipment. It is important that those with responsibility for the design, construction, and maintenance of buildings also have an understanding of the potential adverse health
effects associated with indoor environments, and the impact that design and
operation of the systems can have on those effects.

**Recommendation: Educate those with responsibilities for the de-
sign and maintenance of indoor environments about the magnitude
and severity of diseases caused by indoor allergens.**

Engineers, architects, contractors, and building maintenance personnel
receive limited if any education about the health implications of the design,
construction, and maintenance of buildings. Improved education in these
areas is important to reducing the incidence, prevalence, and severity of
adverse health effects associated with indoor environmental exposures.

**Recommendation: Develop educational processes and accountabil-
ity procedures for architects, engineers, contractors, and building
maintenance personnel with respect to the health implications of
the design, construction, and operation of buildings.**

An interdisciplinary approach to the prevention and control on the adverse health effects associated with indoor exposures, including indoor allergens, is important. Such an approach should improve education in all
areas of expertise and result in reduced health risks for building occupants.

**Recommendation: Develop interdisciplinary educational programs
for health care and building design, construction, and operations
professionals.**

Research Agenda

MAGNITUDE AND DIMENSIONS OF SENSITIZATION AND DISEASE

As outlined in this report, allergic disease constitutes a substantial public health problem in the United States. Data on its exact magnitude and dimensions, however, are incomplete, or lacking, in many cases. Better data regarding the incidence, prevalence, attributable fraction, and cost of allergic diseases are essential to the development and implementation of effective programs of prevention and control. Accurate determinations of the magnitude and dimensions of sensitization and disease caused by regional and local indoor allergens would be useful in this regard.

Research Agenda Item: Determine prevalence rates of sensitization, allergic diseases, and respiratory morbidity caused by regionally and locally relevant indoor allergens and assess the contributions of different allergens to these conditions.

Socioeconomic status seems to contribute to asthma prevalence rates and to indices of disease severity. Similarly, several studies have reported racial differences in the prevalence and severity of asthma in the United States, but such results are inconsistent. In order that these differentials can be translated into effectively targeted public health interventions, additional research is needed to clarify these relationships.

Research Agenda Item: Conduct studies to clarify the relationships that exist between socioeconomic status, race, and cultural environment and the incidence, prevalence, severity, and mortality associated with allergic disease including asthma.

The annual costs (direct and indirect) associated with asthma have been estimated at more than $6.2 billion—an increase of 39 percent since 1985. The size of these costs and the trend towards even greater costs argue for careful attention not only to the effects of allergy in causing asthma but also to the litany of other conditions that are more commonly thought to be associated with "allergies." Accurate assessments of the costs associated with asthma and other allergic diseases would be useful in the development of health policy initiatives and the implementation of appropriate and cost effective interventions.

Research Agenda Item: Determine the economic impact of asthma and other allergic diseases.

Allergic disease occurs when a genetically predisposed or susceptible individual is exposed to an allergen and becomes immunologically sensitized. In the early stages of sensitization, people who are sensitized have not developed symptoms of disease. The magnitude of this group within the population is of interest, however, because it reflects the proportion of the population at greatest risk of developing allergic disease. Additional exposure to the sensitizing allergen leads to the development of an allergic reaction (disease) that can be mild, moderate, or severe, depending on the amount of exposure.

The relationship between exposure, sensitization, and disease, and the potential for a threshold level of exposure below which the risk of sensitization is reduced, is of critical importance in the prevention and control of allergic disease. Epidemiological data would be useful in determining these relationships and in developing and evaluating public health and medical intervention strategies.

Research Agenda Item: Conduct appropriate epidemiological studies of exposure-response relationships of important defined indoor allergens that induce sensitization in humans. Such studies should include a focus on identifying threshold exposures.

Indoor allergens are associated with a wide variety of particles in a broad size range, only some of which are microscopically identifiable, culturable, or detectable with existing immunoassays. Evaluation of indoor allergens requires both air and source sampling, and several different analytical techniques (including microscopy, culture, and immunoassays) must be used to characterize even the well-known allergens. Because of the complexity of the assessment problem, indoor allergens, with a few exceptions, have not been identified and studied.

Research Agenda Item: Encourage and conduct additional research to identify and characterize indoor allergens. The new information should be used to advise patients about avoiding specific allergenic agents.

AGENTS, SOURCES, SOURCE CONTROLS, AND DISEASES

Dust Mite, Cockroach, and Other Arthropoda

Although most studies investigate asthma (because it is common and can be measured objectively), sensitivity to indoor allergens is also very common among patients with other allergic conditions such as chronic rhinitis and atopic dermatitis. Because the important cause of inflammation that is common for all of these diseases is exposure to allergens, avoidance of the exposure should be the primary anti-inflammatory treatment. Developing realistic avoidance protocols for routine use is a challenge that must not take second place to pharmaceutical treatment.

Research Agenda Item: Conduct detailed studies of physical factors, such as sources and emission rates, that influence the levels of exposure to arthropod and other indoor allergens. These studies should include the effects of (a) protocols for reducing exposure and (b) devices advertised as reducing indoor allergen concentrations. More specifically, test the effectiveness of allergen avoidance protocols on the management of allergic asthma and other allergic diseases using protocols that have been demonstrated to reduce exposure by 90 percent or more. Such protocols should be tested under field conditions and should encompass the socioeconomic, cultural, ethnic, and geographic diversity of these problems in the United States.

The more important question now is the effectiveness of avoidance measures for reducing the development of disease (i.e., sensitization) related to arthropod allergen exposure. Specific measures include the use of polished wooden floors and no upholstered furniture in bedrooms and possibly in schoolrooms, keeping pillows and mattresses covered, and using bedding that can be washed regularly in hot water. Other measures, which require further evaluation, include air filtration, chemical treatment of carpets, and dehumidification.

Research Agenda Item: Develop allergen-free products to help reduce the incidence of allergic disease in the general public. The initial objective of this initiative, which should be carried out by the industrial/business sector, would be to provide an aesthetically appealing bedroom with reduced allergen exposure for children under the age of five.

There is considerable evidence that allergen avoidance is an effective means of reducing allergy symptoms. Practical measures to control exposure to mite and other allergens in the house can reduce both the symptoms and the underlying bronchial reactivity. For example, controlled studies of the effects of avoidance measures conducted in the homes of patients who were allergic to dust mites have shown significant improvement in both asthma and bronchial reactivity. The effects of avoidance measures on an individual's quality of life, however, remain to be determined.

Research Agenda Item: Conduct longitudinal studies to determine whether long-term allergen avoidance has a positive effect on quality of life.

Mammals and Birds

One of the difficulties in producing good epidemiological data on allergy is the lack of well-characterized, standardized allergenic extracts for diagnostic purposes. Cat allergens have been well studied, although the role of serum albumin and its overall importance in allergic reactions to cats have yet to be determined. Allergenic extracts that are standardized for the content of *Fel d* I have appeared only recently. Well-characterized, standardized extracts of dog allergen preparations with known concentrations of *Can f* I are not available and should be developed. Similarly, there is a limited understanding of the identity and characteristics of many other mammalian and avian allergens. The lack of standardized extracts is partially responsible for the lack of development of immunochemical assays such as monoclonal antibody-based assays for many mammalian and avian allergens. Methods for measuring airborne allergen concentrations are critical for devising and evaluating control measures.

> **Research Agenda Item: Characterize important allergens from indoor animal sources (e.g., cats, dogs, birds, rodents) more precisely in order to develop standardized allergenic extracts for diagnostic purposes and immunoassays suitable for monitoring exposure.**

Despite an increasing body of knowledge regarding the role of indoor allergen exposure, particularly to mammals, as a cause of asthma, much remains to be learned. The relationship between exposure to indoor pets and the increasing morbidity and mortality of asthma requires further clarification.

> **Research Agenda Item: Determine the relationship between exposure to indoor pets and the incidence, prevalence, and severity of asthma.**

In addition, rodents that infest inner-city dwellings need to be examined as potential risk factors for asthma among individuals exposed to these potent allergens.

> **Research Agenda Item: Explore the possibility that exposure to rodent populations in inner-city areas may be a risk factor for asthma.**

For many allergens, the size of airborne particles and their distribution in the air have not been elucidated and should be studied. Reservoirs of animal and avian allergen exposure and their dissemination through ventilation systems of offices, apartments, and other large buildings likewise require investigation.

> **Research Agenda Item: Investigate the potential role of mammalian- and avian-allergen-contaminated ventilation systems in the development of allergic disease among inhabitants of apartments, offices, and other large buildings.**

Although control measures may reduce airborne concentrations of mammalian and avian allergens, the ability of these approaches to influence symptoms in sensitized patients or to prevent the sensitization of naive individuals requires clarification and study. The use of personal monitoring systems should be useful in making these determinations.

Research Agenda Item: Evaluate the effectiveness of environmental control measures on patient symptoms. This should include assessments of preventing sensitization in the naive individual as well as symptom reduction in those already sensitized.

The epidemiology, diagnostic techniques, and even pathogenesis of hypersensitivity pneumonitis from avian proteins are poorly characterized and require further elucidation. In addition, the risk factors and natural history of the disease are poorly understood and the effects of allergen avoidance remain controversial in the ultimate prognosis of the disease. All of these issues require clarification through further research.

Microbial Allergens

Overall, the fungus-associated allergies have been the least well-studied. Little data is available on the distribution of airborne fungal products, dynamics of human exposure, nature of the allergens, factors influencing the quality of skin test and immunotherapy materials, and the nature of fungus-related allergic disease.

Research Agenda Item: Initiate and conduct studies to determine the relative etiologic importance, geographic distribution, and concentrations of airborne fungus material associated with indoor allergy.

Fungi grow indoors in damp environments such as basements, window sills, shower stall surfaces, and in dust. Fungal spores and other effluents become airborne indoors when disturbed by air movement and normal human activities. The composition of aerosols of fungus-derived particles depends on the abundance and strength of sources as well as dissemination factors, mixing, dilution, and particle removal.

Research Agenda Item: Investigate the dynamics of fungal colonization of indoor reservoirs and emission of allergens from these sources. The results of such research should permit the risks associated with indoor fungal growth to be evaluated.

Exposure to fungal spores (and possibly other fungal antigen-carrying particles) can produce both IgE-mediated disease (e.g., asthma) and hypersensitivity pneumonitis while other allergens (e.g., dust mite, pollen) produce only the IgE-mediated diseases. It is not clear why or under what conditions fungal particles can have this dual effect.

Research Agenda Item: Study the differences between fungal and other allergen-carrying particles that control the development of hypersensitivity pneumonitis as opposed to IgE-mediated asthma.

Chemicals

Many of the protein allergens have long been recognized, but a lengthening list of newly recognized allergenic chemicals is developing. Allergic diseases caused by these chemicals can differ from those caused by protein allergens in terms of symptoms, mechanisms of action, and appropriate treatment. The diseases can differ also in terms of etiology and exposure, i.e., often occurring at the work site. A better understanding of these differences will assist in the formulation of improved measures of prevention and treatment.

Research Agenda Item: Determine the types of allergic diseases caused by reactive allergenic chemicals, their prevalence rates, and the mechanisms responsible for the resulting airway reactions.

A body of knowledge about chemical allergens is available, but many areas have not been well studied. Other chemicals besides those already reported to cause allergic reactions may provoke responses. Thus, as new chemicals are introduced, the list of agents that elicit allergic reactions is likely to grow.

Research Agenda Item: Identify the risk factors, such as a specific immunologic response, that are predictive of the development of chemically induced sensitization or allergic disease, and as soon as possible after their introduction, determine the sensitizing potential of new chemical entities. This knowledge will facilitate the development of primary and secondary preventive strategies.

The allergic rhinitis, conjunctivitis, and asthma that arise from exposure to chemicals appear to be due to classic immunologic reactions. However, late respiratory systemic syndrome (LRSS) and immunologic hemorrhagic pneumonitis occur only in response to chemical exposures and are not the result of response to the usual protein allergens; the mechanisms of immunologic damage in these two cases are not entirely known. The mechanism of non-IgE-mediated isocyanate asthma is also unclear.

Research Agenda Item: Determine the disease mechanisms of chemically induced LRSS, of immunologic hemorrhagic pneumonitis, and of non-IgE-mediated isocyanate asthma. Appropriate in vitro or in vivo models should also be developed.

Low molecular weight reactive allergenic chemicals can cause immunologic sensitization and consequent allergic reactions. At a minimum, hundreds of thousands of U.S. workers are exposed to such chemicals that can form haptens with airway proteins and induce allergic diseases. The goal of reducing the incidence and severity of allergic disease caused by chemical exposure is achievable, although it may not be possible to prevent all such disease.

Research Agenda Item: Determine the number of workers exposed to allergenic chemicals in various industrial and non-industrial settings and the prevalence of allergic disease resulting from such exposure. Populations in close contact with reactive allergenic chemicals and highly potent sensitizers would be logical candidates for study.

For those individuals who develop allergic disease from exposure to chemicals, it is important to determine their long-term prognosis. In particular, if immune responses that are predictive of allergic disease can be identified, and reduced exposure can be shown to result in resolution of disease (and disappearance of immunologic sensitization), then reduced exposure may represent the most practical approach for preventing allergic disease arising from chemical exposure.

Research Agenda Item: Conduct dose-response studies in humans to determine both the relationship between allergen concentration and immunologic response, and a threshold environmental exposure concentration for sensitization.

PLANTS AND PLANT PRODUCTS

Indoor plants are commonly found in offices, schools, and the home. Although most indoor plants do not produce aerosols of allergen-containing particles, as more plants are used indoors, especially in large numbers in office settings, the risk of exposure to plant allergens increases.

Research Agenda Item: Assess the significance of workplace exposures to indoor plants, including the contribution to the overall magnitude of indoor allergic disease.

Latex allergy has recently received substantial attention because of an increasing number of reports of its occurrence and its potential, in certain individuals, to produce life-threatening anaphylactic reactions. In addition to health care workers and manufacturers, children with spina bifida are at increased risk for latex allergy.

Research Agenda Item: Conduct research to further characterize the immune response to natural rubber. This effort should include studies of the incidence and prevalence of natural-rubber-related allergic disease.

MECHANISMS OF IMMUNE FUNCTION

Evidence suggests that the adjuvant effects associated with macrophage activation are related to endotoxin. Additional research is necessary, however, to clarify this phenomenon.

Research Agenda Item: Determine whether bacterial products (such as endotoxin) or fungal products may act as adjuvants in the immune responses to indoor allergens.

The magnitude of allergen exposure appears to be related to the risk of sensitization. Allergen exposure is also related to the risk of developing asthma and the age at which asthma develops. However, genetic and other local host factors are also important in atopy and asthma. Understanding the relative importance of these individual factors and their interaction with each other is essential to understanding the mechanisms of immune function.

Research Agenda Item: Conduct research to identify risk factors other than exposure, and clarify their potential significance relative to indoor allergy. This effort should include an evaluation of the role of genetic and local host factors in allergen sensitization.

MEDICAL TESTING METHODS

Medical History and Diagnosis

In spite of universal agreement on the primary importance of a patient's allergy history, very little space in medical textbooks is devoted to the topic, and no standard exists for collecting appropriate information. A standardized, validated allergy-history questionnaire would be useful in both clinical and research settings.

Research Agenda Item: Develop, test, and validate a standardized allergy-history questionnaire for use in multi-center studies.

Skin Tests

A major unknown in skin testing is the identity of the more prevalent allergens involved in many indoor exposures. Studies to characterize these allergens are important for the development of reliable diagnostic reagents.

Additional research is needed to identify, characterize, and standardize indoor allergenic extracts used for diagnostic testing.

> **Research Agenda Item: Develop standardized, well-defined indoor-allergen reagents for skin tests that can be used in clinical diagnosis and research studies.**

In Vitro Tests

The accuracy of any immunodiagnostic test is highly dependent on the characteristics of the test reagents, in particular, the allergen reagent. Standardization and characterization of allergen reagents used for immunodiagnostic tests are imperative. There are a variety of characterization methods and unitages that could be used. Similarly, the existence and characterization of control antibody, whether polyclonal or monoclonal, would be valuable for standardization and quality control of immunodiagnostic tests. Ideally, minimal standards for quality control should be devised for labs reporting results of tests to detect specific immunologic responses to indoor allergens.

Once the above are instituted, the specificity, sensitivity, and positive predictive value of immunodiagnostic tests could be determined for use in major epidemiological studies or to determine specific immunologic responses of individuals to indoor allergens. In addition, the degree of cross-reactivity of antibody developed in response to a given allergen could be determined by cross-inhibition with other allergens. It would be important to assess these factors prior to embarking on any large epidemiological studies. There are also several unclear aspects of the immunopathogenesis of allergic disease that need elucidation in order to define the role of various tests (e.g., tests of specific cell-mediated immunity in asthma). Further immunopathogenic studies of non-IgE-mediated asthma and cellular studies in all types of immunologic asthma will be required to clarify these issues.

Future studies in the development of in vitro diagnostic tests should include the following:

> **Research Agenda Item: Identify selected allergens of potential research usefulness, and prepare pure reference standards for the development of immunoassays, including those that can be used in large scale epidemiological studies.**

> **Research Agenda Item: Develop and assess immunoassays for new allergens, including low molecular weight allergenic chemicals, that can be used for research and for the diagnosis of allergic disease.**

Pulmonary Function

Bronchial hyperreactivity is a feature of asthma that correlates with clinical severity and does not require repeated measurement. It is unclear, however, whether bronchial hyperreactivity can be correlated with exposure to indoor allergens.

Research Agenda Item: Determine whether changes in bronchial hyperreactivity can be correlated with exposure to indoor allergens. If such a correlation exists, determine how reducing the level of allergens affects bronchial hyperreactivity.

ASSESSING EXPOSURE AND RISK

Some of the issues to be considered when undertaking a risk assessment for aeroallergens have been discussed and an example has been given by using data from the literature on dust mite exposure and the development of sensitization. The data can be described by a linear model and indicates that there is a positive relationship between cumulative exposure to dust mite allergen and the risk of sensitization.

Some residual sensitization (i.e., approximately 10 percent in this example) will occur irrespective of exposure to dust mite allergen according to these estimates. This finding is consistent with the knowledge that other factors may also result in sensitization. For this reason, information on cross-reactivity of allergenic agents in study subjects is desirable, and important to the analysis of potential mechanisms of sensitization.

Research Agenda Item: Determine whether a practical method could be developed to measure concentrations of dust mite allergens that are capable of sensitizing humans.

Variability in the methods used in the multiple study protocols reported to date is high. More uniformity in the collection of exposure data would be useful in the risk assessment process. Although reservoir sampling has yielded meaningful results that assist in remediating exposure, further development of standardized air sampling collection and analytical methods is needed.

Research Agenda Item: Standardize methods of collecting and analyzing indoor allergen samples to facilitate comparative and collaborative studies.

Improved, standardized methods of collecting and analyzing indoor allergen samples would be particularly valuable in establishing the relationship between reservoir samples, personal exposure measures, levels of activity, and the potential for airborne exposure of sufficient magnitude to induce negative health outcomes.

Research Agenda Item: Quantitate the relationship of allergens in reservoirs (and on surfaces) to aerosols and develop monitoring methods

for quantitating airborne-allergen concentrations in personal breathing zones.

Assessment of exposure is a rapidly advancing, complex, and multistep process that entails numerous variables and estimations. Most monitoring, for example, is often based on sampling for indicators rather than the actual allergen. There is a need for developing improved methods for estimating environmental concentrations of aeroallergens and the resultant individual exposures.

Research Agenda Item: Develop appropriate exposure metrics for specific indoor allergens that are analogous to time-weighted averages and permissible-exposure limits for industrial chemicals.

Engineering Control Strategies

Dry vacuum cleaning is traditionally used to remove dirt and debris from the fibrous pile of carpets. Little information is available, however, on the effectiveness of this cleaning method in removing the various types of particles, including specific allergens that may adhere to pile fibers, carpet backing, and other furnishings. In addition, the physical cleaning process itself may be sufficient to disperse fine allergenic particles.

Research Agenda Item: Develop standardized tests for rating the effectiveness of vacuum cleaners in removing allergen-containing particles of known size from carpets, upholstery, drapes, and other materials. The tests should take into account the possible dispersion of particles from carpet caused by the cleaning process itself.

The effectiveness of air cleaning devices and practices depends on variables such as the volume of air that passes through the filter, the particle size of the air contaminant to be removed, and the source emission rate. If the air flow rate through an air cleaning device is low, for example, and the emission rate of the allergen is high, then the beneficial effect of the air cleaner is likely to be nonsignificant.

Research Agenda Item: Develop standardized test procedures for rating the effectiveness of air cleaning devices and other methodologies for removal of known size classes of particles containing allergens. The tests should address the capability of the device or methodology in removing airborne particulates from entire rooms or zones of buildings.

Restricted airflow and dissemination of particulates into occupied spaces are valid reasons for cleaning air supply ducts. Protocols for cleaning air conveyance systems are currently in development by the National Air Duct Cleaners Association. However, the effectiveness of duct cleaning in controlling allergic disease is yet to be determined.

Research Agenda Item: Evaluate the role of duct cleaning in controlling allergic diseases.

As described throughout this report, ambient relative humidity is often considered a major controlling factor for indoor allergens. Control of relative humidity, or water vapor pressure in occupied space and in the HVAC system is an important part of allergen control in both residential and commercial buildings.

Research Agenda Item: Develop a public-use guideline on moisture and allergen control in buildings. The guideline should describe the proper use of vapor retarders and other techniques for moisture control in both naturally and mechanically ventilated buildings.

There are approximately 4 million commercial and 84 million detached residential buildings in the United States. About 75–85 percent of the buildings that will exist in the year 2000 have already been built. Maintenance, operation, renovation, and housekeeping practices affect the useful life span of a building and the quality of the indoor air. Cost effective strategies for source and exposure control are needed to address the problems associated with normal degradation of the HVAC performance that occurs as a building ages.

Research Agenda Item: Determine the relative efficacy of currently recommended environmental control strategies and develop cost-effective strategies for controlling aeroallergens throughout the lifetimes of residences and other buildings.

1

Introduction

Most Americans spend a majority of their time indoors. For many, this can be nearly 24 hours per day, including time in transit (Figure 1-1; NRC, 1981; Spengler and Sexton, 1983). The quality of indoor air, therefore, is at least as important to health as the quality of outdoor air, and in recent years concerns have increased about possible adverse health effects from indoor allergens (i.e., biological or chemical substances that cause allergic reactions). In fact, there is a growing perception that allergy has been underestimated as a public health problem. For example, indoor allergens such as those from dust mites, house pets, cockroaches, and fungi are thought to be responsible for much of the acute asthma in adults under age 50. Using asthma as an example of severe allergic disease, it is apparent that indoor allergens adversely affect the quality of life for a significant fraction (more than 10 percent) of the population (Burge and Platts-Mills, 1991). In addition, asthma was the first-listed diagnosis for more than 450,000 hospitalizations in 1987 (NHLBI, 1991). There is also a rising rate of mortality associated with asthma—especially among African Americans (NHLBI, 1991). Depending on age, African Americans are 3–5 times more likely than Caucasians to die from asthma. In 1988, a total of 4,580 people died from asthma in the United States.

The severity of asthma and the fact that it is often more easily identified than some other allergy-related conditions have led to better statistics on the effects and outcomes than are available for other allergy-related conditions. Nevertheless, most such figures are necessarily expressed as estimates. For example, recent estimates have put the total annual cost to

U.S. society from asthma at more than $6.2 billion (K. B. Weiss et al., 1992b), an increase of 39 percent over the estimated cost of the disease in 1985. Despite the lack of exactitude, the size of these costs and the fact that they represent an increasing trend argue for careful attention not only to the effects of allergy in causing asthma but also to the litany of other conditions that are more commonly thought to be associated with "allergies." Allergic rhinitis (hay fever), for example, is a disease of dramatic proportions in this country; although not life-threatening, it nevertheless is estimated to affect some 35 million Americans (17 percent of the population) at some point in their lives. The direct and indirect costs associated with such a burden of disease are not trivial and point to the importance of serious consideration of the effects of indoor allergens on the public's health.

To a majority of people, the indoor environment has always been considered a refuge from the dangers of the outdoors, offering protection from weather, wild animals, and, in modern times, air pollution—long seen as primarily an outdoor phenomenon. In more recent times, however, reduced ventilation in energy-efficient buildings has stimulated debate and concern about increasing indoor pollutant concentrations. Much of the concern has focused on potential carcinogens, with radon and environmental tobacco smoke being good examples. But other agents and potential health effects are beginning to receive increasing consideration, and among these are the indoor allergens.

Evidence is mounting that allergy has a fundamental, causal relation-

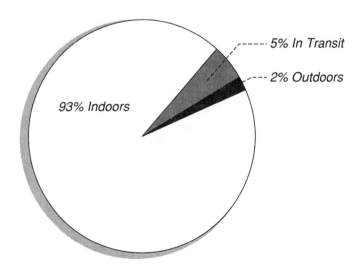

FIGURE 1-1 Amount of time spent indoors, outdoors, and in transit by most Americans. Sources: NRC, 1981; Spengler and Sexton, 1983.

ship to many diseases and that aeroallergens—allergens that are airborne—
are important in disease pathogenesis. For example, allergy to house dust
mites and cats increases the risk of childhood asthma four- to sixfold.

There is a coherence to the body of research on these effects that now
deserves emphasis: Exposure to indoor allergens is common, as are resultant sensitization and allergic disease. One hundred million Americans are
immunologically sensitized to allergens; that is, they have been exposed to
a substance that has raised the sensitivity of their immune system in such a
way that the system will produce an "allergic response" when it encounters
the allergen on subsequent occasions. The number of Americans so sensitized is fourfold more than the number with hypertension; 50 million people
in this country will have a disease related to allergy at some time in their
life.

With increasing levels of interest and concern about the potential health
effects of indoor air, the Environmental Protection Agency's Office of Air
and Radiation, in conjunction with the Agency for Toxic Substances and
Disease Registry, and three of the institutes at the National Institutes of
Health—the National Institute of Allergy and Infectious Diseases, the National Institute of Environmental Health Sciences, and the National Heart,
Lung, and Blood Institute—requested that the Institute of Medicine (IOM)
conduct an independent assessment of the public health significance of indoor allergens. Specifically, the IOM was asked to perform the following
tasks:

1. Examine and characterize what is known about the adverse effects
on human health caused by indoor allergens and chemicals that elicit allergic reactions.

2. Characterize the magnitude of these problems nationally and the populations that are commonly affected.

3. Identify the specific causative agents and their sources.

4. Assess testing methodologies for indoor allergens, including their
applicability and interpretation.

5. Evaluate the adequacy of the knowledge base that is currently available to physicians on this topic and the need for additional information and
research.

The IOM Board on Health Promotion and Disease Prevention responded
to the request by establishing a committee with expertise in allergy and
immunology, epidemiology, industrial hygiene, mycology, pulmonology, general
and family medical practice, engineering, education, and public policy. This
report, the product of the committee's deliberations, reviews published epidemiological reports, describes risk factors, and estimates the health impacts of exposure to indoor allergens. It examines clinical diagnostic practices and the measurement of indoor allergen exposures, and discusses the

need for increased awareness and knowledge of indoor allergens among primary care physicians, allergists, patients, and others.

Because of the large number of people affected by allergic disease, the committee anticipates that this report will be of interest to a broad array of readers (e.g., medical practitioners, public policymakers, members of the public whose health is adversely affected by the indoor environment). Consequently, several sections of background material are presented in this chapter; they should be useful to readers in understanding the rest of the report. The sections below present a brief historical perspective on allergies, the indoor environment, and the twentieth-century innovations that have contributed to current concerns; several concepts and definitions that the committee used as a framework for its discussions are also discussed. The chapter concludes with a brief statement regarding the report's scope and organization.

ALLERGENS AND THE INDOOR ENVIRONMENT

For more than 5,000 years, observers have recorded episodes of disease and demise that today are recognized as allergic reactions. For example, the death of King Menes of Memphis in about 3000 B.C. was the result either of anaphylaxis from the sting of a hornet or of being trampled by a hippopotamus. (A question arises because hornet and hippo share the same ancient Egyptian word.) During the sixteenth, seventeenth, and eighteenth centuries, reactions were noted (mostly to foods) that were surely allergic in nature, and some surprisingly intuitive observations were made regarding cause and effect. During this period, cats, dogs, horses, feathers, and many foods were suspected of causing asthma. In the early nineteenth century, it was recognized that pollen caused "hay fever" and that dust from beaten carpets produced similar symptoms. The first experimental challenges with pollen apparently were done by Kirkman in 1835, followed by Blackley's extensive investigations into both allergy and aerobiology in the 1870s (Blackley, 1873).

More recently, much progress has been made in recognizing specific causes of asthma and hay fever and in defining the mechanisms by which symptoms are elicited. Yet in spite of increasing knowledge of the mechanisms of allergic disease and of the agents that cause sensitization and symptoms, we have modified indoor environments in ways that may contribute significantly to exposure to these agents and to the development of allergic disease. For example, we have adopted central heating on a broad scale, which means that we heat all parts of our houses, even when they are not occupied. By the late 1960s and 1970s it became obvious that much energy was wasted in houses and other buildings owing to excessive heat transfer through surfaces such as walls, ceilings, and windows or through "cracks." This waste inspired a federally sponsored campaign to make houses and other buildings both better insulated and "tighter." Insulation may not directly affect the levels and kinds of allergens in the indoor environment (as long as it remains dry), but decreased ventilation encourages their accumulation and may increase relative humidity, leading to conditions that favor the growth of biological allergen sources (e.g., microbes and mites).

Other twentieth-century amenities have also contributed to increased allergen exposure. The invention and widespread use of the vacuum cleaner have been followed by the practice of installing wall-to-wall carpeting in homes, schools, hospitals, and offices, despite the fact that the vacuum cleaner is probably not effective for removing skin scales, mites, and excess moisture from carpeting. In fact, vacuum cleaning is quite effective in dispersing and suspending allergens and other particles in the air. Heating, ventilating, and air-conditioning (HVAC) systems, which have become widespread in this country, can also create problems in terms of indoor allergens. On the one hand, HVAC systems allow windows to remain closed, filter out outdoor particulate material, and decrease relative humidity, thus lowering the risk of microbial and mite contamination. On the other hand, if air-conditioning units are not maintained properly, they can be sources of microbial contamination and have caused more than one major epidemic of infectious disease (Fraser et al., 1977). In addition to these factors, pets in the home are a major source of allergens and can play a large role in sensitization and allergic disease (Mathison et al., 1982; Ohman et al., 1977; Reed and Swanson, 1987; Swanson et al., 1989). Thus, lifestyle changes have influenced indoor air pollution both positively and negatively. Overall, indoor air quality is better than it was in the Middle Ages, when straw was used as a floor covering, sanitation was nonexistent, and burning fuels created a smoky environment. In modern times, however, homes and buildings still harbor the potential for severe allergic problems.

Over the past 25 years, a steady increase in our knowledge of the biological sources (sensitizing agents) that give rise to immunological sensitization has allowed us to identify specific health risks and to establish methods for measuring some of the indoor agents that are thought to cause hypersensitivity disease (i.e., "allergies"). For most agents, it is clear that exposure creates a risk, but insufficient data are available to propose threshold or risk levels. For a few sensitizing agents, it has been possible to propose specific levels of exposure that constitute a risk for sensitization and symptom development (Platts-Mills and Chapman, 1987). Most research on indoor air quality, however, has focused on the same pollutants that are found in outdoor air (nitrogen oxides, carbon monoxide, etc.) and a few other nonbiological agents that are either relatively easy to measure (e.g., asbestos, radon) or that have stimulated major controversy (e.g., environmental tobacco smoke). Very little research has been directed toward the health effects, prevalence, or relative risks of allergens, toxins, and volatile irritants of biological origin in indoor air.

The major sources of indoor allergens include house dust mites, molds, and animal dander. These and other potential sources of allergens are the focus of this report and are covered in detail in the chapters that follow. Also emphasized are several specific allergic diseases of the upper and lower airways: rhinitis, asthma, and hypersensitivity pneumonitis. Before proceeding to this discussion, however, several concepts and definitions are described that will be useful to an understanding of the material that follows. Readers are also encouraged to consult the glossary, which follows the references at the end of the main text of the volume.

CONCEPTS AND DEFINITIONS

Allergy, generally speaking, is the state of immune hypersensitivity that exists in an individual who has been exposed to an allergen and has responded with an overproduction of certain immune system components such as immunoglobulin E (IgE) antibodies (defined below). More specifically, common use of the term *allergy* usually refers to a *type I* hypersensitivity reaction (the type I reaction is also called the anaphylactic type or immediate type of allergic reaction). It is the common type of allergic reaction caused by IgE antibody and results in the common allergic diseases such as hay fever, allergic asthma, allergic urticaria, and the less common but potentially fatal anaphylactic reaction. About 40 percent of the population have IgE antibodies against environmental allergens, 20 percent have clinical allergic disease, and 10 percent have significant or severe allergic disease.

Gell and Coombs (1968) classified hypersensitivity reactions into four major types, according to the immune mechanisms involved (see Table 1-1). For the purposes of this report (and unless noted otherwise), the term *allergic reaction* means the immediate hypersensitivity response that occurs when a sensitized individual is exposed to an allergen, that is, a type I, IgE-mediated reaction. Although these reactions against indoor allergens are our primary focus, some attention is also given to allergic diseases that operate through different mechanisms, for example, hypersensitivity pneumonitis—a disease that involves IgG antibodies and reactive lymphocytes. Chapter 4 describes the mechanisms of the immune function in technical detail.

There are two major components of the immune system that respond to allergens: *antibodies*, which are specific proteins of the *immunoglobulin E* (IgE) type, and *mast cells*, which carry receptors for IgE and release chemicals such as histamine. The initial phase of the immune response, known as *sensitization*, sets up the mechanisms for an allergic reaction (Figure 1-2).

TABLE 1-1 Four Types of Immunologic Reactions to Indoor Allergens that Cause Pulmonary Disease in Humans

Type of Hypersensitivity Reaction[a]	Immunologic Mediation	Antigen	Method of Study
Type I (anaphylactic)	IgE antibody	Proteins such as animal allergens	Correlation of skin tests or in vitro measurement of antibody with symptoms
Type II (cytolytic)	IgG antibody against cells haptenized with allergenic chemicals	Allergenic chemicals: acid anhydrides and isocyanates	Theoretical concepts under research study
Type III (antigen-antibody complex)	Antigen-antibody complex and inflammatory cells	Proteins such as avian proteins	Precipitating antibody correlated with exposures and symptoms
Type IV (lymphocyte-mediated)	Lymphocyte-mediated reactions in lung against antigen	Avian and microbial proteins, for example	Correlation of clinical disease with lymphocyte activation and cytokine release after antigen exposure

[a]Gell and Coombs's (1968) classification.

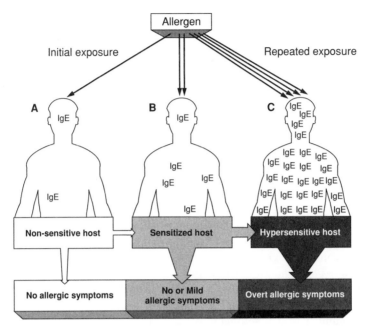

FIGURE 1-2 Hypersensitivity and IgE-mediated allergy. (A) Upon initial exposure to an allergen, there will be no overt manifestations of allergic disease because the patient is nonsensitized. However, this allergen will initiate an immune response that results in the synthesis of IgE and sensitization of the susceptible (atopic) host. (B) Upon subsequent, repeated exposures, this sensitized individual will synthesize increased amounts of IgE, thus becoming hypersensitive, although mild allergic symptoms may or may not be present. (C) From this point onward, reexposure to this specific allergen will provoke the overt manifestations of allergic disease. Source: Fireman and Slavin, 1991.

During sensitization (in the case of airborne allergens, this occurs in the airways or on the skin), the immune system of a susceptible person encounters an allergen in the form of an *antigen* (a substance that stimulates the production of an antibody). The system responds by producing IgE specific to that antigen[1]—that is, an IgE that will recognize and react with that specific antigen when they meet again. During the sensitization process, the antigen-specific IgE that is produced attaches to receptors on mast cells[2]—completing the sensitization process. When the "sensitized individual" is

---

[1] Specific IgE can be assayed in vivo by skin testing or in vitro by such assays as radioallergosorbent test or enzyme-linked immunosorbent assay.

[2] Mast cells are present throughout the body but occur at the highest concentrations in areas that are exposed to the outside world (i.e., the skin and linings of the nose, lungs, and gastrointestinal tract).

exposed again to the allergen, the IgE on the mast cells binds the antigen, triggering the mast cells to release histamine and other chemicals from their cytoplasm. These chemicals interact with surrounding tissues, producing *inflammation* and the typical allergic reaction. IgE antibody production can last for many years, sensitizing the mast cells. Thus, for example, someone who had an allergic reaction to penicillin as a child could still be allergic to the drug as an adult.

Another important term is *atopy.* Atopy is generally defined as the state of having one or more of a defined group of diseases (allergic rhinoconjunctivitis, allergic asthma, and atopic dermatitis) that are caused by the genetic propensity to produce IgE antibodies to environmental allergens (predominantly from pollens, molds, dust mites, animals, and foods) encountered through inhalation, ingestion, and possibly skin contact. Although everyone produces some IgE, people with a genetic predisposition to allergy (i.e., atopy) produce significant quantities of it.

For epidemiological surveys, atopy is sometimes defined differently, that is, solely by the presence of IgE antibodies to one or more allergens— whether or not clinical disease is present. The population identified by this definition is considerably larger than that defined by disease because a sizable number of asymptomatic persons have such antibodies (this condition is known as *asymptomatic atopy*).

Thus, an *atopic individual* is defined differently in different settings and by different researchers as either a person with a particular disease described under atopy or a person with IgE antibody (i.e., positive skin tests). It is obviously important, therefore, that the context and use of the term be well defined in its particular applications and settings. In discussing the published literature in this report, the meaning of the term will correspond to that used in the study being considered.

SCOPE AND ORGANIZATION OF THE REPORT

In addressing the tasks described earlier in the chapter, the committee used the following guidelines to define the scope of its deliberations. First, it focused on *airborne* allergens, although contact allergens (e.g., cosmetics) and food allergens are also in the home (the second major focus). A second major focus in the scope was the residential environment; the broad topic of industrial/occupational allergens was deemed to be beyond the scope of this report and was not covered in depth. Some attention is given, however, to allergens that are found in nonindustrial work sites such as office buildings, hospitals, and schools. Similarly, some attention is given to selected nonbiological chemicals that have been shown to be allergenic and that are found in residential or nonindustrial work sites (e.g., trimellitic anhydride and isocyanates). Primarily, however, the report focuses on airborne allergenic agents of biological origin.

In-depth consideration of multiple chemical sensitivity (MCS) reactions was excluded from the scope of this report because of a lack of data showing immune mechanisms. As stated in a recent NRC report (NRC, 1992a, p. 138), "there is insufficient evidence to ascribe an immune etiology to this disorder." For information on the topic of MCS the reader is directed to the recently published proceedings of a workshop and compilation of individually authored papers (NRC, 1992b).

The remaining seven chapters of this volume present the committee's findings, conclusions, and recommendations on several aspects of the problem of indoor allergens. Chapter 2 discusses the magnitude and dimensions of allergen-caused disease, and Chapter 3 presents a detailed discussion of allergen sources and their distribution. Chapter 4 describes the mechanisms of immune function. A discussion of test methods, with particular attention to their applicability, reliability, and interpretation appears in Chapter 5. Chapter 6 discusses the assessment of exposure and risk. The report concludes with an examination in Chapter 7 of engineering control strategies and the role of education in Chapter 8.

In preparing this report, the committee identified and developed a list of research agenda items and 15 priority recommendations. The recommendations focus primarily on the need to improve awareness and education, and as such are also relatively inexpensive and easy to implement compared to the research agenda items. The research agenda focuses on the longer term, more expensive, and more technical aspects of fundamental research and data collection.

Although cost-benefit analyses were beyond the scope of the committee's charge in conducting this study the committee's findings have been separated into recommendations and research agenda items, as described above, based on relative ease of implementation and estimations of associated costs. Policy development and further priority setting, however, should be based on investigations of the probable costs and benefits of the various strategies for prevention and control. The cost analyses should include both direct and indirect costs (i.e., medical costs as well as lost work), and benefits should include a full assessment of the effects on improving the quality of life and on reducing morbidity and mortality.

All of the committee's recommendations and research agenda items are compiled in the Executive Summary. In the body of the report they are presented in their individually relevant chapters and chapter sections.

2

Magnitude and Dimensions
of Sensitization and Disease Caused
by Indoor Allergens

The magnitude and dimensions of a disease or its antecedents can be measured and described in several ways (Table 2-1). The *prevalence* of a disease, for example, is a measure of its frequency in a population at some specified point in time. The *incidence* of a disease is the number of new cases of the disease that occur during a specified period of time. Exploration of disease incidence can focus public health attention on critical time periods in the development of disease. For example, the observation that allergy often develops by age 5 suggests that interventions are needed for infants and young children.

Incidence and prevalence rates can be determined for the entire population, or they can be specified by gender, age, ethnic group, socioeconomic class, geographical region, or time of year. The increased mortality from asthma among African Americans (Sly, 1988; NHLBI, 1991), for instance, has focused attention on the disease in this segment of the population. Community-based epidemiological studies and national health surveys provide the best available estimates of the overall prevalence of immunologic sensitization and of specific diseases.

Disease *severity*—a measure of the impact of a disease on a patient's life, the intensity of medical care required, and the ultimate outcome—is another indicator of the magnitude and dimensions of a particular disease. Measures of disease severity are often derived from national mortality data and hospital discharge statistics. They are used to evaluate the effectiveness of therapeutic regimens, environmental control strategies, and educational and behavior modification self-management programs.

TABLE 2-1 How Diseases Can Be Measured

Question	Appropriate Measure
How many people have asthma sometime in their life?	Cumulative prevalence
How many people have asthma in the United States today?	Current prevalence
How many people get asthma when they are children?	Age-specific incidence rates
Do children have more asthma than adults?	Comparison of prevalence rates for specific age ranges
Does allergy have a role in asthma?	Comparison of the probability of developing asthma among people with and without skin test reactivity
What proportion of asthma is attributable to allergy?	Attributable fraction
What proportion of asthma in children is attributable to house dust mite exposure?	Attributable fraction

Generally speaking, *mild* disease causes symptoms but only intermittently requires medication and infrequently alters life activities. *Moderately* severe disease may require regular physician visits, regular medication, and lost time from work or school. *Severe* illness may require admission to an intensive care unit, to a regular medical floor, or to a hospital emergency room (in decreasing order of intensity). In the case of allergic disease, death occurs relatively infrequently. Other measures that can be used to indicate the magnitude and dimensions of a particular disease (such as allergy) include quality of life, trends over time, and economic and psychosocial impact.

This chapter uses these and other measures to discuss the magnitude and dimensions of two aspects of allergic disease: sensitization and the specific allergic diseases themselves. Immunologic sensitization is important because it is an indicator of the population at greatest risk of developing allergic disease. The remainder of this chapter discusses the public health significance of immunologic sensitization and allergic disease in these terms.

IMMUNOLOGIC SENSITIZATION

Allergic disease develops through a series of steps that are becoming more clearly understood (Figure 2-1). As described briefly in Chapter 1, allergic disease occurs when a genetically predisposed or susceptible individual is exposed to an allergen and becomes immunologically sensitized. The occurrence of different types of sensitization can be ascertained by skin

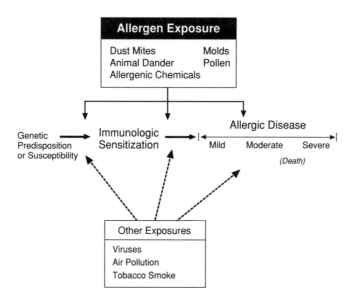

FIGURE 2-1 Development of allergic disease, illustrated schematically. A genetically susceptible individual is exposed to an allergen and becomes immunologically sensitized. At this stage the person is asymptomatic, but the sensitization may be detected by skin tests or laboratory tests. Over time, a proportion of sensitized individuals will develop one of a group of allergic diseases. Exposure to allergen is understood to be a major factor at each stage of the pathogenesis of these diseases.

tests, serologic measurements, or tests of cell function. In the early stages of sensitization, people who are sensitized have not developed symptoms of disease. The magnitude of this group within the population is of interest, however, because it reflects the proportion of the population at greatest risk of developing allergic disease. Additional exposure to the sensitizing allergen leads to the development of an allergic reaction (disease) that can be mild, moderate, or severe, depending on the amount of exposure. Exposure to other substances that might irritate the respiratory tract (e.g., environmental tobacco smoke) can serve to promote the development of allergic reactions and disease.

Skin testing has been the primary diagnostic tool for allergy for over 100 years. Skin test reactivity (i.e., a positive skin test, also known as *atopy*) indicates that an individual has been immunologically sensitized and now has specific immunoglobulin E (IgE) antibody against one or more common allergens. The presence of skin test reactivity indicates an increased risk for one of several diseases including allergic rhinitis and asthma, but at the time skin test reactivity is detected, disease may or may not be present. Table 2-2 shows results from several studies that measured sensitization prevalence in various populations. The results indicate a range from a low of 20 percent to a high of 51 percent in different studies. It is estimated that about 40 percent of people in the United States have been immunologically sensitized as indicated by skin test reactivity to a panel of currently available allergens. The recent International Consensus Report (NHLBI, 1992) claims that about 30 percent of all populations are *skin test positive*.

Early childhood is a common time for sensitization. Barbee and others (1987) showed that 22 percent of a cohort of children less than 5 years old had skin test reactivity to one or more allergens. During an 8-year follow-up, the prevalence increased to 44 percent. Gergen and coworkers showed that 18 percent of Caucasian and 28 percent of African American children showed skin test reactivity when tested between the ages of 6 and 11 (Gergen et al., 1987). In a cohort of genetically susceptible children, 20 percent developed skin test reactivity by age 5; another 20 percent became reactive between ages 5 and 11 (Sporik et al., 1990).

Skin test reactivity increases in prevalence until age 20–45 and then decreases (Figure 2-2; Barbee et al., 1976, 1987; Gergen et al., 1987). In most studies, skin test reactivity is similar for both sexes, although men have more total IgE (Barbee et al., 1976; Burrows et al., 1980; Klink et al., 1990). Although immunologic sensitization is generally stable over time, small seasonal variations have been observed (Peat et al., 1987). Prevalence rates of sensitization also increase somewhat with the use of a greater number of allergens during testing (Barbee et al., 1987).

Table 2-3 and Figure 2-3 show estimates of the prevalence of sensitization in certain population samples to specific allergens as determined by skin test reactivity. Among Australian schoolchildren, for example, sensitization to house dust mite was shown to be most common, followed by grass and weeds, animal danders, and molds (Peat et al., 1987). Similar rates of sensitization have recently been found in middle-class American children. The prevalence of cockroach allergy shown in Figure 2-3 may well be higher in selected populations, such as among residents in cities in which cockroach infestations are widespread. A study of patients visiting an emergency room in Virginia showed that 5 percent of nonasthmatics were sensitized to cockroaches, compared to 33 percent of asthmatics (Pollart et al., 1989). Fifty-four percent of patients referred to a Kansas City clinic for possible allergic disease were sensitized to cockroach allergen (Hulett and Dockhorn, 1979).

Risk Factors

In keeping with a public health approach to prevention and control of allergic disease and the need to improve the understanding of the etiology,

TABLE 2-2 Partial Listing of Published Data on the Prevalence of Skin Test Reactivity

Source	Population	%[a]	Notes
United States			
Barbee et al., 1987	Community sample (N = 1,333; Arizona)	39	Ages 8–60 Rates with retesting 8 years later
Gergen et al., 1987	Nationwide sample (N = 16,204; NHANES[b])	51 20.2	Some people with disease were excluded; the skin test flare was used as the response
Freidhoff et al., 1981	Workplace study (N = 174; Maryland)	24	Lower rates are seen in other workplace studies
Barbee et al., 1976	Community sample (N = 3,012; Arizona)	34	Ages 5–55
Hagy and Settipane, 1969	College freshmen (N = 1,243)	30.9	
Henderson, 1993 (personal communication 1-26-93)	Schoolchildren (N = 225; North Carolina)	40	

Other countries

D. Charpin et al., 1991	Schoolchildren (N = 933; France)	25	Methodologic differences probably account for the lower rates in this study
Sears et al., 1989	Birth cohort; (N = 714; New Zealand)	45.8	
Peat et al., 1987	Schoolchildren (N = 2,363; Australia)	39	
Chan-Yeung et al., 1985	Adult workers (Canada)	22	Lower rates are seen in other workplaces
Cockroft et al., 1984	Canadians	35	
Haahtela et al., 1980	Teenagers (N = 708; Finland)	49	
Woolcock et al., 1978	Australia Community sample (N = 188)	39	
	New Guinea natives (N = 317)	49	

[a]Percentage of population with skin test reactivity.
[b]NHANES, National Health and Nutrition Examination Survey.

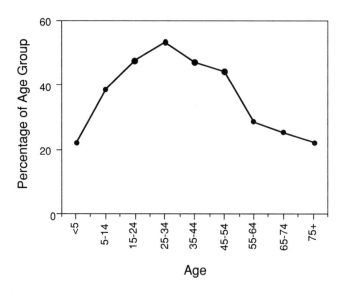

FIGURE 2-2   Changes in skin test reactivity with age. By age 5, approximately 20 percent of children will be sensitized to common aeroallergens. Skin test reactivity increases with age until ages 25–34 and then declines. Source: Barbee et al., 1987.

it is important to identify potential risk factors associated with allergic disease. Risk factors are biological, environmental, and behavioral characteristics that are causally associated with health-related conditions (Lalonde, 1974; Last, 1986). Heredity, for example, is an important biological risk factor in the development of immunologic sensitization and allergy. Infants of parents with allergic disease develop positive skin tests at higher rates than infants in population-based studies. Skin prick test reactivity during the first year of life has been reported as ranging between 30 and 70 percent of at-risk infants—i.e., those who have one or both parents with allergic disease (Zeiger, 1988). In general, if one parent has allergies and the other does not, then the chances are one in three that each of their children will have allergies. If both parents have allergies, it is much more likely (seven in ten) that each of their children will have some manifestation of allergic disease.

Exposure to allergens is an example of an environmental risk factor related to the prevalence of sensitization. Household exposure to elevated levels of dust mite allergen (see Chapter 3) in infancy, for example, has been associated by age 5 with an increased prevalence of positive dust mite skin tests and an increased concentration of dust mite IgE antibody (Zeiger, 1988). Another example is found among people living at high altitudes in Briançon, France, where significantly lower rates of sensitization to house

TABLE 2-3 Partial Listing of Published Prevalence Data on Sensitization (per 100 population) to Some of the Common Aeroallergens

Source	Population	House Dust Mite	Grasses and/or Weed Pollen	Animal Danders	Molds	Notes
Barbee et al., 1976	Community (N = 3,101; United States)	ND	17–24	ND[a]	8	Crude house dust: 9[b]
Peat et al., 1987	Schoolchildren (N = 2,363; Australia)	19–31	18–29	20–25	10–20	
Sears et al., 1989	Birth cohort to age 13 (N = 714; New Zealand)	30.1	32.5	13.3		
D. Charpin et al., 1988a	Adults (France) Marseille (N = 4,008; low altitude)	44.5				
	Briançon (N = 1,055; high altitude)	10				
D. Charpin et al., 1991	Schoolchildren (France) Martigues (N = 693; low altitude)	16.7	8.5	5.6		
	Briançon (N = 240; high altitude)	4.1	21.7	3.3		
Henderson, 1992 (personal communication)	Schoolchildren (North Carolina)	27	Pending	Pending		
Godfrey and Griffiths, 1976	Schoolchildren (N = 303; Southampton, England)	26	24			

[a]ND, no data.
[b]Crude house dust is a skin test reagent that contains unpredictable amounts of antigens such as dust mite, animal dander, and molds. Its use has been largely replaced by purified allergen extracts.

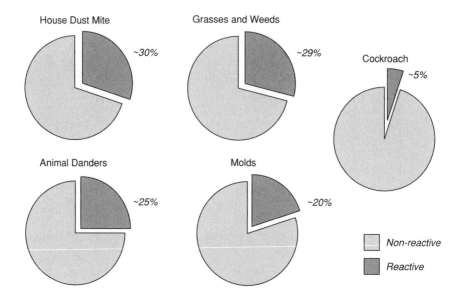

Allergy in School Children

FIGURE 2-3   Estimated prevalence of skin test reactivity to selected aeroallergens among schoolchildren. The presence of skin test reactivity indicates that the child has been sensitized to that aeroallergen by an IgE mechanism. Henderson found lower rates of skin test reactivity to animal danders and molds in their population. (Table 2-3 shows results from additional epidemiological studies.) Sources: Henderson, 1993; Peat et al., 1987; Sears et al., 1989.

dust mite occur than in comparison populations at lower altitudes (Martigues for children and Marseilles for adults) (D. Charpin et al., 1988a, 1991; Vervloet et al., 1979). Analyses of house dust from Briançon found lower levels of house dust mite compared with the levels in dust from low-altitude homes (see Table 2-3). In addition, Sporik and colleagues (1990) demonstrated a trend toward an increasing degree of sensitization among children at age 11 with greater dust mite exposure at age 1. Chapter 6 analyzes these and other studies and suggests that there may be an exposure-response relationship between dust mite exposure and the prevalence of skin test reactivity (sensitization).

Exposure to environmental tobacco smoke is another example of an environmental risk factor in that it appears to be associated with increased skin test reactivity in children (Burrows and Martinez, 1989) and a twofold increase in serum IgE in infants (Zeiger, 1988). Zeiger also discussed the possible effects of exposure to ingestants and microbial agents on the risk of sensitization. Animal studies have shown an increased rate of new IgE sensitization when allergen exposure is interspersed with ozone, nitrogen dioxide, or sulfur dioxide inhalation (Matsumura, 1970a–c; Riedel et al., 1988; Sheppard, 1988b). Diesel fumes have also been shown to act as an adjuvant for IgE sensitization (Muranaka et al., 1986).

The effect of active smoking—a possible behavioral risk factor—on the prevalence of skin test reactivity in adults is uncertain, given the existence of conflicting data. Barbee and colleagues (1987) found lower rates of skin test reactivity among current smokers than either nonsmokers or ex-smokers; ex-smokers had higher rates of skin test reactivity than nonsmokers. The study team interpreted this finding as demonstrating a self-selection process; that is, smokers with skin test reactivity would stop smoking. In contrast, Gergen and coworkers (1987) did not find a relationship between skin test reactivity and smoking status.

Other Types of Sensitization

Immunologic sensitization may also occur through non-IgE mechanisms. Specific IgG in the serum, for example, is a marker of sensitization to allergens known to cause hypersensitivity pneumonitis. Tests of lymphocyte proliferation can indicate sensitization through cellular immune mechanisms (see Chapter 4).

Exposure to allergens in the workplace may result in immunologic sensitization and occupational disease. Table 2-4 lists agents that are associated with causing asthma in occupational settings. The prevalence of sensitization among workers ranges from less than 10 percent for some agents to as much as 80 percent for platinum. Exposures may occur in an industrial or office setting. Although many of the protein allergens have long been recognized, a lengthening list of newly recognized allergenic chemicals is developing (see Chapter 3).

The prevalence of exposure and sensitization to defined, reactive, allergenic chemicals has led to a relatively new but growing category of chemical allergens. Most of these chemicals were initially developed for use in specific industrial processes, but their commercially useful reactive properties often make them react with human proteins as well, a feature that is thought to contribute to their allergenicity. As with other allergens, immunologic sensitization to these chemicals may occur in the absence of symptoms of allergic disease.

DISEASES

Although some people remain asymptomatic despite being immunologically sensitized, many develop one of several diseases including asthma,

TABLE 2-4 Selected Agents Causing Asthma in Selected Occupations

Occupation or Occupational Field	Agent
Animal-related	
Laboratory animal workers, veterinarians	Dander, urine proteins
Food processing	Shellfish, egg proteins, pancreatic enzymes, papain, amylase
Dairy farmers	Storage mites
Poultry farmers	Poultry mites, droppings, feathers
Granary workers	Storage mites, aspergillus, indoor ragweed, grass pollen
Research workers	Locusts
Fish food manufacturing	Midges
Detergent manufacturing	*Bacillus subtilis* enzymes
Silk workers	Silkworm moths and larvae
Plant proteins	
Bakers	Flour
Food processing	Coffee bean dust, meat tenderizer (papain), tea
Farmers	Soybean dust
Shipping workers	Grain dust (molds, insects, grain)
Laxative manufacturing	Ispaghula, psyllium
Sawmill workers, carpenters	Wood dust (western red cedar, oak, mahogany, zebrawood, redwood, Lebanon cedar, African maple, eastern white cedar)
Electric soldering	Colophony (pine resin)
Cotton textile workers	Cotton dust
Nurses	Psyllium, latex
Inorganic chemicals	
Refining	Platinum salts
Plating	Nickel salts
Diamond polishing	Cobalt salts
Stainless steel welding	Chromium salts
Manufacturing	Aluminum fluoride
Beauty shop	Persulfate
Refinery workers	Vanadium
Welding	Stainless steel fumes
Organic chemicals	
Manufacturing	Antibiotics, piperazine, methyldopa, salbutanol, cimetidine
Hospital workers	Disinfectants (sulfathiazole, chloramine, formaldehyde, psyllium, glutaraldehyde)
Anesthesiology	Enflurane
Poultry workers	Aprolium
Fur dyeing	Paraphenylene diamine
Rubber processing	Formaldehyde, ethylenediamine, phthalic anhydride

TABLE 2-4 *Continued*

Occupation or Occupational Field	Agent
Organic chemicals—*cont'd*	
Plastics industry	Toluene diisocyanate, hexamethyl diisocyanate, diphenylmethyl isocyanate, phthalic anhydride, triethylene tetraamines, trimellitic anhydride, hexamethyl tetramine
Automobile painting	Dimethyl ethanolamine toluene diisocyanate
Foundry workers	Furfuryl alcohol resin

SOURCE: NHLBI, 1991

rhinitis, eczema, hypersensitivity pneumonitis, and allergic bronchopulmonary aspergillosis. Figure 2-4 shows the generally accepted range of prevalence for these diseases in the population. Estimates of the magnitude of these diseases are affected by the diagnostic criteria used, as well as different study and data collection methods. Individual diseases are discussed in further detail below and are grouped as follows: first, common diseases clearly related to allergy, then less common diseases clearly related to allergy, and finally, common diseases possibly related to allergy.

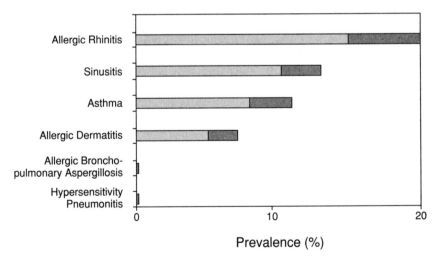

FIGURE 2-4 Estimated range of prevalence of diseases in total U.S. population that are commonly, although not invariably, related to allergy. Darker screen indicates the range in published data. Source: NHLBI, 1991.

Common Diseases Clearly Related to Allergy and IgE Antibody

Three common diseases are clearly related to exposure to indoor allergens: asthma, rhinitis (hay fever), and allergic skin conditions (i.e., eczema and urticaria). Although specific causal genes have not been identified, the genetic predisposition for these diseases is well established. If neither parent has a history of allergy or atopy, a child has only a 0–19 percent chance of having a childhood allergic disease. If one parent has atopy, the risk rises to 31–58 percent; if both parents have atopy, the risk rises still further to 60–100 percent (Zeiger, 1988; Table 2-5). In addition, an earlier age of onset of allergic disease is related to a family history of atopy (Figure 2-5; Smith, 1988). Examples of the effects of these diseases on quality of life are provided in Box 2-1. The economic impact of these conditions was discussed briefly in Chapter 1.

TABLE 2-5 Prediction of Development of Sensitization or Allergic Disease in Childhood Based on Parameters Present During Infancy

Factor	Likelihood of Allergic Disease (%)
Parental atopy history	
Both parents	60–100
One parent	31–58
Neither parent	0–19
Aeroallergen exposure assessed at age 1*	
<10 µg *Der p* I/g of dust	6
>10 µg *Der p* I/g of dust	28
Serum IgE during infancy	
Increased >1 standard deviation above	
geometric mean	75
Normal	5
Illness during infancy	
Recurrent croup	58
Recurrent wheezing	36
Wheezing and blood eosinophilia	75
Wheezing and positive radioallergosorbent test	44
None of these illnesses	13

SOURCE: Zeiger (1988) and Sporik et al. (1990).

*The data on aeroallergen exposure, which were derived from Sporik and colleagues, were obtained from a cohort of newborns, each of whom had one parent with allergic disease. Asthma was defined as active wheezing and bronchial hyperreactivity. The numbers in parentheses show the percentage of all subjects who reported a history of wheezing. The relative risk of active asthma was 4.8 times greater if the child had been exposed to more than 10 µg of *Der p* I/g of dust in infancy. The percentage of sensitization for children exposed to less than 10 µg *Der p* I/g of dust was 31 percent; the percentage for children exposed to more than 10 µg was 56 percent.

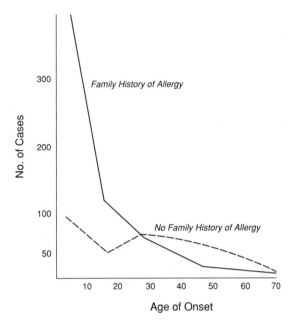

FIGURE 2-5 Relationship between a family history of allergic disease and the time of onset of allergic disease in children. Family history is an extremely important determinant of the risk of allergic disease in children. Source: Smith, 1988.

ALLERGIC ASTHMA

Definition and Diagnosis

Asthma is a lung disease characterized by (1) airway obstruction that is reversible (but not completely so in some patients) either spontaneously or with treatment, (2) airway inflammation, and (3) increased airway responsiveness to a variety of stimuli (NHLBI, 1991). Lung inflammation is often present even when symptoms are absent. People with asthma often experience intermittent wheezing, coughing, chest tightness, and shortness of breath.

The diagnosis of asthma for an individual patient is usually made during a physician office visit, emergency room visit, or hospitalization. The diagnosis is based on one or more of the following: history of episodic symptoms, signs of asthma on physical examination (chiefly wheezing), or pulmonary function test results demonstrating reversible airflow obstruction and bronchial hyperreactivity (see Chapter 5). Bronchial hyperreactivity may be quantified by determining the concentration of methacholine or histamine required to induce a transient drop in lung function. In asthmatics, the concentration of methacholine required to induce a drop in lung function is usually lower than in nonasthmatics.

BOX 2-1   Impact of Allergic Disease on Quality of Life

Allergic diseases can disrupt childhood education and family life and have a negative effect on quality of life.

• Unpredictable absences from work may affect employability.
• Over-the-counter medications commonly purchased to reduce symptoms of allergic rhinitis may make people sleepy; performance impairment has been shown.
• Children with asthma are absent from school far more often than nonasthmatic children.
• Children with asthma may become ill because of school activities that most people accept as routine, such as keeping a hamster in class or workers painting a hallway.
• Allergic dermatitis is characterized by severe paroxysmal itching, which can disrupt work or school activities.
• People with asthma often pay much higher health insurance premium rates (and may be denied insurance).
• People with asthma may be unable to work at jobs that expose them to allergens or irritants.
• The accepted treatment for occupational asthma is removal from exposure. This often requires a job change and may cause job loss.
• Treatment for contact dermatitis is to prevent the exposure. This may require a job change and may cause job loss.

Once a diagnosis of asthma is made, the contribution of allergy to asthma is of great interest. Allergy may be confirmed through either skin testing or measurement of specific IgE antibodies in blood samples, or both. The causal relationship between allergy and asthma is established by the individual's history (see Chapter 5). For example, asthma that occurs in the spring may be due to allergens from tree pollen. People with perennial asthma are often unaware of the allergens that trigger their disease. Current recommendations include environmental modification as an early step in asthma management (NHLBI, 1991).

The diagnosis of asthma in population-based studies has been made in a variety of ways. Diagnostic criteria may be influenced by the local language, customs, and definitions and by physician habits, questionnaires, and study objectives (Kryzanowski et al., 1990). Examples of the epidemiologic definition of asthma include a history of episodic or persistent wheezing, a physician diagnosis of asthma, evidence of reversible airflow obstruction on pulmonary function tests, evidence of bronchial hyperreactivity, or a combination of the above. Airway inflammation, a cardinal feature of asthma, cannot practically be determined in population studies. Differences in diagnostic criteria may explain variable rates of disease. Another source of variation is diagnostic overlap with conditions such as bronchitis (Gregg, 1983; Lebowitz et al., 1990) which can occur in children or adults. Forty to 50 percent of children with asthma also have chronic bronchitis. In adults, 50 to 80 percent of smokers have chronic bronchitis, and only a minority of these have asthma (Snider, 1988).

Research indicates that of those persons with asthma, 90 percent of children, 70 percent of young adults, and 50 percent of older adults also have allergy (Lehrer et al., 1986; Peat et al., 1987; Platts-Mills et al., 1992). These data strongly suggest a role for allergy in the pathogenesis of asthma, particularly in childhood.

Attributable Fraction

Public health officials are often asked to estimate the fraction of asthma in the population that could be attributed to allergy or to exposure to a specific allergen. Several conceptual issues must be understood before making these calculations. Last (1986) states that attributable fractions in the population cannot be estimated from prevalence data but must be calculated from incidence data obtained in cohort studies. This is particularly true for conditions such as cancer where the case-fatality rate is high, and the prevalence in a population grossly underestimates the cumulative incidence rate. It is also true for brief, self-limited illnesses such as influenza where the point prevalence would underestimate cumulative incidence. By contrast, asthma is a chronic disease with a low case-fatality rate, and therefore the prevalence and the cumulative incidence rates are similar. The prevalence of asthma is a reasonable measure of the burden of the disease in a society. Estimates of the fraction of asthma attributable to allergy or to a specific allergen may, therefore, utilize cumulative incidence or prevalence data.

Last (1986) also states that attributable fractions should not be obtained from the distribution of risk ratios obtained in case control studies. Approaches have been developed that use data from case control studies (Schlesselman, 1982), but these should be verified with population-based data. The essential limitation in using case control data to estimate population attributable risk is that it is difficult to know whether the control group is representative of the general population. This concern lessens when the outcome of interest is similar in multiple control groups.

Another problem with calculating attributable fractions is the issue of competing risks. Individuals may be exposed simultaneously to several risk factors, such as house dust mite, cockroach, and cat allergens. The assumption is generally made that exposure to competing risks is uniform in the population, and that the calculations listed below are fairly robust to competing risks. It is possible that the attributable fractions for several risks will sum to more than 1 (i.e., more than 100 percent).

Finally, the risk estimates should not be generalized to the entire population when they are calculated based on studies of a segment of the population, such as children. The reason for this is that measures of risk may systematically change over time. As noted above, the prevalence of skin test reactivity is lower in school children than in young adults.

Studies that are used to calculate attributable fractions should meet several criteria. For example, the exposed and unexposed populations should be alike with respect to demographic variables that could affect disease rates. The study instruments that are used should be the same, and preferably the exposed and unexposed populations should have been studied by the same team.

Finally, calculation of attributable fractions should be specified as referring to the entire population or to only the exposed population.

According to Last (1986) the population attributable risk is calculated as:

$$AF_p = \frac{I_p - I_u}{I_p}$$

where AF_p, the attributable fraction (population), also called the population attributable risk, is the proportion of the disease that can be attributed to exposure to the risk factor; I_p is the incidence in the population; and I_u is the incidence in the unexposed population.

The attributable fraction in the exposed population is calculated as:

$$AF_e = \frac{I_e - I_u}{I_e}$$

where AF_e is the attributable fraction (exposed), i.e., the proportion of cases among the exposed population that can be attributed to exposure to the risk factor of interest; I_e is the incidence rate among the exposed group; and I_u is the incidence rate among the unexposed group.

Table 2-6 presents some estimates of attributable fraction using data from several published studies.

Prevalence

Recent surveys have found prevalence rates of asthma of 8 to 12 percent in the U.S. population—an estimated 20–30 million people (Burrows et al., 1989; Mak et al., 1982; Pattemore et al., 1990; Weiss et al., 1992a). As yet unpublished data suggest that 10 percent of children had been prescribed an inhaler for the treatment of asthma (Platts-Mills, personal communication; Sporik et al., in press). The National Asthma Education Program has estimated that 10 million people in the United States currently have asthma (NHLBI, 1991). Figure 2-6 shows how the definition of asthma can affect estimates of prevalence in epidemiological studies. Additional work is needed to clarify the relationship between asthma and the various definitions of allergy phenotype.

Incidence

About half of the cases of childhood asthma develop before age 5, and most of the rest develop before adolescence. Incident cases are uncommon between ages 20 and 40, with a gradual increase in incidence beginning in the fifth decade of life. Overall, cases of asthma have increased in the past few decades. During the 1920s and 1930s, incident cases were uncommon; a gradual increase began in the 1940s, however, and has continued up to the present. The overall annual incidence of asthma in a Tucson study was 0.46 percent (Lebowitz, 1989).

Impact

Asthma is generally treated in outpatient settings. In 1985, of the 640 million total ambulatory care visits estimated by the National Ambulatory Medical Care Survey (NHLBI, 1991), 6.5 million (1 percent) had asthma as a first-listed diagnosis. From 1970 to 1987, hospital discharge rates for asthma increased nearly threefold. African Americans were more than twice as likely as Caucasians to be hospitalized.

Most recently K. B. Weiss and colleagues (1992b) estimated that 1.8 million people required emergency room services for asthma in 1985, and 48 percent of the visits involved children under age 18. There were an estimated 1.5 million hospital outpatient visits to treat asthma. K. B. Weiss and colleagues (1992b) also estimated the cost of illness related to asthma in 1990 to be $6.2 billion.

Another measure of impact is absenteeism. A WHO/EURO report (1990) reported that a substantial portion of absenteeism from work or school is associated with infectious and allergic episodes caused by exposure to indoor air. Similarly, a 1979 report (NIAID, 1979) stated that asthma was among the leading causes of physician visits, hospitalizations, and workdays lost.

Trends Over Time

Figures 2-7, 2-8, and 2-9 show recent trends over time in asthma prevalence, mortality, and hospitalization rates, respectively. The prevalence rate of asthma in the first half of the century was about 1–2 percent in Caucasian populations in industrialized countries (Gregg, 1989). The first estimate of the prevalence rate of asthma in the United States—in 1928—was less than 1 percent in both children and adults; a 1930 survey reported a prevalence

TABLE 2-6 Some Estimates of Attributable Fractions

Study	Population	Definition of Exposure	I_e	I_u	Attributable Fraction
		The attribution of skin test reactivity to allergen exposure			
House dust mites					
Murray et al., 1985	Children visiting allergy clinic (Canada)	u = residence in dry areas (N = 60) e = residence in humid areas (N = 714)	0.4	0.02	0.95
D. Charpin et al., 1991	Schoolchildren (France)	u = residence in low mite region (N = 240) e = residence in high mite region (N = 693)	0.167	0.041	0.75
D. Charpin et al., 1988a	Adults (France)	u = residence in low mite region (N = 1,055) e = high mite region (N = 4,008)	0.275	0.102	0.63
Sporik et al., 1990	Children of an atopic parent (England)	u = exposure to <10 µg/g of mite (N = 17) e = exposure to >10 µg/g of mite (N = 43)	0.47	0.31	0.34
Grass pollen					
D. Charpin et al., 1991	Schoolchildren (France)	u = residence in low grass region (N = 693) e = residence in high grass region (N = 240)	0.217	0.085	0.61

The attribution of asthma to allergy, defined as any skin test reactivity

Freidhoff et al., 1981	Employees (United States)	e = skin test positive to one or more antigens (N = 28) u = skin test negative (N = 87)	0.24	0.06	0.75
Sporik et al., 1990	Children of an atopic parent (England)	e = skin test positive to one or more antigens u = skin test negative	0.94	0.03	0.97
Sears et al., 1989	Children (New Zealand)	e = skin test positive to one or more antigens (N = 320) u = skin test negative (N = 394)	0.56	0.34	0.39

The attribution of asthma to specific allergen exposure

House dust mites

D. Charpin et al., 1988a	French adults (France)	u = residence in high mite region (N = 4,008) e = residence in low mite region (N = 1,055)	0.041	0.024	0.41
Sporik et al., 1990	Children of an atopic parent (England)	u = exposure to <10 µg/g of mite (N = 17) at age 1 e = exposure to >10 µg/g of mite (N = 43)	0.29	0.06	0.79

NOTE: I_e, incidence in exposed population; I_u, incidence in unexposed population; e, exposed population; and u, unexposed population. The method of data presentation precluded derivation of these estimates for a number of other major population studies.

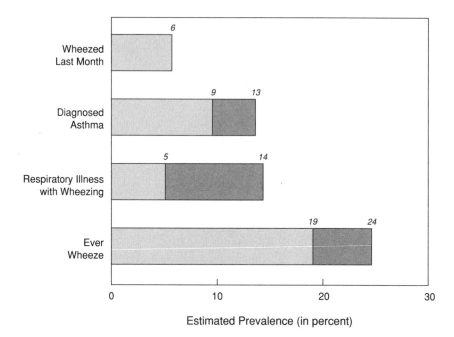

FIGURE 2-6   Effect of case definition on estimates of the prevalence of asthma. Shaded area indicates the range of data.   Sources: Barbee et al., 1985; Britton et al., 1986; Peat et al., 1987.

rate of about 0.5 percent (Barbee et al., 1985).   Over the past 20 years, much higher current and cumulative prevalence rates have been reported from surveys of both children and adults; Figure 2-7 shows the rise in rates for 1980–1987 alone.   Mortality epidemics occurred in Britain and Australasia around 1960 and in Britain and New Zealand in the 1970s.   U.S. surveys in the 1970s showed prevalence rates of between 4 and 6.7 percent (Barbee et al., 1985; NCHS, 1986).   In 1983 Sharma and Balchum stated that 9 million Americans had asthma, 3 million of whom were children.

Recent data also suggest a possible increase in the severity of asthma, using mortality and hospitalization rates as indices of severity (Figures 2-8 and 2-9, respectively; Barbee et al., 1985; Gergen and Weiss, 1992; Gregg, 1983, 1989).   Hospital admissions appear to have increased remarkably in Great Britain and the United States, paralleling the increase in prevalence. The hospital discharge rate with asthma as the first-listed diagnosis rose 43 percent among children less than age 15 from 19.8 to 28.4 discharges per 10,000 population (NHLBI, 1991).   A 5 percent increase in total hospitalizations for asthma occurred between 1987 and 1990 (NCHS, 1992).   These data could reflect changes in the course of the disease or changes in medical practice, including shifting diagnostic criteria or methods (Gregg, 1983, 1989; Newacheck et al., 1986). It should also be noted, however, that except for an increase in mild asthma among older Caucasian children, a significant increase in prevalence rates has not been seen in the National Health Interview Survey (Weitzman et al., 1992).

Asthma death rates have increased substantially in this country over the past 10 years. In 1988, 4,580 people died from asthma in the United States (NHLBI, 1991). During 1979–1984, the death rate for asthma rose from 1.2 per 100,000 people (in 1979) to 1.5 per 100,000 (in 1983 and 1984). Asthma was reported as the underlying cause of 1,674 deaths in the United States in 1977; this statistic increased to 3,564 in 1984 (Sly, 1988), a trend that runs counter to the decreasing death rates seen between 1950 and 1978 (Sly, 1988). Although the first observed rise was concomitant with a change in coding procedures by the National Center for Health Statistics, the rising trend has persisted during a period of uniformity in coding. The increase in death rates has been seen across ages and races but is particularly notable among African Americans (NHLBI, 1991; Sly, 1988). Increases in asthma mortality have been variously attributed to inadequate medical management (especially among minority patients of low socioeconomic status), underutilization of available health care, neglect of personal health, changes in prescribing practices, inappropriate use of medications, and altered environmental fac-

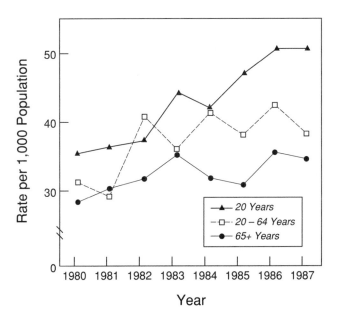

FIGURE 2-7 Trends in Asthma Prevalence. Source: NHLBI, 1991.

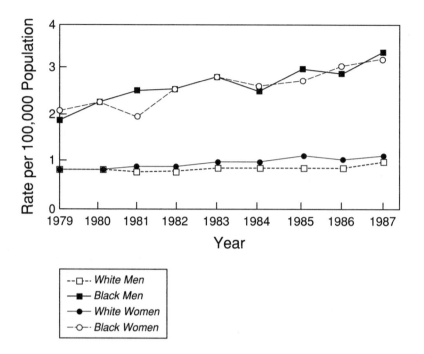

FIGURE 2-8 Trends in asthma mortality, U.S. age-adjusted death rates, 1979–1987. Source: NHLBI, 1991.

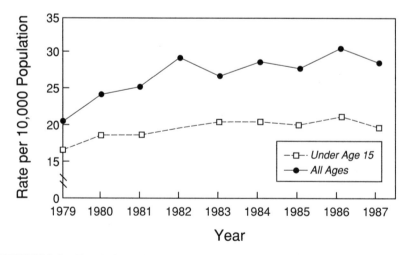

FIGURE 2-9 Trends in asthma hospitalization rates. Source: NHLBI, 1991.

tors, including exposure to new allergens. The New Zealand experience (Sears et al., 1989) indicates that several of these factors are likely to contribute to increased mortality and that broad intervention can reduce these rates (Burrows and Lebowitz, 1992; Lofdahl and Svedmyr, 1991).

Risk factors

Genetic Factors Genetic factors are important in asthma and atopy (Gregg, 1983; Horwood et al., 1985; Lebowitz et al., 1984; Sibbald, 1980; Sibbald et al., 1980). The highest prevalence of asthma occurs among children whose parents have evidence of allergic disease (see Table 2-5; Barbee et al., 1985; Luoma, 1984; Sporik et al., 1990). In addition, the symptom of "persistent wheeze" has been shown to have a familial component. Theoretically, familial aggregation of cases of asthma could be explained either by genetic factors or by common exposure to increased environmental risks. The low prevalence of asthma (as well as low rates of smoking-related lung disease) among Native Americans has been attributed to unspecified genetic factors.

Bronchial Hyperreactivity Bronchial hyperreactivity is both a feature of asthma and a risk factor for its development. The risk of bronchial hyperreactivity increases with increasing skin test reactivity (Burrows and Lebowitz, 1992; Lofdahl and Svedmyr, 1991).

Skin Test Reactivity The presence of skin test reactivity is clearly a strong risk factor for asthma. This relationship has been shown using a variety of definitions of asthma and different study designs. New asthma that develops before age 40 is likely to be associated with allergen skin test reactivity, high total serum IgE, a family history of atopy, and prior symptomatology. Total IgE is still strongly related to asthma prevalence and incidence above age 60 (Burrows et al., 1991). Overall, half of existing cases of asthma have been attributed to allergenic factors.

Recently, the risk of asthma has been related to skin test reactivity to specific allergens. Gergen and Turkeltaub (1992), using data from the second National Health and Nutrition Examination Survey (NHANES II), demonstrated an independent association of asthma with reactivity to an extract of crude house dust (odds ratio, 2.9) and to *alternaria* (odds ratio, 2.3). In a birth cohort of New Zealand children that was monitored up to age 13, sensitivities to house dust mite (odds ratio, 6.7) and to cat dander (odds ratio, 4.2) were highly significant independent risk factors associated with the development of asthma (Sears et al., 1989). Grass sensitivity, although common, was not an independent risk factor (odds ratio, 1.33). *Aspergillus fumigatus* was uncommon (skin test positivity in only 2.4 percent of the sample) but was significantly correlated with asthma (odds ratio, 13.8).

Allergen Exposure   As outlined above, the magnitude of allergen exposure appears to be related to the risk of sensitization. Allergen exposure is also related to the risk of developing asthma and the age at which asthma develops. Sporik and colleagues (1990) demonstrated that exposure to more than 10 μg of dust mite allergen (*Der p* I) per gram of dust was associated with a 4.8 relative risk of asthma by age 11.[1] These investigators also showed that the age at which the first episode of wheezing occurred was inversely related to the level of dust mite exposure assessed when the subject was 1 year old.

Allergen exposure has also been related to disease severity. Voorhorst and coworkers (1967) showed that allergic and atopic asthmatics with sensitivity to house dust mite had clinic visits that correlated well with mite growth curves in their homes, a finding that has been confirmed by others (Andersen and Korsgaard, 1986; Tovey et al., 1981a). Clark and colleagues (1976) found a reduction in peak flows[2] of children with asthma that they associated with house dust when housecleaning increased particle concentrations in the air. O'Hallaren et al., (1991) provided evidence that exposure to the aeroallergen *Alternaria* is a risk factor for sudden respiratory arrest in asthmatics. The O'Hallaren team found that *alternaria* skin test reactivity was associated with a 200-fold risk of respiratory arrest, the timing of which corresponded uniformly with the *alternaria* season in that region.

It has been shown that prolonged allergen avoidance reduces the severity of asthma as indicated by tests of bronchial hyperreactivity (Platts-Mills et al., 1982). For example, asthmatic children who stayed in a sanitorium in Davos, Switzerland, for 1 year showed progressive reductions in bronchial hyperreactivity. It has also been demonstrated that patients with seasonal asthma will lose their bronchial reactivity to histamine after the pollen season.

Hospital emergency room studies have demonstrated an increased risk of allergen sensitization among asthmatics compared with patients being treated for other conditions. The content of allergen in reservoir dust samples from the homes of asthmatics compared with patients with other conditions is also reported to be larger. These studies are covered in detail in Chapter 3.

Geographic location   Rates of asthma in other populations around the world appear to be highly variable (Barbee et al., 1985; Gregg, 1983). Charpin and coworkers (1988b) found no urban–rural difference in asthma

---

[1] That is, a person exposed to this level of dust mite allergen is 4.8 times more likely to develop asthma than a person who has not been so exposed.

[2] Peak flow is a measure of lung function. See the Glossary and Chapter 5 for definition and details.

prevalence rates. Sly found no difference in urban versus rural asthma mortality rates and no variation by geographic region (Sly, 1988). K. B. Weiss and colleagues (1992a), however, found that inner-city asthma was more prevalent than asthma in suburban areas. Moreover, in a comparison of asthmatics being treated at two Delaware hospitals, cat allergy and cat allergen exposure were more common among patients who were treated in a suburban Wilmington hospital, and cockroach allergy and allergen exposure were more common among patients being treated in an urban Wilmington hospital (Gelber et al., in press). This suggests that similarities in urban and rural asthma prevalence rates may mask significant differences in relevant allergen exposures and sensitivities.

Demographics

Age Asthma prevalence rates are highest in earliest childhood, declining to a low at around age 20 and then slowly increasing with age (Barbee et al., 1985; Mak et al., 1982). Between 5 and 14 percent of children will have a respiratory illness with wheezing at some time (Barbee et al., 1985). All wheezing in children is not asthma, however; an infectious disease, bronchiolitis, is also associated with wheezing.

Hospitalization rates for asthma vary with age. NCHS data (1992) indicate that from birth to age 14, 169,000 hospitalizations occurred due to asthma (30.8 hospitalizations per 10,000 population); from age 15 to 44, 86,000 hospitalizations occurred (19.1 per 10,000 population); from age 45 to 64, 86,000 occurred (18.2 per 10,000); and for age 65 and older, 102,000 occurred (32.4 per 10,000 population).

Gender Most studies report a greater prevalence of asthma among boys than among girls, usually by a ratio of 1.5–2 to 1. However, studies in Sweden, Finland, and Arizona have not demonstrated this predominance in males. The increased prevalence of asthma among boys does not appear to be related to genetic or familial factors (Sibbald et al., 1980), but it has been associated with males' greater predilection for infections, especially wheezing lower respiratory infections (Gregg, 1983). S. Weiss and colleagues (1992) report apparent sex differences in the relationship between asthma and lung function development, with males more likely to have asthma but with females experiencing a greater deficit in pulmonary function.

In adults over age 40, new cases of asthma are more likely to occur among women. If the asthma is associated with airflow obstruction (at the time of diagnosis or later), incidence is greater among adult males than among adult females (0.44 percent per year versus 0.39 percent per year; Lebowitz, 1989). Increased rates later in life reflect overlap with "wheezy bronchitis," which is most commonly due to smoking.

Race and Ethnic Group   Several studies have reported racial differences in the prevalence and severity of asthma in the United States, but such results are inconsistent. Schwartz and colleagues (1990) reported an asthma prevalence rate of 7.2 percent among African American children compared with 3 percent among Caucasian children. For adolescents, prevalence is higher among Caucasians (5–6 percent) than among African Americans (3–4 percent); rates for Native Americans are close to zero, and rates among Asians are unknown (Barbee et al., 1985). A study of adults by Di Pede and coworkers (1991) found a prevalence of diagnosed asthma of 4.9 percent among Caucasians (non-Mexican Americans) and a prevalence of 0.9 percent among Mexican Americans. In the same study, Caucasians reported more respiratory symptoms (even within smoking groups) than Mexican Americans. The apparent difference in Caucasian/Mexican American rates may reflect cultural differences in reporting, since the lung function of the two groups is similar for comparable ages, heights, and genders.

Asthma mortality rates are significantly higher among African Americans than among Caucasians (NHLBI, 1991; Sly, 1988). Overall, the death rate from asthma rose from 1.2 per 100,000 population in 1979 to 1.5 in 1983 and 1984. Overall mortality rates among African Americans are two- to threefold higher than among Caucasians and fivefold higher among children. Between 1979 and 1984, the mortality rate rose from 1.8 to 2.5 per 100,000 population among African Americans and from 1.1 to 1.4 per 100,000 among Caucasians.

Socioeconomic Status   Socioeconomic status seems to contribute significantly to asthma prevalence rates and to indices of disease severity. Studies have shown that asthma prevalence rates among children are inversely related to socioeconomic status and residential mobility and are directly related to crowding (Lebowitz, 1977, 1989). Schwartz and colleagues (1990) found that both residence in central cities and low income significantly contributed to asthma prevalence rates. Poverty has also been associated with increased hospitalizations for asthma (NHLBI, 1991). A study of inner-city children in the United States demonstrated a cumulative prevalence rate of asthma of 10.6 percent (Mak et al., 1982). (Inner-city asthma death rates in Chicago were two times greater than those for the United States; in New York City they were three times greater than those for the United States [Evans, 1992; K. B. Weiss et al., 1992a].) The criterion for diagnosis in that study was a positive response to the following question: Have you ever had a condition that causes difficulty in breathing, with wheezing noises in the chest?

Air Pollution   Recent human challenge studies showed a twofold increased sensitivity to allergen in allergic asthmatics following a 1-hour exposure to ozone under conditions typical of summer smog (0.12 parts per million [ppm], 1 hour, at rest; Molfino et al., 1991). However, ozone did not alter the acute response to nasal challenge with allergen when higher ozone exposure levels were used (0.5 ppm ozone, 4 hours, at rest; Bascom et al., 1990). Epidemiological studies point to a contribution by outdoor air pollution to asthma exacerbations. Bates and Sitzo (1987) related increased hospitalizations for asthma to increases in air pollution in the Toronto region. Other studies are well summarized in a recent chapter (Bresnitz and Rest, 1988). Prevalence rates of bronchial responsiveness were associated with specific indoor pollutant exposures only among groups of people of lower socioeconomic status (Quackenboss et al., 1989b).

In summary, asthma is a prevalent disease whose magnitude includes a significant socioeconomic component. Indoor allergens are important at all phases of the disease process, from sensitization to disease onset and severity and prognosis. There is evidence that environmental control strategies can affect the severity of the disease. Nonallergic factors may modify the role of allergens in the disease process.

ALLERGIC RHINITIS

Rhinitis is inflammation of the mucosa (surface cells) of the nose; it causes such symptoms as sneezing, runny nose (rhinorrhea), postnasal drip, and congestion. Allergic rhinitis ("hay fever") is rhinitis caused by IgE-mediated inflammation. The contribution of allergy to this condition is assessed through the individual's medical and environmental history and by skin testing or blood serology to seek specific IgE against suspected allergens.

Other types of rhinitis that can cause similar symptoms include nonallergic rhinitis, vasomotor rhinitis, and infectious rhinitis. These diagnoses are made by the physician on the basis of clinical presentation and are considered when skin testing fails to show evidence for an allergic cause of the rhinitis or when symptom patterns suggest other diagnoses.

Prevalence and Incidence

The prevalence rate of allergic rhinitis is 15–20 percent, although estimated prevalence rates range from 8 to 43 percent (Hagy and Settipane, 1969). Some evidence suggests that the prevalence of allergic rhinitis is rising (Barbee et al., 1987). Incidence rates are highly variable, and are considered untrustworthy at present. Absenteeism and other impacts have not been adequately assessed.

Risk Factors

Allergy A high degree of skin test reactivity to common allergens correlates well with the rate of allergic rhinitis in population studies (Burrows et al., 1976). For example, in the study by Burrows et al. (1976), the fraction of the population with the lowest rate of skin test reactivity had a prevalence of allergic rhinitis of 2 percent; those with intermediate reactivity had a prevalence of 53 percent; and those with the highest skin test reactivity had an 89–100 percent prevalence of allergic rhinitis. In a study conducted in Tucson, allergic rhinitis was similar whether skin test reactivity was positive for perennial allergens (such as house dust mites) or seasonal allergens. Nationwide surveys—specifically, the National Health and Nutrition Examination Survey (NHANES)—also relate seasonal rhinitis symptoms to positive skin tests for seasonal allergens (Gergen and Turkeltaub, 1991). A similar relationship was not demonstrated with perennial rhinitis and dust allergy; however, the NHANES was limited because it did not use now-available allergen extracts that allow for improved sensitivity and specificity. Allergic rhinitis was independently associated with allergy to rye grass, ragweed, and house dust. Another nationwide survey is in progress using additional allergens.

A study in six U.S. cities (Brunekreef et al., 1989) found significant relationships between rates of hay fever and "mold" or "dampness" in homes (odds ratios of 1.57 and 1.26, respectively), although neither finding could be correlated with fungal counts (Su et al., 1989; 1990).

Age The prevalence of allergic rhinitis increases with age, up to the middle years, and then decreases. This same age-related pattern has been seen in studies with widely varying prevalence rates. For children, the National Health Interview Survey (NHIS) developed a prevalence rate of allergic rhinitis without asthma of 5.3 percent for those under age 18. A community-based study of non-Hispanic Caucasians in Tucson showed a prevalence of allergic rhinitis of 29.4 percent in children ages 3–14 (Lebowitz et al., 1975). Rates in other studies have been reported as between 3 and 19 percent (Arbeiter, 1967; Broder et al., 1962; Freeman and Johnson, 1964; Smith and Knowles, 1965). For adults, the NHIS study elicited a rate of 11.2 percent for ages 18–44, whereas the Tucson study found a prevalence of 42.6 percent in those over age 15. This rate was thought to be due in part to the migration of allergic subjects to Tucson. The NHIS study showed a prevalence of 9 percent for ages 45–64 and 5.2 percent for ages 65 and older (NCHS, 1986).

ECZEMA

Eczema (eczematous dermatitis) is a characteristic inflammatory response of the skin to multiple stimuli. There is usually a primary elicitor of the response (such as an allergen) after which many factors may contribute. The unifying feature of eczematous dermatitis is the occurrence of vesicular eruptions, that is, small blisters within the rash. The initial diagnosis is based on the individual's history and the appearance of the skin; occasionally, skin biopsy is used.

Immunologic sensitization is the primary mechanism for two of the major types of eczema, atopic dermatitis and allergic contact dermatitis. These conditions are discussed below.

Atopic Dermatitis

Atopic dermatitis is associated with IgE allergy and, often, with a family history of atopy. The striking feature of the disease is severe, spasmodic itching; it results in "constant and vigorous rubbing, scratching, and even tearing and pounding which many of the tortured patients carry out by day and particularly by night, for periods of months to years" (Sulzberger, 1971, p. 687). The condition is diagnosed through the individual's medical history and physical examination and by identification of IgE allergy using skin tests for specific antibody in the sera.

Prevalence, Incidence, and Natural History Few estimates of prevalence exist for atopic dermatitis. Prevalence rates of 7.7 and 4.8 percent have been reported for children and adults, respectively (Lebowitz, 1975). The incidence of atopic dermatitis varies dramatically with age. Of all people with atopic dermatitis, 60 percent develop the condition before age 1; 30 percent develop it between ages 1 and 5; 10 percent develop it between ages 6 and 20. Of all people with atopic dermatitis, 58 percent have persistent disease for more than 15 years, but most heal by age 60 (NIAID, 1979).

Risk Factors The prevalence of eczema was not related to skin test reactivity in one population-based study (Burrows et al., 1976), but other studies have shown a relationship (Platts-Mills et al., 1991b). Infant eczema is related to cord blood IgE (Halonen et al., 1992).

Allergic Contact Dermatitis

Allergic contact dermatitis is a common skin condition unrelated to IgE allergy that is associated with the cellular immune response. The agents causing this disease are usually molecules that possess one or several reactive chemicals capable of forming stable bonds with tissue proteins and other tissue elements. The agent must penetrate the skin to induce the disease, and it is recognized that primary irritation of the skin favors sensitization.

The mechanism of this dermatitis is delayed-type hypersensitivity, which means that it is an allergic state that is not due to circulating antibodies but to specifically sensitized T-cells. The disease is diagnosed by patch testing, which consists of applying a small quantity of the suspected agent to the surface of the skin and inspecting the site for inflammation 48–72 hours later.

The treatment of contact dermatitis follows three key principles: (1) continued contact with the specific allergen must be avoided by all means, (2) local irritation should be avoided, and (3) local therapy should be employed. (Steroid creams are generally prescribed, with oral steroids used in the most severe cases.)

Prevalence   Contact dermatitis accounts for 20–30 percent of all patients treated in dermatology clinics. Seasonal factors such as the growth of poison ivy alter the rates of dermatitis cases. Most of the allergens associated with contact dermatitis are not aeroallergens; however, there have been instances in which airborne contact allergens have produced contact dermatitis.

Less Common Diseases Clearly Related to Allergy

Allergic bronchopulmonary aspergillosis, hypersensitivity pneumonitis, and humidifier fever are conditions that appear to be less common in the United States than the diseases discussed previously in this chapter. Nevertheless, they merit attention because of their potential severity.

Allergic Bronchopulmonary Aspergillosis

Allergic bronchopulmonary aspergillosis (ABPA) is an intriguing condition characterized by the development of a specific immune response to the *Aspergillus* species of fungi that colonize the central airways. A related disease is allergic *Aspergillus sinusitis*. The diagnostic features of the disease include (1) asthma, (2) a history of infiltrates found by chest radiograph, (3) immediate skin test reactivity to *Aspergillus*, (4) elevated total serum IgE, (5) precipitating antibodies to *Aspergillus fumigatus*, (6) peripheral blood eosinophilia (expected at the time of radiographic infiltrates), (7) elevated serum IgE and IgG to *A. fumigatus*, and (8) proximal (central) bronchiectasis (Greenberger, 1988).

Prevalence

An estimated 10,000 cases of ABPA were prevalent in the United States in 1977; it is thought to be more common in the United Kingdom (NIAID, 1979). Approximately 25 percent of patients with asthma have IgE antibody to *Aspergillus flavus*, and 10 percent have IgG antibody; yet ABPA occurs in only 4 percent of *Aspergillus* skin test-positive individuals (Schwartz et al., 1978). ABPA also occurs in 10 percent of patients with cystic fibrosis (Greenberger, 1988).

HYPERSENSITIVITY PNEUMONITIS

Hypersensitivity pneumonitis is a specific immunologic lung disease characterized by inflammation of the lung parenchyma. The causative agents are numerous and diverse, and the immune pathogenesis includes formation of specific IgG antibody and formation of lung granulomas. The clinical spectrum is also diverse and ranges from recurrent, acute flu-like illnesses to a gradually increasing breathlessness. Hypersensitivity pneumonitis sometimes occurs in sporadic outbreaks, for example, when a building's ventilation system becomes contaminated. Diagnosis is often difficult because it requires a high index of suspicion, and the appropriate laboratory studies (i.e., precipitins) are sometimes unfamiliar to many physicians (see Chapter 5). Table 2-7 lists various causes of hypersensitivity pneumonitis that have been reported in nonindustrial indoor environments.

Prevalence

No prevalence rates are available for the general population. Clinical assessments of the type recommended by Solomon (1990) have not been utilized to obtain rate estimates or population impacts.

Disease Severity

In 1977, it was estimated that 2,000 hospitalizations occurred in the United States in which hypersensitivity pneumonitis was the primary (50 percent) or secondary (50 percent) diagnosis. Treatment consists of corticosteroid therapy and removal from exposure. Failure to institute these measures can result in disease or disability resulting from irreversible fibrosis and respiratory failure (EPA, 1991b).

Risk Factors

Indoor Environment Exposure to allergenic bioaerosols in residential or commercial heating, ventilation, and air-conditioning (HVAC) systems can cause hypersensitivity pneumonitis. Causative agents include thermophilic actinomycetes, *Aspergillus* species, *Aureobasidium* species, and other proteins.

Occupation/Hobby The risk of hypersensitivity pneumonitis increases markedly with exposure to allergens through hobbies and occupations, ranging from 0.5 to 10 percent of exposed populations. For example, the prevalence of hypersensitivity pneumonitis among pigeon breeders ranges from 6 to 15 percent (NIAID, 1979). Farmer's lung, a hypersensitivity pneumonitis that usually arises from exposure to thermophilic actinomycetes, probably occurs in 3–4 percent of exposed populations, with estimates ranging from 0.5 to 10 percent of farmers (NHLBI, 1982; NIAID, 1979). Occurrences among

TABLE 2-7 Hypersensitivity Pneumonitis in Indoor Environments

Disease	Source of Antigen	Probable Allergen
Familial hypersensitivity pneumonitis	Contaminated wood dust in walls	*Bacillus subtilis*
Humidifier lung	Contaminated humidifiers, dehumidifiers, air conditioners	Thermophilic actinomycetes: *Micropolyspora faeni, T. candidus, T. vulgaris, Penicillium* spp., *Cephalosporium* spp. Amebae: *Naegleria gruberi* and *Acanthamoeba castellani*
Cephalosporium hypersensitivity pneumonitis	Contaminated basement (sewage)	*Cephalosporium* spp.
Pigeon breeder's disease	Pigeon droppings	Altered pigeon serum
Laboratory worker's hypersensitivity pneumonitis	Rat fur	Male rat urine
Summer type hypersensitivity pneumonitis	House dust	*Trichosporon cutaneum*

SOURCES: Fink, 1988; Lopez and Salvaggio, 1988; Rose, 1992.

farmers have decreased markedly with changes in methods of harvesting and baling hay.

Other occupations also present opportunities for development of hypersensitivity pneumonitis. Office workers are estimated to develop hypersensitivity pneumonitis at rates of from 1.2 to 4 percent (EPA, 1991b). Banaszak and colleagues (1970) reported that 15 percent of workers in one office displayed pulmonary disease from thermophilic actinomycetes; their exposure had occurred through a contaminated air-conditioning system. Sauna takers disease (caused by *Aureobasidium pullulans*) occurs infrequently. However, lifeguards at an indoor swimming pool reportedly experienced an extremely high rate of attack of a hypersensitivity pneumonitis-like condition; the causative agent of the disease remains unidentified (Rose and King, 1992).

Climate/Season   In Japan, 74 percent of the cases of hypersensitivity pneumonitis reported by hospitals (a total of 835 cases) were associated with predominantly hot, humid climates and occurred in the summer. These cases were attributed to *Trichosporon cutaneum* on the basis of detection of antibody in the blood.

Immunologic Sensitization   In some studies, up to 20 percent of exposed populations will have IgG against the allergen in their serum. Because more than 90 percent of affected individuals will have a specific antibody, these individuals are thought to constitute a subset of the exposed population who are at increased risk of developing the disease. These general figures, however, may vary markedly across individual circumstances.

HUMIDIFIER FEVER

Humidifier fever is an illness with influenza-like symptoms that develops shortly after exposure to aerosols from microbiologically contaminated humidifiers. Recovery can occur within days, even with continuing exposure. It often recurs on the first day of reexposure following a period of no exposure.

Inhalation challenge with extracts from contaminated water can produce symptoms of humidifier fever, but the causative agent is still unknown. Experimental exposure of symptomatic workers to humidifier allergens can induce headache, rhinitis, and lethargy as well as asthma and alveolitis; similar exposure does not cause symptoms in previously unexposed individuals. These findings have led to a presumption that a specific immunologic mechanism is operative. A World Health Organization report, for example, listed endotoxin as the leading suspect (WHO, 1990). Finnegan and others (1987) identified allergens of amebae in contaminated humidifiers but failed to correlate their presence with humidifier fever or with work-related symptoms in a group of 25 workers with humidifier fever and 90 workmate controls.

Prevalence

In general, published estimates are only educated guesses. The U.S. Environmental Protection Agency (EPA, 1991b) states that epidemics in the workplace are rare but that when they do occur, "attack rates are high (30–75 percent)." Based on symptoms, Finnegan and coworkers (1984) estimated a rate of occurrence of 2–3 percent in Great Britain in office buildings with mechanical ventilation. Prevalence rates in the home, however, have not been evaluated (EPA, 1991b).

Common Diseases Possibly Related to Allergy

There are several syndromes in which a role for indoor allergens is possible or suspected but largely undefined. These include chronic sinusitis and bronchitis, sick building syndrome and other nonspecific syndromes, and acute respiratory illnesses.

SINUSITIS

Sinusitis is defined as inflammation of the sinuses, which are four pairs of hollow structures that surround the nasal cavity. Chronic sinusitis is defined by physicians as persistent inflammation of the mucosa of the sinuses lasting for more than 3 months (Slavin, 1989). Symptoms may include facial pressure, nasal stuffiness, hyposmia, prurulent nasal secretions, sore throat, fetid breath, and malaise. Coughing and wheezing occasionally occur.

Prevalence and Severity

Sinusitis is an important cause of morbidity (Slavin, 1989). In 1981, statistics from the U.S. Department of Health and Human Services indicated that 31 million people had chronic sinusitis; data from the National Center for Health Statistics (NCHS, 1986) are similar (13.9 percent, more than 30 million people). Sinusitis is thus more prevalent than arthritis (27 million) and hypertension (25.5 million). In Great Britain, the Department of Health and Social Security estimates that a half million working days are lost in that country each year due to sinusitis (Slavin, 1989).

In 1975, the rate of physician-confirmed sinusitis in Tucson adults was reported to be 29.4 percent (Lebowitz et al., 1975). The difference between this rate and the figures listed above does not reflect decreasing prevalence over time but rather variable case definitions.

Risk Factors

Age The prevalence of chronic sinusitis varies with age. According to estimates from the National Center for Health Statistics, it occurs in 6 percent of people under age 18, 16.4 percent of those between ages 18 and 44, 18.5 percent of those age 45–64, and 15.4 percent of those over age 65 (NCHS, 1986). A similar age pattern was reported in Tucson (Lebowitz et al., 1975).

Allergy Allergic rhinitis is considered a common risk factor for both acute and chronic sinusitis (Slavin, 1989), but the proportion of chronic sinusitis for which it is the dominant factor is unknown. The NHANES data, analyzed by Gergen and Turkeltaub (1992), did not show a relationship between reported sinusitis and skin test reactivity.

Nonallergic factors that predispose an individual to sinusitis are upper respiratory infection, overuse of topical decongestants, hypertrophied adenoids, deviated nasal septum, nasal polyps, nasal tumors, foreign bodies, cigarette smoke, swimming and diving, barotrauma, and dental extractions. Immunodeficiency syndromes, cystic fibrosis, bronchiectasis, and the immotile cilia syndrome can also be associated with chronic sinusitis.

CHRONIC BRONCHITIS

Chronic bronchitis is commonly defined by clinicians as a chronic productive cough, without a medically discernible cause, that is present more than half the time for 2 years (Snider, 1988). Epidemiologists define chronic bronchitis more precisely as a cough productive of phlegm for a total of 3 months per year for at least 2 years in a patient in whom other causes of chronic cough have been excluded (e.g., infection with *Mycobacterium tuberculosis*, carcinoma of the lung, chronic congestive heart failure; Snider, 1988). The major risk factor for chronic bronchitis is cigarette smoking. The prevalence of chronic bronchitis among nonsmokers rises from age 15 to 60, increasing from 7 to 18 percent; prevalence among smokers rises from 40 to 82 percent (Snider, 1988).

Figure 2-10 illustrates the overlap between asthma, chronic bronchitis, and emphysema (Snider, 1988). Asthma, by definition, is characterized by reversible airflow obstruction, although a few patients may develop unremitting airflow obstruction. Patients with chronic bronchitis may have partially reversible airflow obstruction (Snider, 1988). The term *chronic obstructive pulmonary disease* (COPD) is often used by doctors when adult patients have evidence of one or more of three diseases: chronic bronchitis, emphysema, and asthma.

The nonproportional Venn diagram in Figure 2-10 shows subsets of patients with chronic bronchitis, emphysema, and asthma in three overlap-

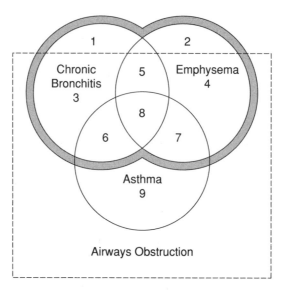

FIGURE 2-10   Relationship between asthma, chronic bronchitis, emphysema, and allergy. Numbers refer to subsets described in text. Source: Snider, 1988.

ping circles. Subsets of patients lying within the rectangle have obstruction of their airways. Patients with asthma, subset 9, are defined as having completely reversible airways obstruction and lie entirely within the rectangle; their diagnosis is unequivocal. Patients in subsets 6 and 7 have reversible airways obstruction with chronic productive cough or emphysema, respectively. Patients in subset 8 have features of all three disorders. It may be difficult to be certain whether patients in subsets 6 and 8 indeed have asthma or whether they have developed bronchial hyperreactivity as a complication of chronic bronchitis or emphysema; the history helps. Patients in subset 3 have chronic productive cough with airways obstruction but no emphysema; it is not known how large this subset is, since data from epidemiologic studies using the computer tomography scan, the most sensitive in vivo imaging technique for diagnosing or excluding emphysema, are not available. It is much easier to identify in the chest radiography patients with emphysema who do not have chronic bronchitis (subset 4). Most patients in subsets 1 and 2 do not have airways obstruction as determined by the FEV_1 (forced expiratory volume exhaled in 1 second) but have clinical or radiographic features of chronic bronchitis or emphysema, respectively. Because COPD, when defined as a process, does not have airways obstruction as a defining characteristic, and because pure asthma is not included in the term COPD, patient subsets 1–8 are included within the area outlined by the shaded band that denotes COPD (Snider, 1988).

There is a recognized association between bronchial hyperresponsiveness, atopy (skin test reactivity), and COPD. Studies have demonstrated that the risk of bronchial hyperresponsiveness is related to skin test reactivity (Cockroft et al., 1984; Lebowitz et al., 1991, Peat et al., 1987; Sears et al., 1989). The index rises with the number of positive skin tests and the magnitude of the skin test reaction to each allergen. Therefore, an individual with many strongly positive skin tests to many allergens is much more likely to show bronchial hyperresponsiveness than an individual with no skin test reactivity. The association between atopy, bronchial hyperresponsiveness, and COPD may be explained by three alternative models (Sparrow et al., 1988). First, cigarette smoke may cause inflammation and mucosal damage, resulting in three unrelated by-products: atopy, bronchial hyperresponsiveness, and COPD. Second, when a person has atopy and therefore bronchial hyperresponsiveness, exposure to cigarette smoke leads to COPD. Third, cigarette smoke, atopy, and bronchial hyperresponsiveness are independent factors that contribute to the development of COPD.

SICK BUILDING SYNDROME

Sick building syndrome (also known as tight building syndrome, closed building syndrome, and new building syndrome) is a term given to nonspecific building-related illness. Figure 2-11 shows the relationship between these and other diseases and conditions that may occur in indoor environments and the general approach to their evaluation and management. Sick building syndrome describes a constellation of symptoms including mucosal irritation, fatigue, headache, and occasionally, lower respiratory symptoms and nausea. Patients or workers report that symptoms increase with the amount of time spent in certain buildings and tend to improve when they leave that building. Symptom prevalence rates associated with indoor environments vary tremendously, from less than 5 percent to as much as 50 percent.

For the majority of cases of sick building syndrome, the cause is unknown. A contribution by allergy has been considered unlikely since atopy or specific sensitivity to indoor allergens often is not found. Nevertheless, people with allergic disease frequently are the individuals who are most affected when an indoor air quality problem is occurring. A definitive conclusion that indoor allergens are not related to sick building syndrome awaits further study, particularly with respect to fungal allergens.

SPECIFIC BUILDING-RELATED ILLNESS

Specific building-related illness is defined as illness caused by identifiable toxic, infectious, or allergenic agents, which can be detected by appro-

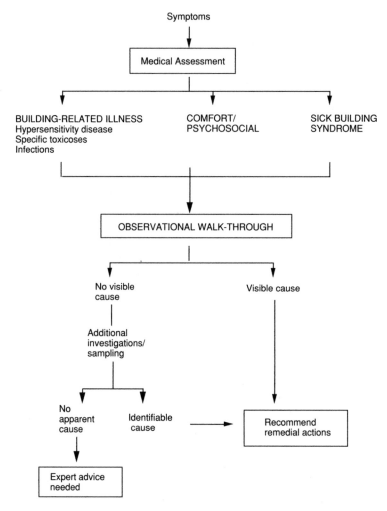

FIGURE 2-11   Approach to the evaluation of symptoms occurring in indoor environments.

priate clinical laboratory tests in patients or by identification of the source in the building (Samet, 1990). Hypersensitivity pneumonitis, humidifier fever (see above), and infection with *Legionella* species are included in this category of illness.   Prevalence rates for specific syndromes are largely unknown and estimates vary tremendously among buildings.

Some researchers believe that there is a set of nonspecific symptoms that are distinguishable, that are different from the other conditions described above, and that are due to indoor allergens (Burge, 1990). Fungus-related illnesses, for example, have been identified or suggested in some instances (Anderson and Korsgaard, 1986; Finnegan et al., 1984; Morey, 1988). Further information can be obtained from a review and an editorial by Kreiss (1988, 1990).

Acute Respiratory Illnesses

Assessments of disease magnitude often depend on a subject's recalling the occurrence of a disease and recognizing its relationship to allergy. For example, questionnaires will ask, "Have you had hay fever?" For an affirmative answer, the subject must recognize that his or her recurrent nasal symptoms are in fact due to an allergic nasal condition. Diary studies, that ask subjects to record symptoms on a daily basis, are another way of assessing the burden of respiratory symptoms in the population. In general, they suggest that respiratory symptoms are quite common, more so than recall studies would indicate. In a diary study of Manhattan (New York) residents, it was determined that the rates of symptoms were much greater when diaries were used than when subjects were asked to recall their health status over the past several years (Lebowitz et al., 1972a,b). The study of 1,707 residents assessed respiratory symptoms for 1,168 person-years of observation. One or more symptoms occurred at least one-quarter of the time among 22 percent of subjects in the study and at least half the time among 6.5 percent. The symptoms that were most often reported included common cold and rhinitis (8 percent of all person-days each), cough (5 percent), headache (2.45 percent), eye irritation (1.8 percent), and chest whistling/wheezing (0.77 percent). The investigator in this study has speculated that reported colds and other infections could have been mislabeled and that an unknown portion of these symptoms could be attributable to allergic diseases.

Data from the NHIS (NCHS, 1986) indicate that in 1985 there were 87.1 acute respiratory conditions per 100 persons. Of these, 46.4 percent were categorized as influenza, 35 percent as common colds, and 11.7 percent as other acute respiratory illnesses. The remainder were labeled as acute bronchitis (3 percent), pneumonia (1 percent), and other (2 percent).

CONCLUSIONS AND RECOMMENDATIONS

As outlined in this chapter, allergic disease constitutes a substantial public health problem in the United States. Data on its exact magnitude and dimensions, however, are incomplete, or lacking, in many cases. Better data regarding the incidence, prevalence, attributable fraction, and cost of allergic diseases are essential to the development of effective programs of prevention and control. Accurate determinations of the magnitude and dimensions of sensitization and disease caused by regional and local indoor allergens would be useful in this regard. The following recommendations for research and data collection will be useful in accurately determining the magnitude and dimensions of sensitization and disease caused by indoor allergens.

Research Agenda Item: Determine prevalence rates of sensitization, allergic diseases, and respiratory morbidity caused by regionally and locally relevant indoor allergens and assess the contributions of different allergens to these conditions.

Socioeconomic status seems to contribute to asthma prevalence rates and to indices of disease severity. Similarly, several studies have reported racial differences in the prevalence and severity of asthma in the United States, but such results are inconsistent. In order that these differentials can be translated into effectively targeted public health interventions, additional research is needed to clarify these relationships.

Research Agenda Item: Conduct studies to clarify the relationships that exist between socioeconomic status, race, and cultural environment and the incidence, prevalence, severity, and mortality associated with allergic disease including asthma.

The annual costs (direct and indirect) associated with asthma have been estimated at more than $6.2 billion—an increase of 39 percent since 1985. The size of these costs and the trend towards even greater costs argue for careful attention not only to the effects of allergy in causing asthma but also to the litany of other conditions that are more commonly thought to be associated with "allergies." Accurate assessments of the costs associated with asthma and other allergic diseases would be useful in the development of health policy initiatives and the implementation of appropriate and cost effective interventions.

Research Agenda Item: Determine the economic impact of asthma and other allergic diseases.

Allergic disease occurs when a genetically predisposed or susceptible individual is exposed to an allergen and becomes immunologically sensitized. In the early stages of sensitization, people who are sensitized have not developed symptoms of disease. The magnitude of this group within the population is of interest, however, because it reflects the proportion of the population at greatest risk of developing allergic disease. Additional exposure to the sensitizing allergen leads to the development of an allergic reaction (disease) that can be mild, moderate, or severe, depending on the amount of exposure.

The relationship between exposure, sensitization, and disease, and the potential for a threshold level of exposure below which the risk of sensitization is reduced, is of critical importance in the prevention and control of allergic disease. Epidemiological data would be useful in determining these relationships and in developing and evaluating public health and medical intervention strategies.

> **Research Agenda Item: Conduct appropriate epidemiological studies of exposure-response relationships of important defined indoor allergens that induce sensitization in humans. Such studies should include a focus on identifying threshold exposures.**

Indoor allergens are associated with a wide variety of particles in a broad size range, only some of which are microscopically identifiable, culturable, or detectable with existing immunoassays. Evaluation of indoor allergens requires both air and source sampling, and several different analytical techniques (including microscopy, culture, and immunoassays) must be used to characterize even the well-known allergens. Because of the complexity of the assessment problem, indoor allergens, with a few exceptions, have not been identified and studied.

> **Research Agenda Item: Encourage and conduct additional research to identify and characterize indoor allergens. The new information should be used to advise patients about avoiding specific allergenic agents.**

3

Agents, Sources, Source Controls, and Diseases

The indoor environment contains many allergens that can be airborne. They derive from various organic and inorganic sources and may be inhaled as particles, vapors, or gases. Chemically, most are proteins, but low-molecular-weight (LMW) reactive chemicals in industrial settings have also caused allergy. Indoor allergens can be derived from the outside, from the structure or furnishings of a building, or from the humans, animals, and plants within it. Similarly, microbiological aerosols can originate in outside air or in sources within the building.

Biological sources of allergens are surprisingly diverse: they range from domestic animals that shed allergen-containing particles to such sources as food substances dropped on the floor, fungi growing on walls or under carpets, plant materials brought into the house, microorganisms within the air-conditioning system, and a variety of arthropods (in particular, the house dust mite) that may grow within the structure of the house or in the furniture (Table 3-1). Homes, apartment buildings, schools, offices, hospitals, stores, and factories each have unique features that affect the types and quantities of allergens that are present. The major allergenic protein molecules—and in some cases, even the allergenic epitopes—have been identified and characterized in the case of house dust mites, cats, dogs, and certain fungi.

DUST MITE, COCKROACH, AND OTHER ARTHROPODA

House dust (called house dust because most studies have focused on houses—but it also occurs in schools, offices, and other buildings) is made

86

up of fibers from carpets and furniture, grit and sand particles, human skin scales, and food debris. This mixture is combined with allergens from domestic animals, insects, and a variety of microscopic arthropods, bacteria, algae, and fungi growing within the house and other buildings. In addition, air coming into these buildings can carry pollen and molds from outside that then become part of the dust. Because the occupants of a house are inevitably exposed to house dust, any source of foreign protein in the dust is a potential cause of sensitization. Skin testing of patients with asthma or rhinitis initially utilized extracts made from dust collected from their own house (i.e., autologous dust; Kern, 1921); now, however, commercial extracts of house dust are widely used for diagnosis and immunotherapy. In 1980 the U.S. Food and Drug Administration (FDA) estimated that at least 10 million injections of "house dust" extract were administered annually in the United States.

Until 1967, house dust allergenicity was attributed to animal dander, insects, and fungi (Spivacke and Grove, 1925; Vannier and Campbell, 1961). In that year, however, Voorhorst and his colleagues in the Netherlands observed large numbers of mites in dust samples and demonstrated that dust

TABLE 3-1 Biological Sources of Allergens in Houses

Acarids	Fungi*
Dust mites	Inside
Dermatophagoides pteronyssinus	Multiple species including
Dermatophagoides farinae	*Penicillium, Aspergillus, Rhizopus,*
Euroglyphus maynei	*Cladosporium* (growing on surfaces or
Blomia tropicalis	wood)
Storage mites	Outside
Spiders	Entry with incoming air, multiple species
Insects	**Pollens**
Cockroaches	Derived from outside
Blattella germanica (German)	
Periplanetta americana (American)	
Blatta orientalis (Oriental)	
Other	**Sundry**
Crickets, flies, beetles,	Horsehair in furniture, kapok
fleas, moths, midges	Food dropped by inhabitants
Domestic Animals	**Rodents**
Cats, dogs, ferrets, skunks, horses,	Wild: mice, rats
rabbits, pigs	Pets: mice, gerbils, guinea pigs

*Fungal spores may contain very little allergen and may require germination to produce significant exposure.

mites of the genus *Dermatophagoides* were a major source of house dust "atopen" (Voorhorst et al., 1967). They also developed techniques for growing mites in culture, which made it possible to produce extracts commercially for skin testing. Most patients with positive skin tests to house dust have specific immunoglobulin E (IgE) antibodies to dust mite allergens (Johansson et al., 1971). Sensitization thus can be detected either by skin tests or measurement of serum IgE antibodies.

Dust mite sensitivity was found to be strongly associated with asthma by J. M. Smith and colleagues (1969) and Miyamoto and coworkers (1968). Indeed, in some countries (e.g., Brazil, Australia, New Zealand, Japan, the Netherlands, Denmark, and England), sensitivity to dust mites appears to be so common among young asthmatics that other sources of indoor allergens are relatively unimportant (see Arruda et al., 1991; Clarke and Aldons, 1979; Sears et al., 1989; and Sporik et al., 1990). In dry climates, however, such as in northern Sweden and central Canada, and in high-altitude areas (e.g., Colorado), mite growth is poor and domestic animals predominate as the major source of indoor allergens. Humidity enhances the growth of mites in carpets, mattresses, and other household items (Korsgaard, 1983a). In some inner-city areas, cockroach debris or rodent urine may be the dominant sources of allergens in house dust (Bernton et al., 1972; Hulett and Dockhorn, 1979; Kang et al., 1979; Twarog et al., 1976). Many different protein sources thus contribute to house dust allergenicity (see Table 3-1).

Heavy exposure to house dust can give rise to sneezing in anyone, and it has been suggested that endotoxins or other substances in dust can be directly toxic. The association between exposure to house dust and diseases such as asthma, chronic rhinitis, and atopic dermatitis, however, has been shown only in individuals who have developed hypersensitivity. The symptoms produced by house dust allergens in sensitized (i.e., allergic) individuals include asthma, perennial rhinitis, and atopic dermatitis. For each disease the symptoms range from severe to very mild; moreover, some individuals, despite their having IgE antibodies, suffer no discernible symptoms. In some cases, the correlation between exposure to a specific indoor allergen and symptoms is obvious; certainly, many individuals who are allergic to cats experience the rapid onset of symptoms on exposure to cat allergens. In contrast, most symptoms related to exposure to house dust are nonspecific and not temporally related to exposure. Thus, in general, it is not possible to distinguish the role of different specific indoor allergen sources solely on the basis of an individual's medical history. Indeed, many patients with asthma are not aware of any other symptoms that would be recognized as allergic. Because their histories are not specific and in many cases exposure to house dust allergens is perennial, understanding the relationship between exposure and disease has required both measurement of exposure and documentation of sensitization.

Dust Mites as a Source of Indoor Allergens

Many different species of dust mites have been found in house dust, but the predominant ones in most parts of the world belong to the family Pyroglyphidae: *Dermatophagoides pteronyssinus, D. farinae,* and *Euroglyphus maynei.* In Florida, Central America, and Brazil (see Hughes, 1976; Van Bronswijk, 1981; Voorhorst et al., 1967; and Wharton, 1976), several species of storage mites and *Blomia tropicalis* are important sources of allergens. It is probably best to reserve the term *dust mites* for pyroglyphid mites and to use the term *domestic mites* to cover any species of mites that are found in houses (Platts-Mills and de Weck, 1989).

Mites are eight-legged and sightless, and they live on skin scales and other debris. They absorb water through a hygroscopic substance extruded from their leg joints and are thus entirely dependent on ambient humidity. In addition, they have a narrow optimal growth temperature range of between 65° and 80° F. As humidity falls, mites will withdraw from surfaces, but even in very dry conditions it may take months for mites in sofas, carpets, or mattresses to die or for allergen levels to fall (Arlian et al., 1982; Platts-Mills et al., 1987).

Mites excrete partially digested food and digestive enzymes as a fecal particle surrounded by a peritrophic membrane (Tovey et al., 1981a). Large quantities of fecal particles are found in mite cultures, and they are a major form of the mite allergen in house dust. The peritrophic membrane probably keeps the particles intact; however, chitin is not waterproof; consequently, allergens elute from fecal particles quite rapidly (Tovey et al., 1981b). Mite fecal pellets are similar to pollen grains in size (10–35 μm in diameter), in the quantity of allergen they carry (i.e., ~0.2 ng), and in their rapid release of proteins.

Dust mites are approximately 0.3 mm in length. Moving mites can be seen by light microscopy, but the great majority of mites in dust are dead and are therefore difficult to identify without separating them from other dust particles (Arlian et al., 1982; Wharton, 1976).

Dust Mite Allergens

The first mite allergen to be purified, *D. pteronyssinus* allergen I (or *Der p* I; Chapman and Platts-Mills, 1980), is a 24,000-MW glycoprotein that has been sequenced and cloned; it has sequence homology with papain and functional enzymatic activity (Chua et al., 1988). In 1984, high-affinity monoclonal antibodies to *Der p* I were reported, opening the door to the development of assay systems that would improve the sensitivity and specificity of measurement (Chapman et al., 1984). A second major allergen (MW 15,000) was first identified in 1985 and has now been fully defined,

TABLE 3-2 Defined Indoor Allergens

Allergen	Molecular Weight (kDa)	Sequence	Function	Monoclonal Antibodies[a]	Assay
House Dust Mite					
Group I					
Der p I	25	cDNA	Cysteine protease	++	ELISA
Der f I	25	cDNA		++	ELISA
Eur m I	25	Nucleotide		+	—
Group II					
Der p II	14	cDNA	Unknown	++	ELISA
Der f II	14	N-terminal		++	ELISA
Group III					
Der p III	29	N-terminal	Serine proteases	−	RIA[b]
Der f III	29	N-terminal		++	RIA[b]
Cat—Felis domesticus					
Fel d I	35	cDNA	Unknown	++	ELISA
Albumin	68	—	—	±	RIA[b]
Dog—Canis familiaris					
Can f I	27	—	Unknown	++	ELISA[b]
Cockroach					
Blattella germanica					
Bla g I	20–25	—	Unknown	+	ELISA[b]
Bla g II	36	N-terminal	Unknown	++	ELISA
Periplanetta americana					
Per a I	20–25	—	Unknown	−	—
Rodent					
Mus musculus	19	Protein sequence	alpha-2U-globulin	+	ELISA[b]
Rattus norvegicus	19	cDNA		+	RIA[b]

NOTE: kDa, kilodaltons; cDNA, complementary DNA; ELISA, enzyme-linked immunosorbent assay; and RIA, radioimmunoassay.

[a]"++" Indicates that more than one epitope has been defined.
[b]Assays in research use.

cloned, and named *Der p* II (Chua et al., 1990; Heymann et al., 1989; Lind, 1985).

There are homologous cross-reacting allergens produced by *D. farinae* (Dandeau et al., 1982; Haida et al., 1985; Heymann et al., 1986; Yuuki et al., 1990); see Table 3-2 for current groupings of allergens. The group II mite allergens show greater than 90 percent sequence homology and are very strongly cross-reacting. They are also relatively stable in relation to heat and pH, but they have no enzymatic activity and their function in the mites is not known. Researchers assume that the group I proteins are digestive enzymes because they are found in glands surrounding the gut and are present in high concentrations in fecal pellets (Tovey, 1982). Currently, simple, sensitive monoclonal antibody-based assays are widely used for quantitating group I allergens (Horn and Lind, 1987; Luczynska et al., 1989). Assays for group II (and also group III) are in development.

COMMERCIALLY AVAILABLE ALLERGEN EXTRACTS

House dust extracts for use either in skin testing or for in vitro assays of IgE antibodies are made from vacuum cleaner bags collected outside the pollen seasons from houses without animals. Other sources for dust (e.g., schools, offices) are not commercially available. The quantity of dust mite allergen in commercial house dust extracts varies from 0.05 to 2.0 µg of *Der p* I/ml.[1] Commercial dust mite extracts can be made from either whole mite culture or from isolated mite bodies. At present, the FDA requires that mite extracts be made from isolated mites. Mites are photophobic; they can be separated by using a light source to "drive" them out of the culture material. Horse dander or human shavings contained in the same media that are used for growing mites may expose recipients of the extract to these proteins. Extracts made from bodies of *D. pteronyssinus* typically contain 30 µg of *Der p* I and 20 µg of *Der p* II.

Insects as a Source of Indoor Allergens

Many insects have been identified as sources of inhalant allergens in case reports or small outbreaks; these include moths, crickets, locusts, beetles, nimitti flies, lake flies, and houseflies (Ito et al., 1986; Kay et al., 1978; Kino et al., 1987; Koshte et al., 1989). Yet the only insect that has been repeatedly recognized as a common source of indoor allergens is the cockroach (Bernton et al., 1972; Kang et al., 1991; Pollart et al., 1989, 1991).

---

[1] Cat allergen (*Fel d* I; see later discussion) in house dust extract varies from less than 0.01 to 2 µg of *Fel d* I/ml; in general, there is no detectable cockroach allergen and little in the way of fungal allergens.

Sources of insect allergens are diverse and may include skin scales from moths and hemoglobin from lake flies or river flies (Mazur et al., 1987). For domestic cockroaches, fecal material and saliva may contribute to the allergen reservoir, and large quantities of allergen can be washed off the outside of the roach.

DOMESTIC COCKROACHES

The German cockroach *Blattella germanica* is very common in crowded cities, in the southern United States, and in tropical countries of the world. As early as 1964, Bernton and his colleagues recognized that many patients with asthma who sought treatment at indigent care clinics had skin tests positive for cockroach (Bernton et al., 1964). Positive skin tests have been reported in several urban clinic populations including Boston, New York, Kansas City, Detroit, Chicago, and Washington, D.C. (Call et al., 1992; Hulett and Dockhorn, 1979; Kang et al., 1979). Subsequently, Kang and her colleagues reported positive bronchial provocation and good responses to immunotherapy with cockroach extract (Kang et al., 1979, 1991). In most suburban clinics, few or no patients have positive skin tests to cockroach extracts.

A case-control study of emergency room patients confirmed the significant association of cockroach sensitivity with asthma (Pollart et al., 1989). An unpublished study on cockroach allergen levels in houses in different parts of a town showed that the correlation between cockroach sensitization and asthma was restricted to that part of the town in which cockroach allergen was present in the houses (Gelber et al., in press). Researchers have identified two cockroach allergens, *Bla g* I (MW ~30 kilodaltons [kDa]) and *Bla g* II (MW 36 kDa; see Table 3-2), and have developed monoclonal antibodies and assays specific for these allergens (Pollart et al., 1991). Details regarding the sources of these allergens, their cross-reactivity with those derived from *Periplaneta americana*, and the nature of the particles that become airborne are not well established (Swanson et al., 1985). Cockroach allergen can be found throughout the house, but the highest levels are generally found in kitchens.

Further work is needed to define the nature of insect allergens, the nature of the particles that become airborne, and their role as indoor allergens.

Measuring Exposure to House Dust Allergens

The major outdoor allergens are components of well-defined particles (i.e., pollen grains or fungal spores) that are disseminated by wind and that can be identified microscopically. In contrast, indoor allergens come from a variety of particles that are not naturally airborne and that cannot be identified microscopically. Thus, evaluation of airborne indoor allergens depends on sensitive immunoassays and requires a method for collecting particles. This can be done either with a filter or with a multistage impactor (Solomon and Matthews, 1988; see also the discussion in Chapter 6).

In 1981, Tovey and colleagues (1981b) showed that fecal particles were a major form in which the allergen *Der p* I becomes airborne and that very little or no (i.e., <1 ng per cubic meter of air) mite airborne allergen was detected in undisturbed rooms (de Blay et al., 1991b; Platts-Mills et al., 1986; Swanson et al., 1989; Yasueda et al., 1989). Furthermore, airborne mite allergen falls rapidly after disturbance. These results support the view that mite allergen is predominantly airborne on particles that are larger than or equal to 10 μm in diameter. The levels found in the air during disturbance depend critically on the form of the disturbance and vary from 5 to 200 ng of *Der p* I/m^3. Assuming that airborne *Der p* I is carried predominantly on fecal particles, it is possible to estimate the number of particles that become airborne and to an estimate of the number of particles that could enter the lung, since the mean allergen content of the particles is known. Chapter 6 discusses methods of assessing exposure and risk.

Thresholds: The Relationship Between Exposure, Sensitization, and Disease

Voorhorst and his colleagues (1967) found that dust from the houses of symptomatic allergic patients generally had more than 500 mites/g of dust. During the 1980s, further data accumulated demonstrating a dose-response relationship between exposure to mite allergens (or mites) and both sensitization and asthma (Bernton et al., 1972; Kang et al., 1991; Pollart et al., 1989, 1991). From these results, it also appeared that there were levels of exposure (or thresholds) below which the risk of sensitization or asthma was much less. This finding notwithstanding, the results suggest that in areas in which all houses contain high levels of mite allergen, sensitivity to mites is a major risk factor, not only for wheezing but for hospitalization of children with asthma.

Fewer data are available on the levels of exposure associated with sensitization or disease for allergens other than dust mite. However, there are data about the levels of cat allergen present in house dust. Dust from all houses with a cat contains at least 8 μg of *Fel d* I/g (the levels range as high as 1.5 mg of *Fel d* I/g). In houses without a cat, levels vary from less than 0.2 μg/g to 80 μg/g; it is thought that this allergen is transported into the houses on the clothes of inhabitants. Levels of cat allergen of less than 1 μg of *Fel d* I/g of dust appear not to give rise to sensitization or disease.

For cockroach allergen the rarity of sensitization among suburban patients suggests that the levels found in suburban houses (i.e., less than 1 unit of *Bla g* II/g of dust) are insufficient to sensitize individuals. By contrast, the levels found in inner-city houses (i.e., more than 10 units of *Bla g* II/g of dust) are clearly sufficient to induce sensitization and appear to be associated with asthma (Gelber et al., 1992; Pollart et al., 1991). There is a general misconception that levels of allergen will be (or should be) higher in the houses of patients with allergic disease than in the homes of nonallergic individuals living in the same area. It is equally likely, however, that levels of exposure are similar for individuals with and without allergic disease in a given region and that differences in individual responses are a function of individual susceptibility. It is important to understand the actual findings. (Chapter 2 presents a brief discussion of exposure to allergens as a risk factor for sensitization.)

Because the common allergens are thought to cause or exacerbate asthma by the inhaled route, measurement of inhaled allergen might seem to be the best method for determining exposure (Price et al., 1990; Swanson et al., 1985; Tovey et al., 1981b). There are several reasons, however, why current threshold levels for indoor allergen exposure are based on measurements of allergen in dust collected from carpets, mattresses, sofas, and other such items. First, the quantities that become airborne (commonly, 5–50 ng/cubic meter of air) are too small to measure in epidemiological surveys. Second, the relevance of airborne levels depends on particle size, which is technically difficult to determine. Third, the quantities of these allergens that become airborne in a house depend critically on domestic disturbance. Thus, at present, overwhelming practical reasons exist for basing threshold levels on the measurement of a representative allergen in "reservoir" dust. An index of exposure using these measurements of reservoir dust assumes that they are positively correlated with inhaled exposure. Chapter 6 addresses issues related to assessing exposure and risk and presents a risk assessment for sensitization related to dust mite exposure as an example.

Reducing Exposure to Dust Mites

Reducing exposure to so-called "trigger factors," i.e., factors that trigger an allergic response, has been a standard part of the treatment of allergic disease for many years, and for many years it was normal practice to recommend avoidance measures to patients who had skin tests that were positive for house dust. This practice was strongly supported by the experiments of Storm van Leeuwen (Storm van Leeuwen et al., 1927) and Rost (1932), who demonstrated benefits to patients with asthma and atopic dermatitis, respectively, from living in a "climate chamber."

Until recently, there has been only limited clinical use of avoidance measures in treating allergic diseases associated with dust mite sensitivity, in part because the control measures that were originally proposed were not effective in controlling mite allergens (Burr et al., 1980; Korsgaard, 1982). In addition, several general medical tests have suggested that avoidance measures should be considered in patients who have a certain typical history. As discussed earlier, however, many allergic patients are not aware of an association between dust exposure and their symptoms (particularly the association between dust and asthma or atopic dermatitis). Today, there is considerable evidence that full avoidance (i.e., 95 percent reduction of mite allergen) can be achieved and can reduce both symptoms and bronchial reactivity. For example, moving patients to a hospital room or sanatorium has been consistently effective (Dorward et al., 1988; Ehnert et al., 1991; Platts-Mills et al., 1982; Warner and Boner, 1988); these units generally have very low levels of mites (i.e., less than 20 mites/g of dust) and mite allergen (less than 0.4 µg of *Der p* I/g). Recently, four controlled studies of the effects of avoidance measures conducted in the homes of patients have found significant improvement in both asthma symptoms and bronchial hyperreactivity (Dorward et al., 1988; Ehnert et al., 1991; Murray and Ferguson, 1983; Walshaw and Evans, 1986).

Avoidance measures can be divided into those for use in the bedroom and those for use in the rest of the house (Box 3-1). In the bedroom, the following have been shown to be effective: covering mattresses and pillows with impermeable covers; washing bedding at 130°F once per week

---

BOX 3-1 Avoidance Measures for Mite Allergen

A. Bedrooms
- Cover mattresses and pillows with impermeable covers
- Wash bedding regularly at 130° F
- Remove carpets, stuffed animals, and clutter from bedrooms
- Vacuum clean weekly (wearing a mask)[*]

B. Rest of the House
- Minimize carpet and upholstered furniture; do not use either in basements
- Reduce humidity below 45 percent relative humidity or 6 g/kg
- Treat carpets with benzyl benzoate or tannic acid

[*]There is a temporary increase in potential exposure to allergens associated with the vacuuming process. The net potential for exposure should be reduced by vacuuming, however, and is considerably less than the cumulative effects of not vacuuming. Wearing a mask while vacuuming should help reduce exposure while vacuuming.

(Miller et al., 1992; Owen et al., 1990); and removing carpets (although Rose and colleagues [1992] have shown that the use of acaricides or tannic acid treatment can also reduce mite allergen). Other control strategies for the bedroom are designed to eliminate sites in which mites can grow and to reduce dust collectors to make cleaning easier. The recent report from NHLBI (1991) is an excellent source of information regarding allergen avoidance. (See also Box 8-1 in Chapter 8.)

Three different approaches are possible for control of mite growth in the rest of the house:

1. Design the house with polished floors and wooden, vinyl, or leather furniture to eliminate sites where mites can grow.

2. Maintain indoor relative humidity at below 50 percent (absolute humidity below 6 g/kg). Korsgaard (1983a) has shown that in some areas of the world this can be accomplished by simply increasing ventilation. In other areas, it would be necessary to use air conditioning during the humid months.

3. Use acaricides to treat carpets or furniture, including pyrethroids (D. Charpin et al., 1990b), natamycin (an antifungal), pirimiphos methyl (Mitchell et al., 1985), and benzyl benzoate (Bischoff et al., 1990). In each case the chemicals are effective in killing the mites, although methods for applying the agents may present problems (de Saint-Georges-Gridelet et al., 1988; Platts-Mills et al., 1992).

Several different chemical treatments (as in approach 3 above) have been shown to achieve 90 percent reduction in allergen levels for a month or more. In addition, 1 or 3 percent solutions of tannic acid have been recommended for denaturing mite allergens (Green et al., 1989). Again, this method achieves a 90 percent reduction of mite allergen, but because tannic acid does not kill mites, the effect is temporary (i.e., approximately 6 weeks to 2 months). Carpets fitted onto unventilated floors—for example, in basements or on the ground floor of a house built on a concrete slab—are particularly difficult to treat. Under these circumstances water can accumulate either because of condensation onto the cold surface of the concrete or because of leakage (either domestic or rainwater from outside). In either case, once the carpet is wet, it will stay wet and become an excellent environment for the growth of fungi and mites.

Avoidance Measures for Other Allergens

For most other allergens, only case reports are available as guidance regarding the clinical effectiveness of avoidance measures. Removal is certainly the logical approach to management for most domestic animals or rodents (Wood et al., 1989); if sensitivity to insects (cockroach or others) is demonstrated, eradication, or moving, is the logical approach in such circumstances. At present, there is little evidence about the effectiveness of eradication measures in reducing insect allergen levels. The recent development of assays for cockroach allergens, however, means that it is now possible to evaluate avoidance measures.

The situation is much more confusing for indoor fungi. No simple assays exist for indoor fungal allergens, and the relationship between spore counts and allergen exposure is not clear. Many spores are not viable, and some allergens may only be expressed once the spore germinates (Arruda et al., 1990; Lehrer et al., 1986). It will be difficult to evaluate proposed procedures to reduce exposure to fungal allergens until accurate assays are available. Even then, evaluation of exposure will not be as simple as for mite or cat allergens. (See Chapter 6 for a discussion of monitoring for indoor fungal allergens.) It is often difficult to tell whether fungal spores come from inside or outside the house, because many species can grow in either environment. Nonetheless, it is reasonable to recommend controlling humidity, removing sites for fungal growth, avoiding basements, and cleaning surfaces with fungicides.

Grass pollen can become a major component of house dust and has been found at high levels in dust from the houses of grass pollen-sensitive patients who present for treatment with asthma (Pollart et al., 1988). Filtering incoming air or keeping doors and windows closed can help control the entry of pollens, although other problems may be created (see Chapter 7).

Conclusions and Recommendations

With most Americans spending the great majority of their time indoors—and most of that in their own houses—it is not surprising that the bulk of inhaled foreign protein is associated with indoor air. The evidence shows that a large proportion of asthmatics are allergic to indoor allergens and that several changes in the way we live indoors may have affected the levels of these allergens. These changes include increased mean temperatures, reduced ventilation (with consequent increased humidity), fitted carpets, and cool-wash detergents which have led to water temperatures for washing bedding that do not kill mites.

Once identified, reducing exposure to allergenic "trigger factors" has been a standard part of the treatment of allergic disease for many years. Since approximately half of existing cases of asthma have been attributed to allergenic factors, it is reasonable to expect that asthmatics who require more than occasional treatment might also have allergies that induce their asthma.

Recommendation: Provide appropriate allergy evaluation of asthmatics who require more than occasional treatment. Where allergic

factors are present, ensure that these patients are given specific, practical information about how to reduce their exposure to arthropod and other allergens.

Although most studies investigate asthma (because it is common and can be measured objectively), sensitivity to indoor allergens is also very common among patients with other allergic conditions such as chronic rhinitis and atopic dermatitis. Because the important cause of inflammation that is common for all of these diseases is exposure to allergens, avoidance of the exposure should be the primary anti-inflammatory treatment. Developing realistic avoidance protocols for routine use is a challenge that must not take second place to pharmaceutical treatment.

Research Agenda Item: Conduct detailed studies of physical factors, such as sources and emission rates, that influence the levels of exposure to arthropod and other indoor allergens. These studies should include the effects of (a) protocols for reducing exposure and (b) devices advertised as reducing indoor allergen concentrations. More specifically, test the effectiveness of allergen avoidance protocols on the management of allergic asthma and other allergic diseases using protocols that have been demonstrated to reduce exposure by 90 percent or more. Such protocols should be tested under field conditions and should encompass the socioeconomic, cultural, ethnic, and geographic diversity of these problems in the United States.

The more important question now is the effectiveness of avoidance measures for reducing the development of disease (i.e., sensitization). Specific measures include the use of polished wooden floors and no upholstered furniture in bedrooms and possibly in schoolrooms, keeping pillows and mattresses covered, and using bedding that can be washed regularly in hot water. Other measures, which require further evaluation, include air filtration, chemical treatment of carpets, and dehumidification.

Research Agenda Item: Develop allergen-free products to help reduce the incidence of allergic disease in the general public. The initial objective of this initiative, which should be carried out by the industrial/business sector, would be to provide an aesthetically appealing bedroom with reduced allergen exposure for children under the age of five.

There is considerable evidence that allergen avoidance is an effective means of reducing allergy symptoms. Practical measures to control exposure to mite and other allergens in the house can reduce both the symptoms and the underlying bronchial reactivity. For example, controlled studies of the effects of avoidance measures conducted in the homes of patients who were allergic to dust mites have shown significant improvement in both asthma and bronchial reactivity. The effects of avoidance measures on an individual's quality of life, however, remain to be determined.

Research Agenda Item: Conduct longitudinal studies to determine whether long-term allergen avoidance has a positive effect on quality of life.

MAMMALS AND BIRDS

Allergic reactions to mammals and birds are frequent and have been recognized as such for many years. An estimated 100 million domestic animals reside in the United States, the most common being cats and dogs (Knysak, 1989), and from one-third to one-half of homes in the United States have a domestic pet. Although a large number of individuals in the population experience allergic reactions to various animals, the exact prevalence of this common problem has not been determined. Skin test reactions to animal extracts are frequent, but the relationship between sensitization as demonstrated by a positive skin test and the frequency of clinical symptoms is unknown. Nonetheless, exposure to domestic pets, particularly cats and dogs, accounts for most of the allergic diseases caused by mammals.

In addition to exposures to animals kept as pets, individuals may be exposed to mammal and bird allergens in occupational settings. Bird allergens can cause not only allergic rhinitis and asthma but also hypersensitivity pneumonitis. Exposure to rodents among laboratory and animal care facility workers often results in sensitization and allergic symptoms. Farmers, veterinarians, animal health care workers, and zookeepers are also at risk. The prevalence of allergic diseases as a result of these occupational exposures is largely unknown.

Animals and birds are not only a source of allergens but may also serve as vectors, introducing outdoor allergens indoors. For example, pets may carry inhalant allergens such as pollens and fungal spores on their coats.

Cats

Cats are among the most common household pets in urban areas. A survey of 16,204 individuals ages 6 to 74 in the United States showed that 2.3 percent had positive skin prick tests to cat dander extracts (Gergen et al., 1987). Thus, extrapolating from these data, from 6 to 10 million Americans are potentially allergic to cats (Luczynska et al., 1990; Ohman and Sundin, 1987). Approximately one-third of the 2.3 percent of people with positive cat dander reactions have cats in their homes.

Cat allergy is especially apt to cause asthma, and the presence of IgE antibody to cat allergen is a significant risk factor for acute attacks of asthma in patients who seek treatment in emergency rooms (Luczynska et al., 1990). Individual sensitivity to these animals may be exquisite, so that even brief exposure to cats can precipitate severe asthma episodes.

The most important cat allergens are *Fel d* I and cat albumin. *Fel d* I is the major cat allergen (Ohman et al., 1973), and it is distributed throughout all breeds of cats. Approximately 80 percent or more of cat-allergic patients have IgE antibody to *Fel d* I (Duffort et al., 1991). In recent years, the complete amino acid sequence of *Fel d* I has been determined, the genes encoding the protein have been identified, and several allergenic epitopes have been mapped (Morgenstern et al., 1991).

Commercially available cat extract reagents for skin testing have variable quantities of *Fel d* I and serum albumin. They are prepared from dander (hair with epithelial scrapings) or whole pelts. The concentration of *Fel d* I is 10 times greater at the root than at the tip of the hair (C. Charpin et al., 1991). It is produced primarily in the sebaceous gland and to a lesser extent in the basal squamous epithelial cells of the skin. It appears to be stored mainly on the surface of the epidermis and the fur. Because saliva also contains *Fel d* I, licking and grooming may spread the allergen on the hair. The mean daily production of *Fel d* I from cats is approximately 3-7 μg (Dabrowski et al., 1990); however, there is day-to-day and diurnal variability in *Fel d* I shedding from individual cats. Washing the cat provides only temporary relief: the original concentration of *Fel d* I reappears within 1 month.

Cat-allergic patients develop symptoms rapidly upon entering a house with a cat, which suggests that the allergen is constantly airborne. Indoor air sampling measurements in fact have confirmed this hypothesis. In addition, studies have determined the quantity of *Fel d* I that will produce a 20 percent decrease in FEV_1 (forced expiratory volume at 1 second) in cat-sensitive patients who undergo bronchoprovocation testing. A room containing two live cats has a concentration of *Fel d* I in this range (Van Metre et al., 1986); however, many cat-allergic individuals will be symptomatic in the presence of even one cat. In a 1988 study by Wood and colleagues, this allergen was found in virtually 100 percent of homes sampled for its presence in settled household dust (Wood et al., 1988). The median level of *Fel d* I in the dust samples was 90 ng per gram of dust (range: 2–130,000 ng *Fed d* I/g of dust).

Although there are significant differences in allergen content in homes with and without pets, many homes that normally are without a cat nonetheless contain surprisingly high levels of *Fel d* I in dust (Wood et al., 1988). These levels vary from less than 2 to 7,500 ng/g of *Fel d* I/g; it is thought that the allergen is transported into houses on the clothes of inhabitants.

Levels of cat allergen as low as 2 μg *Fel d* I/g of dust, commonly observed in houses without cats, may be a risk factor for sensitization to *Fel d* I (Wood et al., 1988).

The size range of the particles that contain cat allergen is broad. Some are of fairly large aerodynamic size (greater than 10 μm in diameter); in contrast, many particles that carry *Fel d* I are less than 2.5 μm in diameter and are therefore respirable and capable of penetrating deeply into the lung (Findlay et al., 1983; Luczynska et al., 1990). Luczynska and coworkers (1990) estimated that 10–60 percent of airborne *Fel d* I in domestic houses is associated with small, respirable particles of less than 2.5 μm in diameter. Modest levels of disturbance of dust such as that created by an electric fan can dramatically increase the airborne concentration of cat allergens. During vigorous vacuum cleaning of rooms with a single cat, levels of up to 212 ng of *Fel d* I/m^3 have been observed (de Blay et al., 1991b). Swanson and colleagues (1985) found that bedspreads on which cats slept contained 2,000–4,000 μg *of Fel d* I. Therefore, beds, stuffed furniture on which cats sleep, and carpets constitute continuing reservoirs for the contamination of the environment. The adherent properties of *Fel d* I may be an important aspect of their contaminating capability; thus, walls and other surfaces can also contribute to the reservoir of cat allergen (Wood et al., 1992). Finally, because voided urine from male cats contains *Fel d* I activity, litter boxes are a source of environmental exposure.

CONTROL MEASURES

The following are several animal allergen avoidance measures for the home:

1. Remove the animal from the home. Once a cat is removed from the environment, it takes at least 20 weeks from the time of removal for the levels of *Fel d* I in settled dust to fall to that found in homes without cats (Wood et al., 1989).

2. Where it is not possible to remove the animal, confine it to certain restricted areas. Animals should be kept out of the bedroom. Washing the animal can also be beneficial.

3. Steam cleaning of carpets has no benefit over regular vacuuming. However, aggressive measures, including carpet removal, can substantially reduce cat allergen content in settled dust more rapidly.

As an alternative to removing the cat from the environment, control measures with the cat in situ can be employed (de Blay et al., 1991a). For example, in an uncarpeted room, a combination of vacuum cleaning, air filtration with a high-efficiency room air cleaner, and washing of the cat can reduce the airborne allergen exposure level by 90 percent or greater. It should be noted, however, that the effects of these measures on allergic symptoms have not been evaluated.

Dogs

Dogs are among the most abundant of household pets. The prevalence of clinically significant allergic diseases caused by dogs appears to be less than that for cats; nevertheless, dog sensitivity can be a major cause of allergic symptoms, including asthma, in some persons. In a survey of U.S. citizens ages 6 to 74 who were unselected for allergy, 2.3 percent of the 16,204 individuals surveyed exhibited positive skin prick tests to dog extracts (Gergen et al., 1987). This frequency of sensitization is similar to that reported for cats, although other investigators have found a higher prevalence of cat sensitivity in their study populations (Sears et al., 1989). Differences in reported sensitivity may be due in part to less well-characterized and standardized allergic extracts used in skin testing for the presence of dog sensitivity. Moreover, in many parts of the country, dogs are kept out of doors; those that are kept indoors may be washed at frequent intervals.

The major dog allergens are *Can f* I and dog serum albumin; *Can f* I has been purified (de Groot et al., 1991; Schou et al., 1991b). *Can f* I can be obtained from hair collected by brushing dogs and from dog saliva; very little is available from urine and feces. The allergen has been found in varying degrees among nine dog breeds and among individual dogs (de Groot et al., 1991). In approximately 15 percent of dog-sensitive patients, there are significant differences among the skin test responses to different dog breeds, indicating breed-specific allergens (Lindgren et al., 1988). However, there is no evidence to suggest that there are "nonallergenic" breeds.

In a study by Lind and others (1987), 63 percent of indoor air samples from 43 homes in the Baltimore, Maryland area indicated the presence of dog allergens. Dog allergen has also been found in buildings, such as schools, in which dogs are not kept, which suggests that the allergen can be transported on clothing. Dust samples from 103 households across the United States contained a range of from 10 to 10,000 µg of *Can f* I/g of dust in those homes with a dog and from less than 0.3 to 23 µg of *Can f* I/g of dust in homes without dogs (Schou et al., 1991a). *Can f* I is a relatively stable molecule and may persist in dust for extended periods of time. Effective control measures to reduce exposure to dog allergens have not been investigated.

Rodents

Rodent allergens are a significant occupational cause of allergic rhinitis and asthma among workers exposed to laboratory animals. In addition, certain rodents, such as hamsters, gerbils, and guinea pigs, may contribute to household allergen levels because they have become increasingly popular household pets. At least 35,000 individuals are exposed to rodent allergens because they work in scientific investigations or breed and care for rodents. In the United States, approximately 11–15 percent of laboratory workers are allergic to rodents (Slovak, 1987); many who become allergic are forced to seek alternative employment. Fifty-five percent of them are sensitive to two or more species; the majority (37–75 percent) are sensitive to rats, mice, and rabbits, whereas 24–33 percent are sensitive to guinea pigs. Patients with atopy may develop symptoms more rapidly than those without it. Moreover, patients with a family history of allergy and positive skin tests to other environmental allergens may be at risk for the development of asthma (Sjostedt and Willers, 1989). Hypersensitivity pneumonitis resulting from exposure to rodent proteins is rare.

Two major allergens have been found in rat urine, *Rat n* I (α-2-euglobulin) and prealbumin. Skin testing for both allergens shows an equal degree of frequency in sensitized patients. Euglobulin is a major component of pubertal male rat urine but is barely detectable in the urine of females (Knysak, 1989).

In rat rooms and during disturbance of rat litter, a large portion of the urinary allergen is airborne; particles are approximately 7 μm in diameter. Air sampling measurements during certain laboratory activities have indicated that feeding or cleaning of cages results in levels of 20 ng of *Rat n* I/m^3; injections or handling of the animals results in levels of 13 ng/m^3. Lower levels (3.1 ng/m^3) are found during surgery or during sacrifice of the animal (Eggleston et al., 1989).

The male rat is capable of releasing high levels (up to 20 ng/min) of airborne urinary allergens (Swanson et al., 1990). Personal sampling devices on 12 sensitized subjects working with rats revealed allergen concentrations that ranged from less than 1.5 to 310 ng/m^3; even higher levels were noted during cage cleaning. These subjects experienced nasal symptoms with exposure, and researchers obtained direct evidence for an allergic reaction by finding histamine in their nasal lavage fluids. Five of the 12 subjects also had lower respiratory tract responses with decreases in pulmonary function (Eggleston et al., 1990).

Two major allergens, *Mus m* I and *Ag 3*, have been identified as related to mice. *Mus m* I is a urinary prealbumin; it has been molecularly cloned, and its amino acid sequence was found to share 80 percent homology with *Rat n* I. Research has noted biochemical differences between urinary proteins from different inbred strains of mice, which may explain why some mouse laboratory workers become sensitized to specific strains. The *Ag 3* allergen is found primarily in hair follicles and the stratum corneum, implying that it is found predominantly on the hair and skin of mice. *Mus m* I has been identified in dust from the walls and air-conditioning filters of rooms in which mice are housed. Airborne allergen concentrations ranging from 1.8 to 825 ng/m^3 of air have been detected; concentrations vary with both the number of mice and the degree of work activity in the room (Twiggs et al., 1982). Mouse urinary allergenic protein has been found in air samplings from inner-city dwellings (Swanson et al., 1985), but the effects of this exposure on the health of inhabitants are not known.

Guinea pig allergens are present primarily in the urine. At least two major allergenic components have been identified. Air sampling studies may show varying concentrations of the urinary protein in the air, depending on the number of animals. Most allergenic activity in the air is associated with particles in two size ranges: those less than 0.8 μm in diameter and those greater than 5 μm (Swanson et al., 1984).

Occasionally, hamsters and gerbils may account for allergic symptoms. Positive skin tests to hamster dander extracts have been reported among laboratory workers, and gerbil extracts prepared from serum and hair/epithelium have produced positive skin tests in sensitized patients. Allergens from these rodents have not been characterized further. Even less fully characterized are rabbit allergens. The major known allergen is obtained from fur extracts and saliva (Warner and Longbottom, 1991).

CONTROL MEASURES

Control measures for laboratory workers include the use of protective clothing and appropriate respiratory protection. Source control measures involve placing filter caps on animal cages, increasing the frequency of air exchange (including 100 percent exhaust of the animal room with no recirculation), locating exhaust ducts at floor level, increasing the frequency of removal of animal waste and bedding, and using high-efficiency filters to provide filtered laminar flow air. Because of the high rate of production of rat urinary protein allergens, very high air exchange rates are required to substantially reduce allergen levels in rooms that house a large number of animals (Swanson et al., 1990).

Farm Animals

The prevalence of sensitivity to cattle is unknown; however, two major allergens from cattle with molecular weights of 20 and 22 kDa have been identified (Ylönen et al., 1992). These allergens are derived from cow hair and dander, but the 20-kDa allergen is also found in urine. Immunochemical assays for quantitating cattle allergens in cow-shed air samples have shown concentrations ranging from 137 ng/m^3 to 19.8 μg/m^3 (Virtanen et al., 1986). In addition to cattle epithelium, other potential allergens in barns include fungi and mites (Campbell et al., 1989).

From 8 to 19 percent of allergic patients react to horse allergens on intracutaneous testing (Solomon and Matthews, 1988); however, clinical sensitivity is not as frequent. Horse-sensitive patients may react to extracts of mule and donkey allergens as well. Allergy to horses is less common now than it was in the early years of this century, but significant exposure still occurs among agricultural workers, racetrack and stable attendants, and avocational riders. Allergens have been identified in horsehair and horse dandruff (Lowenstein et al., 1976).

Many workers in swine confinement areas experience respiratory symptoms that are not mediated by IgE. It is believed that most of these symptoms are due to the irritant effects of nitrogenous waste products from the animals or to endotoxins from gram-negative organisms in feces. Nevertheless, occupational asthma caused by allergy to pig urine has been reported (Matson et al., 1983).

Wild Animals

Occasionally, veterinarians and zookeepers exposed to wild animals may develop allergies. For example, large cats (lions and tigers) have the *Fel d* I allergen; in addition, allergy to deer and elk has been reported. Allergenic sources from deer include hair, dander, saliva, serum, and urine (Gillespie et al., 1985). Asthma triggered by exposure to monkeys has been reported in primate centers. The allergenic sources of that condition appear to be hair and dander, and possibly saliva (Petry et al., 1985).

Birds

Positive skin test reactions to feather extracts but not fresh feathers are common, a finding that may be explained by contamination from dust mites. The most common sources of exposure to feathers are pillows, comforters, quilts, down-filled clothing, and feather beds. About 20 percent of budgerigar and canary fanciers with allergic rhinitis and asthma have IgE antibodies to allergens found in extracts of the feathers of these birds (Solomon and Matthews, 1988).

Another avian source of allergens is chickens. About 10 percent of workers in egg-processing plants develop asthma (A. B. Smith et al., 1990). The allergens are aerosolized in liquid form or occasionally as dried airborne egg protein.

Avian proteins may cause hypersensitivity pneumonitis (also called extrinsic allergic alveolitis). Several specific terms are in use, including bird fancier's, breeder's, or handler's lung or pneumonitis. Parakeets, budgerigars, pigeons, turkeys, and chickens have been implicated as causes of these conditions. An estimated 5–20 percent of the 250,000 pigeon breeders in the United States may develop pigeon breeder's lung (Christensen et al., 1975). The hypersensitivity pneumonitis seen in these cases is caused by allergy to proteins in the droppings and sera of the birds; exposure occurs through inhalation of these dried materials. Potentially, contamination of ventilation or air-conditioning systems by bird (especially pigeon) droppings could expose a building's occupants to these allergens, but large-scale outbreaks of hypersensitivity pneumonitis from such a cause have not been reported.

Pigeon serum gamma globulin appears to be the major allergen in pigeon breeder's disease, and ongoing allergen exposure even to minimal levels results in impaired lung function (Rose and King, 1992). The diagnosis of the disease is based on clinical presentation and the presence of IgG precipitating antibodies to the allergen in question. Patients with pigeon breeder's disease may have positive skin tests and IgE antibodies to pigeon serum as well; however, crude extracts often are nonspecifically irritating and therefore not useful for diagnosis. Positive in vitro lymphocyte transformation tests with pigeon allergens can be demonstrated more consistently in symptomatic individuals than in asymptomatic but exposed individuals (Fink, 1988).

Conclusions and Recommendations

One of the difficulties in producing good epidemiological data on allergy is the lack of well-characterized, standardized allergenic extracts for diagnostic purposes. Cat allergens have been well studied, although the role of serum albumin and its overall importance in allergic reactions to cats have yet to be determined. Allergenic extracts that are standardized for the content of *Fel d* I have appeared only recently. Well-characterized, standardized extracts of dog allergen preparations with known concentrations of *Can f* I are not available and should be developed. Similarly, there is a limited understanding of the identity and characteristics of many other mammalian and avian allergens. The lack of standardized extracts is partially responsible for the lack of development of immunochemical assays such as monoclonal antibody-based assays for many mammalian and avian allergens. Methods for measuring airborne allergen concentrations are critical for devising and evaluating control measures.

Research Agenda Item: Characterize important allergens from indoor animal sources (e.g., cats, dogs, birds, rodents) more precisely in order to develop standardized allergenic extracts for diagnostic purposes and immunoassays suitable for monitoring exposure.

Despite an increasing body of knowledge regarding the role of indoor allergen exposure, particularly to mammals, as a cause of asthma, much remains to be learned. The relationship between exposure to indoor pets and the increasing morbidity and mortality of asthma requires further clarification.

Research Agenda Item: Determine the relationship between exposure to indoor pets and the incidence, prevalence, and severity of asthma.

In addition, rodents that infest inner-city dwellings need to be examined as potential risk factors for asthma among individuals exposed to these potent allergens.

Research Agenda Item: Explore the possibility that exposure to rodent populations in inner-city areas may be a risk factor for asthma.

For many allergens, the size of airborne particles and their distribution in the air have not been elucidated and should be studied. Reservoirs of animal and avian allergen exposure and their dissemination through ventilation systems of offices, apartments, and other large buildings likewise require investigation.

Research Agenda Item: Investigate the potential role of mammalian- and avian-allergen-contaminated ventilation systems in the development of allergic disease among inhabitants of apartments, offices, and other large buildings.

Although control measures may reduce airborne concentrations of mammalian and avian allergens, the ability of these approaches to influence symptoms in sensitized patients or to prevent the sensitization of naive individuals requires clarification and study. The use of personal monitoring systems should be useful in making these determinations.

Research Agenda Item: Evaluate the effectiveness of environmental control measures on patient symptoms. This should include assessments of preventing sensitization in the naive individual as well as symptom reduction in those already sensitized.

The epidemiology, diagnostic techniques, and even pathogenesis of hypersensitivity pneumonitis from avian proteins are poorly characterized and require further elucidation. In addition, the risk factors and natural history of the disease are poorly understood and the effects of allergen avoidance remain controversial in the ultimate prognosis of the disease. All of these issues require clarification through further research.

MICROBIAL ALLERGENS

There are many kinds of microorganisms: viruses, bacteria, fungi, slime molds, algae, and protozoa. With the exception of viruses, each will be discussed below in relation to allergic disease. Viruses will not be considered here because they are obligate intracellular parasites and do not cause allergic environmental disease, although viral infections may exacerbate existing conditions.

The Fungi

Fungi are eukaryotic organisms that have rigid, chitinous cell walls and occupy reservoirs ranging from continuously wet to minimally moist. Fungi compete with bacteria for nutrients and oxygen. They can grow at a low pH level and can kill or inhibit many bacteria by excreting toxic metabolites (antibiotics).

Most organic material can be degraded by at least one fungus, provided that sufficient water is present. The fungi digest their food externally, excreting enzymes into the environment. Some fungi change their pattern of enzyme excretion with a change in food source or with changing environmental conditions.

Fungi reproduce both asexually and sexually and often produce spores by one or both of these kinds of life cycles. The fungal spores are allergenic; that is, responsible for causing allergic responses. Many fungal spores resist drying; they also require specific triggers that initiate germination only under favorable circumstances. Fungi are divided into classes (*Oomycetes*, *Zygomycetes*, *Ascomycetes*, and *Basidiomycetes*) on the basis of the ways in which their sexual spores are produced. A large group of fungi either has lost the ability to reproduce sexually or the sexual stages have yet to be discovered. These species are classed as *Hyphomycetes* (asexual spores are produced on naked hyphae) and *Coelomycetes* (asexual spores are produced within fruiting bodies).

SOURCES AND DISSEMINATORS

Fungi are ubiquitous outdoors, growing on living and dead plants, animals, and other microorganisms. Field crops such as corn, wheat, and soybeans are well colonized with fungi during active growth and especially at harvest (Burge et al., 1991). One group of *Basidiomycetes* (e.g., members of the genera *Coprinus*, *Pleurotus*, *Merulius*, and *Ganoderma*) live in the soil or as saprophytes or parasites on plants. These fungi produce macroscopic fruiting bodies that elevate spores into the air (Adams et al., 1968; Cutten et al., 1988; Koivikko and Savolainen, 1988). There are also many fungi that can invade living plant tissue (plant pathogens such as the rusts, smuts, and powdery and downy mildews). Fungal growth and spore dissemination depend on available substrates, season, climate, and human activity such as agriculture.

Stored organic material such as compost, silage, hay, or grain supports the growth of fungi that favor low oxygen levels, dryness, and heat (e.g., *Aspergillus fumigatus, A. flavus, Penicillium* spp.; Campbell et al., 1989). Their spores become airborne when the moldy stored material is handled.

Spores can penetrate interior environments from outdoors either with ventilation air, or on the surfaces of people, animals, or objects (Pasanen et al., 1989; Su et al., 1992). Air conditioning and mechanical ventilation and filtration, however, allow doors and windows to remain closed, thus preventing entry of most outdoor fungal aerosols (Hirsch et al., 1978; Pan et al., 1992).

Indoors, fungi grow in wet environments such as basements, window sills, and shower stall surfaces (Kapyla, 1985; Verhoeff et al., 1990b). Species of *Aspergillus* and *Penicillium* often are dominant indoors (Brunekreef et al., 1990). Some fungi (e.g., *Aspergillus versicolor, A. flavus, Wallemia sebi*) prefer environments that are relatively dry and will grow at a water activity level of 0.65 (i.e., on substrates that contain 65 percent water) (Kendrick, 1985). The humidity at which hygroscopic materials in dust (including human skin scales) absorb enough water from the air to support the growth of these xerophilic ("dry-loving") fungi is unknown, but it is probably close to from 60 to 70 percent. Other residential substrates that have been attractive to fungi include urea formaldehyde foam insulation, wicker baskets used as plant containers, and carpeting installed in bathrooms (Bisset, 1987; Kozak et al., 1980b).

Fungal spores and other effluents become airborne indoors when disturbed by air movement and normal human activities (O'Rourke et al., 1990). Contaminated air conditioners and humidifiers can actively spray spores, fragments, and dissolved allergens into the air (Baur et al., 1988; Burge et al., 1980; Kumar et al., 1990). Humidifiers that cause water to evaporate into the air are less likely to produce measurable aerosols (Burge et al., 1980), although exposure to such aerosols probably occurs during cleaning.

NATURE OF THE AEROSOLS

The composition of aerosols of fungus-derived particles depends on the abundance and strength of sources, as well as on dissemination factors, mixing, dilution, and particle removal. Natural aerosols are almost always composed of mixed species. In agricultural situations and in indoor environments with actively disseminating reservoirs, aerosols may be monospecific (i.e., containing particles derived from a single fungus species), increasing the risk of exposure to species-specific toxins or allergens and thereby increasing the risk of hypersensitivity pneumonitis. Massive exposure to toxin-producing fungi can occur when moldy organic material is handled. Such exposure can cause exacerbation of allergic disease as well as direct toxic effects such as immunosuppression and cancer (Baxter et al., 1981).

Dose-response data for fungal allergens are unavailable. Standardized protocols for the collection of fungal aerosols are not in wide use, and some of the current methods for quantitation may be unreliable (see Chapter 6). Some studies have reported concentrations of measured viable fungal units (i.e., colony-forming units) in the air of homes that vary over several orders of magnitude both within individual homes, between homes in one community, and between communities (Beaumont et al., 1985; Brunekreef et al., 1990; Su et al., 1992; Verhoeff et al., 1990b).

DISEASES

Of the many different kinds of microorganisms, the fungi are most often associated with allergic disease. Airborne fungal allergens have been implicated in allergic rhinitis/conjunctivitis, allergic asthma, and hypersensitivity pneumonitis. Certain fungi grow saprophytically in the mucous lining of the lungs of patients with allergic bronchopulmonary fungosis or aspergillosis and in the sinuses of people with allergic fungal sinusitis. In addition, conditions favorable to fungal growth correlate positively with respiratory symptom rates as determined by questionnaires (Beaumont et al., 1985; Brunekreef et al., 1989, 1990; Dales et al., 1990; Dekker et al., 1991; Platt et al., 1989; Strachan et al., 1990).

All fungi probably produce allergens that will cause disease with appropriate exposure, although skin test rates vary with allergen sources and the populations chosen for study (Cutten et al., 1988; Giannini et al., 1975; Tarlo et al., 1988). Among people referred for assessment of respiratory atopy, from 1 to 10 percent have positive skin prick tests to one or more fungal allergens (Beaumont et al., 1985). In atopic populations, the percentage of responders can be as high as 27 percent (O'Neil et al., 1988). Skin test reaction rates to fungi in atopic asthmatic patients have been as high as 70 percent (Lopez et al., 1976).

Investigators have measured precipitating IgG antibodies that are specific for soluble allergens of a number of different species of *Aspergillus* as well as *Penicillium, Paecilomyces* (Dykewicz et al., 1988), *Pleurotus ostreatus* (Noster et al., 1976), and *Leucogyrophana pinastri* (Stone et al., 1989). These antibodies are related to exposure to high levels of small fungal particles, but diseases such as allergic bronchopulmonary aspergillosis (ABPA) and allergic fungal sinusitis (AFS) require additional host factors that are currently unknown. In ABPA, exposure to allergens released from active fungal surface growth in the lung stimulates the production of both IgE and IgG. The relationship between exposure to airborne spores and either initiation of ABPA or the status of ABPA patients is unclear.

NATURE OF THE ALLERGENS

Fungal allergens may be a structural part of the microbial cell, or they may be produced by the cells and released into the environment. The fungal allergens that have been isolated thus far are water-soluble glycoproteins, some of which are enzymes (Baldo and Baker, 1988; MacDonald et al., 1989); some may also be high-molecular-weight carbohydrates (Savolainen et al., 1990). Only a few have been partially characterized (Aukrust and Borch, 1979; Horner et al., 1988, 1989; Pazur et al., 1990; Savolainen et al., 1990).

Crude fungal extracts are complex mixtures of soluble materials from mycelial and spore walls, cytoplasm, and metabolites. These extracts are produced from fungi grown in a liquid medium for 5–15 days; the mixture is then blended and filtered. Sometimes the fungal growth is removed from the liquid by filtration and subsequently ground, dried, and extracted. Residual culture medium is sometimes used as a second kind of preparation (Kauffman et al., 1984).

Batch-to-batch variability in fungal extracts is often greater than variability among different strains, species, or even genera (Burge et al., 1989; Savolainen et al., 1989). For most of the mushrooms and other macrofungi, field collections are usually used to produce allergen extracts. Preliminary studies of the comparative allergen content of spores, fruiting body tissue, mycelium, and spent culture medium demonstrate both similarities and differences. Variability in allergen content (determined by radioallergosorbent tests inhibition; see Chapter 6) has been observed in the same kinds of mushrooms collected from different sites and, to a lesser degree, from the same site at different times (Liengswangwong et al., 1987). Studies on cross-reactivity of allergens extracted from different taxa of fungi (Baldo and Baker, 1988; De Zubiria et al., 1990; O'Neil et al., 1988; Shen et al., 1990; Weissman, 1987) have not generally documented batch and strain variability within each species.

There are few reports of experimental human challenges with fungal allergens. Licorish and others (1985) provoked immediate and delayed asthma using *Alternaria* whole-spore challenges, but they produced only an immediate response with spore extracts. Lopez and coworkers (1989) induced positive bronchial challenges with basidiospores. For some fungi, possibly because the actual allergens are enzymes associated with germination, it may be necessary for a living unit to begin growth on the respiratory tract mucosa before allergen exposure occurs (Savolainen et al., 1990).

The Bacteria

Bacteria are prokaryotic microorganisms that lack many of the subcellular structures found in fungal, plant, and animal cells. They are usually single celled and reproduce by simple division. Most bacteria are saprophytes and require a source of complex carbon compounds (i.e., nonliving organic material); they decay substrates both aerobically and anaerobically. In general, bacteria require more water for active growth than the fungi and are usually the dominant organisms in water reservoirs with a pH of greater than 7. Some bacteria form spores that are extremely resistant to environmental stressors.

Many organic substrates can be degraded by bacteria, which form biofilms on surfaces that are continuously wet. Biofilm ecosystems also support the growth of algae, protozoa, and fungi. The bacteria are classified by cell shape, staining properties, spore production, metabolic characteristics, and the human diseases they cause.

SOURCES AND DISSEMINATORS

Bacteria occupy a wide range of reservoirs both outdoors and indoors. Gram-negative bacteria often predominate in outdoor reservoirs on living leaf surfaces and are able to survive at least brief periods of transit in the air. A wide variety of bacteria can be found in soil and in natural bodies of water. Some *Bacillus* species and the thermophilic actinomycetes (e.g., *Faenia rectivirtigula, Thermoactinomyces* spp.) will grow only at temperatures between 45° and 60° C. They are found mainly in environments that have become warm from insolation, geothermal conditions, or self-heating.

Outdoor bacteria become airborne with the disturbance of substrates—for example, with the movement of air or rain or with human activity, especially activities related to farming and refuse handling. Indoor reservoirs that allow growth and dissemination of allergenic bacterial aerosols include water-containing appliances such as portable humidifiers, large humidification systems, organic material stored or accumulating indoors, cooling fluids in machining plants, and other such moist areas. *Bacillus* species tend to accumulate in house dust. Thermophilic organisms occupy indoor reservoirs such as humidifiers attached to heating systems, refrigerator drip pans, evaporative cooler media, clothes dryer exhausts, and other such places characterized by organic material, water, and warm temperatures. Air movements, inadvertent human activity, and activities that allow direct handling of contaminated material (e.g., in removal or cleaning procedures) are common dissemination factors for indoor bacteria.

NATURE OF THE AEROSOLS

Similar to the fungi, the composition of bacterial aerosols depends on the abundance and strength of sources, on dissemination factors, and on factors that act directly on the aerosol such as mixing, dilution, and particle removal. Outdoor bacterial aerosols usually are dominated by gram-negative leaf surface bacteria such as *Pseudomonas* species (Nevalainen et al., 1990). Aerosols near such sources as cooling towers may contain more exotic organisms such as *Legionella pneumophila*, the agent of Legionnaires' disease. Indoors, where environmental sources are absent, the bacterial aerosol consists primarily of gram-positive cocci that inhabit human skin and respiratory tract secretions (Nevalainen et al., 1990). When gram-negative rods become dominant in indoor air, it can be assumed that they have been emitted from an environmental (rather than human) reservoir. Concentrations of bacteria that constitute a significant risk for sensitizing or provoking human allergic reactions are unknown.

DISEASES

Exposure to bacterial allergens has been associated with work-related asthma, hypersensitivity pneumonitis, humidifier fever, and a disease resembling allergic bronchopulmonary fungosis. Symptoms classified as humidifier fever have been attributed to gram-negative bacterial aerosols, although it is not clear whether the disease results from exposure to endotoxin alone or from exposure to the adjuvant characteristics of endotoxin acting in conjunction with other allergens (Hood, 1989; Polla et al., 1988). Endotoxin causes some of the symptoms related to infections (i.e., fever, chills) and can cause the same symptoms when large quantities of the organisms (or large amounts of toxin) are inhaled.

Like the fungi, bacteria secrete enzymes that can act as allergens, and these enzymes are being found in an increasingly broader range of products and locations. For example, *Bacillus* species are being used to produce proteases that are added to laundry detergents for stain removal. When initially introduced, aerosols of these enzymes were implicated in outbreaks of hypersensitivity pneumonitis among workers who manufactured these detergents. In this unusual case, a threshold limit value was established in relation to the risk of an allergic disease, and ventilation controls that now allow enzyme-containing products to be manufactured without apparent risk of hypersensitivity pneumonitis have been introduced. Enzymes and spores from gram-positive bacilli and the thermophilic actinomycetes have been implicated in epidemics of hypersensitivity pneumonitis and work-related asthma (I. L. Bernstein, 1972; Dolovich and Little, 1972; C. L. Johnson et al., 1980; Perelmutter et al., 1972; Wiberg et al., 1972).

NATURE OF THE ALLERGENS

Crude extracts of the thermophilic actinomycetes commonly associated with hypersensitivity pneumonitis are produced in a manner similar to that described for the fungi. They are used in double-diffusion assays to evaluate the presence or absence of specific IgG antibodies (precipitins).

Protozoa

The protozoa are microscopic animals that occupy reservoirs similar to those of the bacteria. Intact protozoa are generally too large to remain in the air for long periods of time, although occasionally they cause infection of the eye and brain when introduced into the eye or respiratory tract in large droplets (e.g., those produced by hot tubs). Some protozoa in indoor water reservoirs excrete allergenic material that can become airborne if droplets form. *Acanthamoeba polyphaga* and *Naegleria gruberi* allergens have been associated with humidifier fever and work-related asthma (Finnegan et al., 1987).

Algae

The algae are plantlike organisms with rigid, cellulosic cell walls; for the most part, they live in aquatic environments. Vegetative cells of the microscopic algae can be relatively abundant in outdoor air near bodies of water that support luxuriant algal growth (Schlicting, 1969). They have been reported to cause IgE-mediated allergy, but this question has not been well studied (Mittal et al., 1979).

Slime Molds

The slime molds (*Myxomycetes*) are microorganisms that do not fit well into such classifications as "plant" or "animal." They are motile amebae during part of their life cycle, but at another juncture they become immobile and produce spores that are indistinguishable from those of some fungi. These small but visible organisms occupy niches in the environment that are similar to those of some fungi (e.g., on damp organic material), and their spores form a small part of the outdoor aerosol. There is some evidence that these spores are sensitizing (Giannini et al., 1975; Santilli et al., 1990).

Environmental Control

Environmental control strategies are intended to reduce airborne concentrations of allergens. Several current strategies and methods are designed either to prevent or remove the contamination of indoor air with microbial allergens. The discussion below addresses these issues briefly; additional information is presented in Chapter 7.

PREVENTION

To prevent the contamination of indoor air by microbial aerosols, the penetration of outdoor aerosols must be reduced, and growth in indoor reservoirs must be eliminated. Keeping indoor environments physically separated from outdoors (by keeping doors and windows closed) and using mechanical ventilation and air conditioning are effective ways to control penetration. Water must not be allowed to accumulate, particularly, in ventilation systems; in addition, airborne water vapor must be kept to a minimum. Relative humidity should be maintained below 60 percent to prevent absorption of water by hygroscopic materials and to avoid condensation on cool surfaces. The effectiveness and utility of biocides (used either on surfaces or incorporated into fabrics or paint) have not been clearly established (see Chapter 7).

SOURCE CONTROL

Vacuum cleaning removes some fungus spores from carpeting, but it probably also reintroduces them into the air, either through the action of the beater in the cleaner or through the bag. High-efficiency particulate arresting (HEPA) filters or release of air and dirt directly to the outdoors, as is the case in central vacuum systems, will reduce such contamination. Wet cleaning of carpeting probably removes some microorganisms (Wassenaar, 1988a), but unless drying is rapid, the added water may spur the growth of those that remain. Water reservoirs associated with portable humidifiers can be cleaned each day and treated with a biocide every third day to maintain relatively low bacterial levels in the reservoir. However, the biocides must be removed before the humidifier is operated.

Biocides are usually used in commercial systems (sumps of machining fluid, spray humidification systems, etc.). For example, aldehydes and quaternary amine compounds have been used to control fungal growth (Kapyla, 1985). These kinds of compounds are clearly irritating, may also be sensitizing, and can also enter the air conveyance system. Reducing the amount of biocide in order to minimize risk to the occupants, however, can result in concentrations that are too weak to prevent the growth of all organisms. Biocide usage can also result in changes in the kinds of organisms in reservoirs rather than in a significant decrease in the total number of organisms. The multiple risks of exposure to fungi and biocides must be carefully balanced.

DUCT CLEANING

Duct cleaning involves loosening and removing dirt in ventilation system ductwork. Whether this practice has any effect on exposure to respiratory allergens (either positive or negative) has not been adequately investigated. Chapter 7 discusses duct cleaning in greater detail.

AEROSOL CONTROL

Aerosol control methods may offer some relief in the presence of continuing sources of specific microbial pollutants. Local exhaust that removes pollutants before they can circulate through room air is one of the best options. Machining fluid and bacterial enzyme aerosols have been controlled in this way. Circulating room air through either a central or console air cleaner can reduce particle levels to a steady state that depends on source strengths, dissemination rates, and the rate at which particles are removed by the cleaner. If a HEPA filter is used, all particles larger than 6 μm in size will be removed from the air that passes through the filter. If active sources (e.g., large dust mite populations, active fungus growth, cats) are present, it is likely that dispersal rates will exceed removal rates to a degree and that a steady state will eventually be achieved.

Other modes of aerosol removal depend on electrostatic precipitation and on devices that charge particles so that they become attached to environmental surfaces. The latter devices are not advisable for use with allergens because disturbance of the collecting surfaces (e.g., carpeting, upholstered furniture, walls) will reaerosolize the particles.

Conclusions and Recommendations

Overall, the fungus-associated allergies have been the least well-studied. Little data is available on the distribution of airborne fungal products, dynamics of human exposure, nature of the allergens, factors influencing the quality of skin test and immunotherapy materials, and the basic nature of fungus-related allergic disease.

Research Agenda Item: Initiate and conduct studies to determine the relative etiologic importance, geographic distribution, and concentrations of airborne fungus material associated with indoor allergy.

Fungi grow indoors in damp environments such as basements, window sills, shower stall surfaces, and in dust. Fungal spores and other effluents become airborne indoors when disturbed by air movement and normal human activities. The composition of aerosols of fungus-derived particles depends on the abundance and strength of sources as well as dissemination factors, mixing, dilution, and particle removal.

Research Agenda Item: Investigate the dynamics of fungal colonization of indoor reservoirs and emission of allergens from these sources. The results of such research should permit the risks associated with indoor fungal growth to be evaluated.

Exposure to fungal spores (and possibly other fungal antigen-carrying particles) can produce both IgE-mediated disease (e.g., asthma) and hypersensitivity pneumonitis while other allergens (e.g., dust mite, pollen) produce only the IgE-mediated diseases. It is not clear why or under what conditions fungal particles can have this dual effect.

Research Agenda Item: Study the differences between fungal and other allergen-carrying particles that control the development of hypersensitivity pneumonitis as opposed to IgE-mediated asthma.

CHEMICALS

Low-molecular-weight (LMW) chemical agents have been found to cause immunologic disease primarily in the industrial setting but not generally in the office, school, or residential setting. Nevertheless, a variety of household products may contain immunogenic agents such as reactive chemical anhydrides in epoxy resins and isocyanates in bathtub refinishing agents. In addition, chemical allergy in the industrial setting serves as a model for improving our understanding of allergy mechanisms. More than 150 LMW chemical agents have been reported to cause allergic reactions such as asthma and hypersensitivity pneumonitis (Table 3-3; Butcher et al., 1989; Grammer et al., 1989). As industrialization increases and new agents are introduced, the number of chemicals that cause such allergic reactions is likely to increase. Allergic contact dermatitis to these and other chemicals is another type of allergic response found in industrial settings and an important cause of occupational disease; however, it will not be discussed here.

It has been estimated that in industrialized countries 2 percent of asthma is occupationally related (Salvaggio, 1979). The prevalence of occupational asthma varies with the particular chemical. For example, more than half of all workers exposed to platinum salts became sensitized (Cromwell et al., 1979). Among workers exposed to trimellitic anhydride (TMA), approximately 20 percent developed sensitization (Zeiss et al., 1983), whereas approximately 5 percent of toluene diisocyanate (TDI)-exposed workers developed positive inhalation challenge at levels of less than 20 parts per billion (NIOSH, 1978).

For some industries, studies have estimated the prevalence of allergic

TABLE 3-3 Examples of Chemical Allergens and Indoor Sources

Agent	Source	Reference
Antibiotics	Hospitals, pharmaceutical facilities	Davies et al., 1974, 1983
Other drugs: piperazine, alpha-methyldopa, cimetidine, endofluorane anesthetic	Hospitals, pharmaceutical facilities	Butcher et al., 1989
Metal salts: nickel, platinum, chromates	Platinum processing and nickel or chromium plating plants	Block and Chan-Yeung, 1982; Cromwell et al., 1979; Dolovich et al., 1984; Malo et al., 1982; McConnell et al., 1973; Novey et al., 1983; Pepys et al., 1972, 1979; Pickering, 1972
Anhydrides	Facilities that manufacture curing agents, plasticizers, or anti-corrosive coatings	D. I. Bernstein et al., 1982, 1984; Grammer et al., 1987; Topping et al., 1986; Zeiss et al., 1977
Isocyanates	Facilities that produce or apply paints, surface coatings, or poly-urethane foam	Baur and Fruhmann, 1981; Butcher et al., 1977, 1979, 1980
Ethylenediamine	Buildings in which shellac or lacquer are used	Gelfand, 1983; Lam and Chan-Yeung, 1980
Azo-dyes	Facilities that manufacture or use dyes	Park et al., 1991; Slovak, 1981

reactions to a given chemical among workers. In other instances, estimates have been developed for the number of workers exposed to chemicals such as TMA (20,000 workers) and TDI (50,000–100,000 workers). However, for many chemicals and many industries the number of exposed workers and the prevalence of allergic disease are not known and have not been studied. In many cases, there have been high employee turnover rates in jobs in which workers may develop allergic reactions to chemicals. For example, high turnover rates have been found among platinum workers as a result of respiratory sensitization (Roberts, 1951). In another study of the electronics industry, many of the workers who left their jobs cited respiratory disease as the reason (Perks et al., 1979). Moreover, occupational diseases are generally underreported (I. L. Bernstein, 1981; NRC, 1987a). Although the annual incidence of work-related disease is believed to be approximately 20 per 100 population, one study (Discher et al., 1975) found that only 2 percent of illnesses were actually reported in employers' logs.

Allergic Diseases Caused by LMW Chemicals

The diseases caused by allergic reactions to LMW chemicals are similar to those caused by other, larger (i.e., high-molecular-weight, or HMW) allergens. More specifically, allergic diseases related to LMW chemicals include allergic rhinitis and conjunctivitis, hypersensitivity pneumonitis, asthma, late respiratory systemic syndrome (LRSS), and hemorrhagic pneumonitis. Scientists assume that chemicals act as allergens by forming haptens (covalently coupling) at multiple sites on the surface of a host carrier protein, which could be in the serum, airway, epithelium, or blood cells. Allergic rhinitis or allergic conjunctivitis, or both, may occur as a result. Allergic asthma can be immediate in onset, delayed, or both (Fink, 1982). The hypersensitivity pneumonitis that occurs as a result of exposure to chemicals in the workplace is generally of the acute type. LRSS is a related disease characterized by cough, chills, fever, and myalgias 4 to 12 hours after exposure (Zeiss et al., 1977). Workers with LRSS have high levels of antibody against TMA conjugated with human proteins such as human serum albumin (TM-HSA).

Another disease, hemorrhagic pneumonitis, is caused by immunologic reactions to chemicals such as TMA (Zeiss et al., 1977) and TDI (Table 3-4). For example, after significant exposure in a TMA-sensitized individual, a hemorrhagic pneumonitis and anemia known as pulmonary disease anemia (PDA) syndrome may occur (Patterson et al., 1978). These workers have very high levels of antibody against TM-HSA and very high levels of TMA exposure, usually from hot fumes. The anemia is likely to be an immune-mediated hemolytic type, probably because reactive chemicals like TMA couple easily with cells (they react readily with cell surface proteins). This process results in type II immunologically mediated cytotoxicity, a condition that cannot occur with complete allergens such as foreign proteins because they cannot react covalently with cell surface proteins (Patterson et al., 1979). There have also been reports of hemorrhagic pneumonitis without anemia caused by TDI (Patterson et al., 1990).

Chemical Agents That Cause Allergic Disease

A variety of pharmacologic agents have caused asthma among hospital and pharmaceutical workers when airborne dust is inhaled. Numerous anti-

TABLE 3-4 Characteristics of Syndromes Related to Inhalation of TMA

Characteristic	Rhinitis and Immediate-Type Asthma	LRSS	Pulmonary Disease-Anemia	Irritant Syndromes
Latent period (duration of exposure prior to onset of symptoms)	Weeks to months of work exposure	Weeks to months of work exposure	Weeks to months of work exposure	Occurs on *first* high-work dose exposure
Onset of symptoms after work exposure	Immediate (minutes)	4–12 hours	Progressive with further work exposure	Variable, depending on exposure
Degree (type) of exposure	TMA dust (mild) or fumes	TMA dust (moderate) or fumes	TMA fumes (heavy)	Fumes or dust (great)
Skin tests (TM-HSA)	Positive immediate-type	Negative	Not done	Negative
Total antibody to TM-HSA	Present	High	Very high	Variable
IgE to TM-HSA	Present	Absent	Absent	Absent

NOTE: TMA, trimellitic anhydride; LRSS, late respiratory systemic syndrome; TM-HSA, trimellitic anhydride-human serum albumin.

SOURCE: Zeiss et al., 1982.

biotics, including penicillin, sulfa, and spiramycin, are known to induce specific IgE, positive skin tests, and asthma (Davies and Pepys, 1975; Davies et al., 1974). Other pharmacologic agents including cimetidine and alpha-methyldopa can cause asthma on an immunologic basis, that is, as a result of an antigen being recognized by specific antibody or sensitized cells (Butcher et al., 1989).

Metal salts of nickel, platinum, and chromates can cause rhinitis, conjunctivitis, or asthma (Block and Chan-Yeung, 1982; Cromwell et al., 1979; Dolovich et al., 1984; Malo et al., 1982; McConnell et al., 1973; Novey et al., 1983; Pepys et al., 1972, 1979; Pickering, 1972). Positive skin tests, specific IgE, and positive bronchial challenges have all been reported. Exposure occurs in processing or plating facilities. Some investigators believe that sensitization to platinum is virtually universal, given a large enough exposure.

Acid anhydrides are used as curing agents in the manufacture of epoxy resins. Exposure may occur in a variety of industries including those that manufacture curing agents, plasticizers, or anticorrosive coating agents. In addition to allergic rhinitis, conjunctivitis, and asthma, two other allergic reactions or diseases, LRSS and PDA, described above, may result from TMA. Other anhydrides, including phthalic anhydride and tetrachlorophthalic anhydride, have also caused asthma (D. I. Bernstein et al., 1982; Topping et al., 1986).

Isocyanates are used to produce a number of products including paints, surface coatings, and polyurethane foam. They are also found in some home improvement products such as refinishing agents. In contrast to people who react to TMA, many individuals affected by isocyanates do not have specific IgE or positive skin tests (I. L. Bernstein, 1982). Isocyanate asthma is a major cause of LMW chemical-induced asthma, but to date, the mechanism (i.e., allergy versus nonimmunologic sensitivity) is unknown. In addition to asthma, reports have linked isocyanates with hypersensitivity pneumonitis and an immunologically mediated hemorrhagic pneumonitis (Patterson et al., 1990).

Research has shown that ethylenediamine induces asthma in individuals exposed to shellac or lacquer (Gelfand, 1983; Lam and Chan-Yeung, 1980). Positive skin tests and positive bronchial responses, both immediate and delayed, have been reported. Azo-dyes, such as azodicarbonamide, can also cause asthma (Park et al., 1991; Slovak, 1981). These studies describe positive skin tests and changes in pulmonary functions after a work shift. Exposure to such chemicals may occur in plants that manufacture or weigh dyes.

Formaldehyde is a chemical that is often found at very low levels in homes, offices, and schools and at higher levels in workplaces that use the substance. Asthma is sometimes reported following gaseous formaldehyde exposure; bronchial provocation studies are usually negative but on occasion they may be positive (Hendrick and Lane, 1977; Nordman et al., 1985). Two studies that investigated immunologic sensitization reported negative challenges even in those individuals with specific IgE to formaldehyde (Dykewicz et al., 1991; Grammer et al., 1992). Thus, immunologically mediated asthma resulting from formaldehyde exposure has yet to be proved. Other volatile organic chemicals such as toluene and turpentine can act as irritants but are not specific sensitizers.

Exposure and Risk

Various air sampling and personal monitoring techniques are used to measure chemical exposures (Eller, 1984), but they are not without limitations. For example, intermittent samples may not reflect an individual's actual exposure because of variations in the exposure levels of the chemical. In addition, only a few allergenic chemicals such as TDI and TMA have threshold limit values (TLVs) set by the American Conference of Governmental Industrial Hygienists (ACGIH, 1986) or permissible exposure limits (PELs) set by the Occupational Safety and Health Administration (OSHA; CFR, 1991).

TLVs and PELs are generally established to help prevent chemical toxicity among workers. Thus, they may have no relevance to levels of chemicals that may sensitize an individual or provoke allergic responses and that may be many orders of magnitude below toxic levels. Very few studies report threshold levels for human exposure to chemicals that elicit allergic responses, and those that do describe exposure concentrations present them only as estimates. A Japanese study of 41 workers exposed to two enzymes and three antibiotics showed that the incidence of occupational allergy was correlated with the frequency and concentration of exposure to allergens (Chida, 1986). In other studies, approximately 50 TMA workers were evaluated in a facility that reduced worker exposure over time by improved ventilation, work practices, and respiratory protection. The levels of antibody in workers decreased with decreasing exposure to TMA (Boxer et al., 1987; Grammer et al., 1991b). In a study of isocyanate workers, the group with the highest exposure had the highest prevalence of positive antibody (Grammer et al., 1991a). In a study of 500 workers at a TMA manufacturing facility, five categories of exposure were identified (Zeiss et al., 1992). Only workers in the highest exposure categories developed specific antibody and allergic reactions to TMA.

In contrast to human studies, estimates of exposure in animal models are considerably more accurate. Studies have reported a concentration-dependent immunologic response in a guinea pig model of TDI asthma (Karol, 1983) and a threshold concentration and concentration-immunologic response relationship in a rat model of immunologic TMA disease (Zeiss et al., 1989). It is likely that such relationships also exist in humans, but there are no data to illuminate such linkages. (Quantitative exposure data from animal experiments may not necessarily translate to humans.)

Other than exposure, risk factors for the development of sensitization to a given chemical have not been defined. In studies of risk factors for development of sensitization to chemicals such as TDI, atopy and airway hyperreactivity were either not predictive or only weakly so (I. L. Bernstein, 1982; Chester and Schwartz, 1979; Nicholas, 1983).

Control by Avoidance and Exposure Reduction

Although some of the control measures used in industrial settings may not be directly applicable to the indoor air environment, the model of reduced exposure that results in reduced sensitization rates is applicable to indoor aeroallergens. Prospective studies of TMA workers (Grammer et al., 1991b; Zeiss et al., 1983, 1992) have reported that serial immunologic studies are useful in predicting which individuals are likely to develop immunologically mediated disease. With careful monitoring, those workers who develop specific antibody can be removed from exposure at the onset of any allergic symptoms. Alternatively, if the development of disease seemed very likely, the worker could be relocated at the onset of serologic positivity. In TMA workers, there is evidence that development of specific antibody is predictive of allergic disease, but this finding has not been confirmed in a definitive manner in populations of workers exposed to other chemical allergens.

The timing of removal from exposure relative to disease onset is important. Some data suggest that early removal of workers who develop asthma as a result of chemical exposure will allow most of them to return to normal pulmonary function. In contrast, asthma tends to persist among workers who have had the disease for several years before they are removed from exposure (Chan-Yeung, 1990). This is especially true for workers who already have abnormal pulmonary function.

With respect to reducing exposure, there is evidence that decreasing airborne concentrations of a chemical such as TMA reduces disease prevalence (Boxer et al., 1987; Chan-Yeung, 1990). As outlined earlier, other evidence in animals and in humans suggests the existence of environmental exposure concentration thresholds and environmental concentration exposure-immunologic response relationships. If these thresholds and relationships could be defined for chemical allergens, reducing exposure could be the best approach to preventing allergic disease caused by these substances. Exposure reduction measures would include improved ventilation, work practices, and protective equipment.

Conclusions and Recommendations

Many of the protein allergens have long been recognized, but a lengthening list of newly recognized allergenic chemicals is developing. Allergic diseases caused by these chemicals can differ from those caused by protein allergens in terms of symptoms, mechanisms of action, and appropriate treatment. The diseases can differ also in terms of etiology and exposure, i.e., often occurring at the work site. A better understanding of these differences will assist in the formulation of improved measures of prevention and treatment.

Research Agenda Item: Determine the types of allergic diseases caused by reactive allergenic chemicals, their prevalence rates, and the mechanisms responsible for the resulting airway reactions.

A body of knowledge about chemical allergens is available, but many areas have not been well studied. Other chemicals besides those already reported to cause allergic reactions may provoke responses. Thus, as new chemicals are introduced, the list of agents that elicit allergic reactions is likely to grow.

Research Agenda Item: Identify the risk factors, such as a specific immunologic response, that are predictive of the development of chemically induced sensitization or allergic disease, and as soon as possible after their introduction, determine the sensitizing potential of new chemical entities. This knowledge will facilitate the development of primary and secondary preventive strategies.

The allergic rhinitis, conjunctivitis, and asthma that arise from exposure to chemicals appear to be due to classic immunologic reactions. However, late respiratory systemic syndrome (LRSS) and immunologic hemorrhagic pneumonitis occur only in response to chemical exposures and are not the result of response to the usual protein allergens; the mechanisms of immunologic damage in these two cases are not entirely known. The mechanism of non-IgE-mediated isocyanate asthma is also unclear.

Research Agenda Item: Determine the disease mechanisms of chemically induced LRSS, of immunologic hemorrhagic pneumonitis, and of non-IgE-mediated isocyanate asthma. Appropriate in vitro or in vivo models should also be developed.

LMW reactive allergenic chemicals can cause immunologic sensitization and consequent allergic reactions. At a minimum, hundreds of thousands of U.S. workers are exposed to chemicals that can form haptens with airway proteins and induce allergic diseases. The goal of reducing the incidence and severity of allergic disease caused by chemical exposure is achievable, although it may not be possible to prevent all such disease.

Research Agenda Item: Determine the number of workers exposed to allergenic chemicals in various industrial and non-industrial settings and the prevalence of allergic disease resulting from such exposure. Populations in close contact with reactive allergenic chemicals and highly potent sensitizers would be logical candidates for study.

For those individuals who develop allergic disease from exposure to chemicals, it is important to determine their long-term prognosis. In particular, if immune responses that are predictive of allergic disease can be identified, and reduced exposure can be shown to result in resolution of disease (and disappearance of immunologic sensitization), then reduced exposure may represent the most practical approach for preventing allergic disease arising from chemical exposure.

Research Agenda Item: Conduct dose-response studies in humans to determine both the relationship between allergen concentration and immunologic response, and a threshold environmental exposure concentration for sensitization.

In addition to studies of the threshold concentrations necessary for sensitization, thresholds for elicitation of allergic reactions to chemicals once sensitization has occurred also require study. Such thresholds exist but vary markedly from individual to individual, as shown by bronchoprovocation tests performed with high-molecular-weight allergens. This is probably also the case with chemical allergens, but the issue has not been systematically studied. If threshold concentration levels do exist but are highly variable, and in some cases very low, the only practical way to manage sensitized individuals is to terminate the exposure.

PLANTS AND PLANT PRODUCTS

It is well known that plants produce substances, materials, or products that are allergenic in humans. The best understood are pollen from trees, grasses, and weeds and the oils or resins from the leaves of poison ivy and poison oak. Airborne pollen can produce allergic rhinitis or asthma, or both, in the susceptible atopic population. The oils from poison ivy can produce allergic contact dermatitis; susceptibility is not limited to atopic individuals. Pollen production is seasonal and varies in quantity, depending on geographic location and climatic conditions. Local pollen production also varies annually, depending on weather conditions.

Outdoor pollen can enter the indoors with ventilation air and can also be transported indoors on people and their clothing as well as on pets. Clothing that was previously hung outdoors to dry is another source of pollen. Studies have shown that indoor pollen concentrations can be quite high during the pollen production season (O'Rourke et al., 1990; Platts-Mills et al., 1987; Pollart et al., 1988). Reports of indoor pollen concentrations have ranged from 0 to 5.5 million pollen grains per gram of house dust (O'Rourke and Lebowitz, 1984). Elevated concentrations of pollen in indoor air are found under open window conditions (in one case 600 grains per cubic meter in a room) (O'Rourke et al., 1989). These high airborne pollen concentrations are rare, but 30 pollen grains per cubic meter are not uncommon. Since people spend a majority of their time indoors (NAS, 1981; Quackenboss et al., 1991a,b), over 20 pollen grains can be inhaled in a typical household each day (on average, over all seasons). During the spring, when average outdoor pollen concentrations approach 400 grains per day, indoor exposure can approach 40–80 grains per day (O'Rourke, 1989).

Indoor plants are commonly found in office or school environments and in the home. Most are grown for their green foliage and accommodate low light or a lack of direct sunlight. As such, most do not flower in these environments and therefore are not pollen sources. Indeed, most plants grown indoors are not highly allergenic. Nevertheless, as more plants are used indoors, especially in large numbers in office settings, those considered not allergenic or only slightly allergenic may need to be reexamined. For example, Axelsson and colleagues (1987b, 1991) report that the leaves of *Ficus benjamina* (weeping fig) can produce airborne IgE-mediated rhinitis and asthma. They estimate that the risk of sensitization among truly atopic individuals is 6 percent. Hausen and Schulz (1988) report on a woman who developed conjunctivitis, rhinitis, and asthma from the nectar secretions of an ornamental plant, *Abutilon striatum* (flowering maple). Ford and co-workers (1986) report the development of IgE antibodies to pollen allergens from *Parietaria judaica,* an outdoor allergenic member of the nettle family found in the Mediterranean region but not likely to be found indoors.

Most recent reports in the literature regarding allergic reactions to indoor plants involve contact dermatitis produced by airborne allergens. Plants provoking such reactions include *Allium* (garlic), chlorophora (iroki; Fernandez de Corres et al., 1984, 1985), *Chrysanthemum*, citrus, *Coleus* (Van Hecke et al., 1991), common ferns (Geller-Bernstein et al., 1987), *Compositae* (daisy), *Frullania* (Pecegueiro and Brandao, 1985), lichens, *Lilium* (lily), *Pelargonium* (geranium), *Philodendron, Pinus* (pine), *Platycodon grandiflorum* (balloon flower; Nagano et al., 1982), *Primula* (primrose), *Typha latifolia* (cattail), and *Umbelliferae* (family name for parsley, carrots, coriander, etc.). Because many of these studies report the effects of mixtures of houseplants and garden plants, it is difficult to determine which house or indoor plants are truly allergenic.

The types of plants grown in the home will frequently differ from those grown in the office or workplace, in that the home gardener may well experiment with different types of plants grown in different locations at various times of the year (e.g., growing herbs in an indoor window box with a southern exposure). Similarly, home gardeners with a greenhouse may grow a wide variety of allergenic plants. Although there is only limited knowledge of the extent of allergic disease from allergenic indoor plants, it seems logical to assume that if increased use is made of indoor plants that are pollen producers, atopic individuals may find indoor environments as unpleasant as the outdoors during the traditional pollen season. In addition, indoor blooming patterns are sometimes manipulated to be different from outdoor "normal" seasonal patterns.

Plant Products

In addition to the plants themselves, plant materials such as *Psyllium* and latex can be brought indoors in a variety of consumer products. *Psyllium* is a grass from India used as a fiber and bulk supplement for bowel control. There are reported cases of severe anaphylactic reactions among workers who produce this product, as well as among those who use it. Airborne exposure to the product can also produce allergic reactions among susceptible individuals during use in the home.

Plant sources of allergens that have been shown to produce asthma in selected occupationally-exposed populations are presented in Chapter 2 (Table 2-4). Some of these plant materials can be present in residential and other indoor environments because of the activities of the occupants. Whether they will pose a hazard to the occupant depends on many factors including the amount of airborne exposures. Dried flowers are another example of a potentially allergenic product that can be brought indoors.

LATEX

Latex allergy has recently received substantial attention because of increasing reports of its occurrence and its potential, in certain individuals, to produce life-threatening anaphylactic reactions. The latex (or sap) of the *Havea brasiliensis* plant is the source of natural rubber (*cis*-1,4-polyisoprene). Although rubber production yields a product that is 93–95 percent polyisoprene, the final product may be as much as 2–3 percent protein by weight (Windholz et al., 1983). The protein component of the latex contains allergens that are responsible for numerous recent reports of latex allergy. Exposure occurs by direct contact or by inhalation of dust or powder that is often used for packaging. Patient reactions to latex have ranged from contact urticaria to systemic anaphylaxis. Persons at increased risk include patients with congenital problems of the spinal cord (e.g., myelomeningocele), patients with recurrent bladder catheterizations, or any patient with a history of rash, swelling, or itching after blowing up balloons, wearing rubber gloves, or using other latex-containing products. On rare occasions, latex condoms can cause allergic reactions.

Most cases of allergy to latex are mediated by IgE antibodies. However, because of varying source materials, the heterogeneity of immune response, and a variety of test methods, the identities of the specific protein allergens remain to be determined (Jones et al., 1992).

Case reports of latex allergy began to appear in the literature in 1979 (Nutter, 1979). Most early reports were of contact urticaria (Forstrom, 1980; Meding and Fregert, 1984; Nutter, 1979); reports of allergic rhinitis (Carrillo et al., 1986), anaphylaxis (Axelsson et al., 1987a; Slater, 1989; Turjanmaa et al., 1984), and asthma (Seaton et al., 1988) followed. Occupational asthma related to latex hypersensitivity has also been described; the response has been attributed to inhalation exposure to cured latex during the inspection and packaging of finished gloves (Tarlo et al., 1990). Fisher (1987) has also reported contact urticaria and anaphylactoid reaction as a result of exposure to cornstarch surgical glove powder. Swanson and colleagues (1992) support the conclusion that dust from latex gloves is a significant occupational aeroallergen, reporting 28 medical center employees diagnosed with rhinitis or asthma caused by exposure to dust from latex gloves.

In addition to health care workers and manufacturers, children with spina bifida are at increased risk for latex allergy, and credible evidence supports an IgE-mediated mechanism (Slater, 1989; Slater et al., 1990a, 1991; Spanner et al., 1989; Turjanmaa, 1987). One prospective study reported that 5 out of 12 spina bifida patients (41 percent) have IgE antibody specific for rubber proteins (Slater et al., 1990a). Another report suggested that IgE titers to latex allergen might be due to parenteral exposure to latex surgeon's gloves during primary closure of the meningomyelocele and that early initial exposure and frequent reexposures may predispose children with spina bifida to rubber allergy (Slater et al., 1990b).

Recent episodes of fatal and life-threatening anaphylaxis have made it increasingly urgent to identify the specific allergen(s) responsible for these reactions (Kelly et al., 1991). Jones and others (1992) and Turjanmaa and colleagues (1988) have reported large variations in the latex allergen content of gloves from different manufacturers; powder-free gloves were significantly less allergenic in this survey. Slater and Chhabra (1992) reported that all patients with latex-specific IgE had antibodies to a 14-kDa peptide present in an extract made from nonammoniated latex; many sera recognized a 20-kDa peptide as well. They concluded that current data are insufficient to identify definitively the major allergen(s) and suggested that studies of latex extract would be useful in further characterizing the immune response to natural rubber.

With recent increases in the production and use of latex gloves and other rubber products, clinical sensitivity may be more common than in previous years, and the regulatory, research, and medical communities are responding accordingly. A current regulatory review by the FDA may result in relabeling of latex products—including latex gloves, condoms, catheters, dental dams, and enema kits—to highlight the risks of latex hypersensitivity. The FDA has already issued a Medical Alert (MDA91-1, March 29, 1991) containing recommendations to health professionals regarding the use of latex products. The American College of Allergy and Immunology has also issued interim guidelines on latex allergy.

As noted by Slater (1989), and others, and described in the medical alert and guidelines, patients with a history of rubber-induced allergic reactions, as with all life-threatening allergy, should practice avoidance as the main form of treatment. Health care workers should use nonrubber gloves when treating these patients, and appropriate care should be taken to avoid exposing latex-sensitive patients to either direct or aerosolized contact (e.g., from the cornstarch dust used in packaging latex gloves). It has also been suggested that latex allergen may be carried on syringe needles from the rubber stoppers of multiuse vials (Silverman, 1989).

Finally, it is important that sensitive individuals be recognized prior to surgery so that proper precautions can be taken to avoid latex exposure and minimize the potential for experiencing the associated adverse reactions. In addition to allergen avoidance strategies, some authors believe that latex-sensitive patients should be premedicated according to protocols for the prevention of anaphylactic reactions in surgery (Bielory and Kaliner, 1985; Greenberger et al., 1985; Lasser et al., 1987).

CONCLUSIONS AND RECOMMENDATIONS

Indoor plants are commonly found in offices, schools, and the home. Although most indoor plants do not produce aerosols of allergen-containing particles, as more plants are used indoors, especially in large numbers in office settings, the risk of exposure to plant allergens increases.

Research Agenda Item: Assess the significance of workplace exposures to indoor plants, including the contribution to the overall magnitude of indoor allergic disease.

Latex allergy has recently received substantial attention because of increasing reports of its occurrence and its potential, in certain individuals, to produce life-threatening anaphylactic reactions. In addition to health care workers and manufacturers, children with spina bifida are at increased risk for latex allergy.

> **Research Agenda Item:** **Conduct research to further characterize the immune response to natural rubber. This effort should include studies of the incidence and prevalence of natural-rubber-related allergic disease.**

4

Mechanisms of Immune Function

For some years after the discovery of antibodies, the immune response was thought to be purely protective. It was soon discovered, however, that the same mechanism that protected against infectious microbes also could be activated by innocuous substances, such as milk proteins, to cause immune responses that were potentially dangerous.

When studying the toxicity of extracts of sea anemones, early twentieth-century investigators observed that dogs given a second injection, several weeks after the first, often became acutely ill and died within a few minutes. This response was called anaphylaxis (Gr., *ana* = against; *phylaxis* = protection), implying incorrectly that it represented an increase in susceptibility to a toxic substance rather than the expected increase in resistance. (Earlier discoveries of anaphylaxis were overlooked, unfortunately, as sometimes happens when valid observations are ignored until they can be accommodated within a conceptual framework.) Almost simultaneously, other observers noted similar responses in guinea pigs to widely spaced injections of nontoxic protein antigens. With the increasing use of horse and rabbit antisera to treat various infectious diseases in man, various adverse reactions due to the immune response soon became commonplace.

In an attempt to organize a chaotic set of observations, the term allergy (Gr., "altered action") was introduced to cover any altered response to a substance induced by a previous exposure to it. Increased resistance, called immunity, and increased susceptibility to reactions were called hypersensitivity or allergy and were then regarded as opposite forms of immunizations. Through usage, however, "allergy" and "hypersensitivity" have become synonymous; both terms refer to the altered state induced by an antigen, in which pathological reactions can be subsequently elicited by that antigen or by a structurally similar substance. Within the context of the allergic response the antigen is usually referred to as the "allergen," and the immunized individual, previously called immune, is called sensitive, hypersensitive, or allergic.

Allergic responses were originally divided into two classes, immediate and delayed, on the basis of the lag in their appearance—several minutes after the administration of antigen in the former, and several hours or even a few days in the latter. These terms are still used, but they are now endowed with a different meaning. Not only the reactions that appear within minutes, but also some of the more slowly evolving ones, are mediated by freely diffusible antibody molecules. To emphasize this common feature, both are now called *immediate type* responses (although "immediate" is not to be taken literally). In contrast, the *delayed type* are those slowly evolving responses that are mediated by specifically reactive ("sensitized") T lymphocytes rather than by freely diffusible antibody molecules; hence they are also called *cell-mediated hypersensitivity.* They constitute part of a larger group of reactions, called *cell-mediated immunity,* in which similar mechanisms are also involved in resistance to many infectious agents and to neoplastic cells.

Chapters 1 and 2 of this report provide general descriptions of concepts and definitions related to sensitization, hypersensitivity, and IgE-mediated allergy. To recapitulate in summary, allergic disease occurs when a susceptible individual is exposed to an allergen and becomes sensitized. Additional exposure to the sensitizing allergen leads to the development of an allergic reaction through the release of histamine and other chemicals from mast cells. Exposure to other substances (e.g., environmental tobacco smoke) may serve to promote the development of allergic reactions and disease.

The typical allergic reaction such as hay fever begins when allergenic proteins from substances such as grass pollen, are inhaled by a sensitive individual and pass through the nasal mucosa. The inhaled allergens then interact with IgE antibodies on the surface of mast cells in the mucosa to cause the release of histamine and other bioactive mediators. These mediators, in turn, cause vasodilation of blood vessels and stimulate secretory cells. The result is the clinical symptomatology and manifestations of nasal congestion, rhinorrhea, and conjunctivitis commonly associated with allergies.

As discussed previously in this report, nearly 20 percent of the population suffer to a greater or lesser degree with allergies involving localized anaphylactic reactions to extrinsic allergens such as grass pollen, animal dander, mites in house dust, and so on. Contact of the allergen with cell-bound IgE in the bronchial tree results in asthma, in the nasal mucosa results in hay fever, and in the conjunctival tissues results in allergic conjunctivitis.

The past 10 to 15 years have brought increased recognition of the importance of indoor allergens—in particular, of protein allergens—in generating allergic responses and asthma (Burge and Platts-Mills, 1991a; Burrows et al., 1989; Platts-Mills et al., 1991a). In part, this recognition is based on an awareness of the clinical significance of the increased time spent by modern humans indoors (Spengler and Sexton, 1983). This increased exposure has led to assessments of the potential contribution of proteins inhaled in the indoor environment to the pathogenesis of several allergic conditions, including allergic rhinitis, sinusitis, allergic conjunctivitis, asthma, and hypersensitivity pneumonitis. This chapter addresses the basic mechanisms of cellular, tissue, and airway responses to inhaled proteins. It seeks to use these responses as a framework to plan control strategies based on patterns of exposure and epidemiology.

RESPONSE OF AIRWAY CELLS AND TISSUES TO INDOOR ALLERGENS

This section discusses the resident cells and tissues that are involved in airway host responses; the discussion is organized on the basis of the structure of the upper and lower airways. A second topic is the function of immune and inflammatory cells that may infiltrate resident cells and tissues of these airways in response to an immune reaction directed against inhaled protein allergens. Table 4-1 identifies the cells of the airway and highlights their effector function (i.e., the action they take against a target cell). The

TABLE 4-1 Structural and Inflammatory Characteristics of Cells of the Airway

Cell	Effector Function
Epithelial	APC, cytokines, mediators
Endothelial	APC, cytokines, mediators
Smooth muscle	Bronchial tone, cytokines, mediators
Fibroblasts	Cytokines, mediators, growth factors
Macrophages	APC, cytokines, mediators
Dendritic cells	APC, cytokines, mediators
T cells	Cytokines, specific response (TCR)
B cells	APC, Ig production, mediators
Mast cells	Cytokines, mediators, specific responses (IgE)
Eosinophils	Cationic proteins, cytokines, mediators
Neutrophils	Enzymes, cytokines, mediators
Neuroepithelial cells	Neuropeptides

NOTE: APC, antigen-presenting cells; Ig, immunoglobulin; and TCR, T cell receptor.

final portion of this section discusses the various mediators of inflammation and hyperreactivity in airways and the ways in which these mediators may influence the airway response to inhaled allergens.

Epithelium and Other Airway Structural Tissues and Cells

EPITHELIAL CELLS

The resident cells and tissues of the upper and lower airways include the epithelial covering of the upper airways as well as the lining of the bifurcations of the bronchi and extending to the respiratory bronchioles. The epithelial cells that cover these structures are ciliated and known as epithelial brush-like cells (or type III epithelial cells). These bronchial brush cells have surface cilia that beat material and proteins caught in mucus toward the upper airway and out of the lower and upper airway system. A similar cell is the alveolar brush cell, which is found in the alveolar space (Hunninghake et al., 1979).

In addition to the type III alveolar epithelial or brush cells, two other types of epithelial cells line the bronchial mucosa or overlay the alveolar structure. Known as type I and type II alveolar epithelial cells, these have various and diverse functions. Both the bronchial brush cell with columnar epithelium (the type III epithelial cell noted above) and type I epithelial cells are end-stage differentiated cells that cover the vast majority of the respiratory epithelium. Type II alveolar epithelial cells are capable of dividing and may give rise to both type I and type III cells through their proliferative activity (Hunninghake et al., 1979). They are the source of phospholipid-rich surfactant, which maintains the integrity of the alveolar spaces; in addition, they may produce factors that are involved in nonspecific host defense (Hunninghake et al., 1979). Goblet cells or mucus-producing cells found in the bronchi are also prominent; such cells can produce materials, in particular, mucus, that blanket the bronchial brush cells and enhance their functioning.

SUBEPITHELIAL STRUCTURES

Below the overlay of epithelial cells in the respiratory mucosa is a rich network of subepithelial structures. Obviously, blood vessels are present, which contain both endothelium and smooth muscle; also to be found are other resident cells that are involved in maintaining ground structure. These include fibroblasts; mast cells in a resting or basal, nonactivated state; nerve cells; and a rich biochemical network of ground substances including collagen, fibronectin, and other structural proteins and proteoglycans that contribute to the integrity of the airways. Within the bronchi, structural integrity is further maintained by bronchial smooth muscle; the larger bronchi and the trachea are supported as well by cartilaginous rings. All of the cells that make up the tissues of the upper and lower airway—epithelial cells, fibroblasts, cartilage cells, smooth muscle cells, blood vessels comprising endothelial and vascular smooth muscle cells, nervous tissue, and resident nonactivated inflammatory cells, such as mast cells—can and do respond to inhaled materials with nonspecific host defense (Hunninghake et al., 1979). Moreover, once such a response has been evoked, there may be a further influx of immune and inflammatory cells to the airway.

Immune and Inflammatory Cells in the Airway

A wide variety of immune and inflammatory cells are involved in airway responses to inhaled proteins in the bronchi as well as in the upper and lower airways. Specific responses are initiated by interacting sets of cells composed of macrophages and other significant antigen-presenting cells, as well as lymphocytes—in particular, T helper lymphocytes (although other forms of T lymphocytes and B cells are also involved in this response). Figure 4-1 shows the interaction of allergen with relevant immune and in-

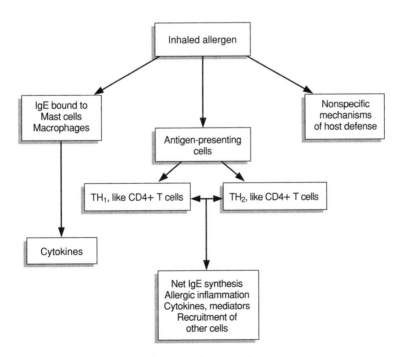

FIGURE 4-1 Immune cells and their interactions with allergens.

flammatory cells in the airway. The discussion below of the response to inhaled protein first deals with the role of macrophages and T and B cells and then describes mast cells, eosinophils, and neutrophils, other cells that are significant in airway host defense.

MACROPHAGES

In terms of number and presumably function, the alveolar and airway macrophages are the predominant airway inflammatory and immune effector cells. Airway macrophages are derived from precursor blood monocytes that migrate to the airways and differentiate there (Unanue and Cerottini, 1989). (The precursors of blood monocytes are monoblasts and promonocytes, which are found in bone marrow.) There are other cells in body tissue that are similar to airway and/or alveolar macrophages in distribution and differentiation—for example, Kupffer cells in the liver and osteoclasts of bone.

FUNCTION

The function of macrophages is to provide nonspecific as well as specific host defense in various tissues; these tissue macrophages can be resting or activated, depending on local conditions. The ability of macrophages to "process" extracellular proteins, including inhaled proteins taken from ambient inspired air and leached from particles, makes them primary airway defenders; they not only eliminate inhaled proteins but also act as allergen-presenting and allergen-processing cells (Unanue and Cerottini, 1989). In addition, macrophages secrete diverse bioactive molecules that influence the physiology of many tissues; in the upper and lower airways they are critical to physiologic function. The nature of these proteins and of other host defense factors involved in the generation of responses by macrophages will be covered in more detail elsewhere in this chapter; macrophages are important in the maintenance of airway tissue.

ANTIGEN-PRESENTING CELL FUNCTIONS

The antigen-presenting cell (APC) is involved in immune and inflammatory reactions in the airway. Antigen presentation is mediated by the expression of specific proteins on the surface of the airway macrophage, which functions in the actual presentation and recognition of protein allergens and other APC antigens—by helper T cells in particular. The response to both inhaled and soluble proteins clearly depends on the CD4+ subtype of helper T cells (see the discussion below); thus, the role of the class II major histocompatibility complex (MHC) macrophages in presenting protein fragments to these T cells is critical (Unanue and Cerottini, 1989).

ANTIGEN PROCESSING

Over the past few years a great deal has been learned about the actual molecular configuration and mechanism of protein processing, which occurs through the subcellular organelles of antigen-presenting macrophages, and its role in generating immunogenic complexes composed of fragments of protein allergens and class II MHC molecules. The identification in 1987 of a crystalline structure for MHC molecules included an antigen fragment-binding pocket within the three-dimensional structure of the molecules (Brown et al., 1988). This discovery led to the postulation of a new model for the recognition by T cells of protein fragments derived from whole allergens or proteins (Brown et al., 1988).

In this process, protein allergens are taken up by the plasma membranes of antigen-presenting macrophages into a phagocytic vacuole that eventually fuses with endosomes. Next, enzymes associated with a nine-subunit subcellular organelle called a proteosome degrade these foreign proteins to peptides. These peptides can then associate with the binding groove of class II MHC that are molecules expressed on the inner membrane of the endocytic vacuole (Martin and Goodman, 1990). The process thus generates a complex composed of a fragment of protein anywhere between 9 and 20 amino acids in length; the complex is bound to the binding groove of a class II MHC molecule. This complex, when redisplayed on the surface plasma membrane of the antigen-presenting macrophage, allows a specific T cell receptor with binding specificity for this particular antigenic fragment (or T cell epitope) to bind to the APC and initiate an immune response.

In addition to this critical protein-processing/antigen-presenting function of macrophages and other APCs, these cells also generate a number of important substances critical to host defense in a nonspecific manner. These substances include cytokines such as interleukin-1 (IL-1), tumor necrosis factor, interferons of the alpha and gamma categories, protein-derived regulatory mediators such as fibronectin, transforming growth factor β, and platelet-derived growth factor (Martin and Goodman, 1990). A number of bioactive lipids are also synthesized by macrophages and APCs, including leukotrienes (LTB4 and LTC4), prostaglandin (PG) E2, and other arachidonic acid metabolites including PGI_2, and hydroxyeicosatetraenoic acids (HETES) and other products of cellular arachidonate (Rosenwasser, 1986). Another bioactive lipid produced by macrophages is platelet-activating factor (PAF). Complement proteins and enzymes including collagenase, elastase, and cathepsins are also major secretory products of airway macrophages and play a significant role in host airway defense (Borish et al., 1991).

OTHER APCs IN THE AIRWAY

In addition to alveolar macrophages, other subsets of cells within the airway can perform some of the major immunological and host defense functions of macrophages. These other classes of antigen-presenting cells include dendritic cells (Borish et al., 1991), which are found in lymphoid tissue and in the subepithelial areas of normal and asthmatic airways; activated B lymphocytes; epithelial cells; fibroblasts; smooth muscle cells; and endothelium. Under the proper conditions, these cells function in vitro as APCs, similar to macrophages. Whether that is their primary role in vivo, however, has not been adequately established.

T LYMPHOCYTES

The major cell type with which antigen-presenting macrophages and other APCs interact in the airway is the CD4+ subset of T cells. Table 4-2 lists the major classes and subpopulations of T cells and notes their major functions and the immune complexes to which they bind (i.e., their specificity). T lymphocytes are prominent among the various populations of airway cells; their importance lies in their major role in the recognition of inhaled proteins (Robinson et al., 1992). T lymphocytes arise from bone marrow-derived precursors selected by thymic maturation through a complicated selection and deletion process that occurs during T cell maturation. The T cells that emerge during this process populate peripheral lymphoid tissues. The majority of such cells display unique combinations of cell surface proteins; it is these proteins that differentiate and function within the mature T cell compartment. Most mature T cells—in fact, the vast majority of T cells that have been selected through the thymus—express on their surface the protein markers CD2 and CD3.

T CELL RECEPTORS

On allergen-specific T cells, the presence of the CD3 marker is associated with coexpression of specific T cell receptors that are developed through molecular recombination events. These events can encompass the association and recombination of various gene segments involved in specific recognition of protein antigens in conjunction with self class II MHC molecules. Hence, the vast majority of CD3+ T cells coexpress an α and a β chain for a T cell receptor. Two or three distinct genetic segments encode these chains, which confer on the cell the specific capability to bind to a protein antigen and class II MHC molecule complex on APCs.

The T cell receptor α chain comprises a variable region, a joining region, and a constant region; the β chain is composed of a variable region, a

TABLE 4-2   Classification of T Lymphocytes

CD2/CD3 Mature T cell	Specificity	Function
CD8 TCRαβ	Class I MHC and peptide	Cytotoxicity Immunoregulation
CD4 (TH1) TCRαβ	Class II MHC and peptide (allergen)	Delayed-type hypersensitivity and host defense IL-2, IL-3, IFNγ, lymphotoxin, GM-CSF
CD4 (TH2) TCRαβ	Class II MHC and peptide (allergen)	Antibody synthesis (IgE) IL-4, IL-5, IL-9, IL-10, IL-3, GM-CSF
CD4⁻, CD8⁻ TCRγδ	Unknown CD1, heat shock proteins	Epithelial defense

NOTE: CD, cluster designate for all surface proteins; TCR, T cell receptor proteins αβ or γδ; MHC, major histocompatibility complex; TH, T helper cells; IL interleukin; IFN, interferon; GM-CSF, granulocyte macrophage colony-stimulating factor; and IgE, immunoglobulin E.

diversity region, a joining region, and a constant region. These seven genetic segments make up the two proteins that are involved directly in recognizing inhaled foreign proteins in the airway. The CD2-CD3+ T cells can be further divided into subcategories on the basis of other cell surface proteins, which indicate the differentiation of these cells according to specific functions. For example, the ability of a particular subset of T cells to help B cells produce antibodies or to help macrophages and other accessory cells express delayed-type hypersensitivity is critically dependent on the coexpression of the CD4 marker on the T cell in conjunction with the CD2 and CD3 markers for T cell delineation.

THE CD4 SUBPOPULATION

The CD4 marker has received a great deal of attention because it is the major receptor for the human immunodeficiency virus (HIV), whose coprotein, gp120, binds to the CD4 protein. The CD4 protein is therefore an important biological marker of T cells, and it makes T cells a target for retroviral infection and eventual destruction in HIV disease. The CD4 marker is important in another regard as well: it acts as a cofactor in recognizing immunogenic complexes of protein antigens in conjunction with class II MHC molecules. Thus, the subset of CD4+ T cells is the helper cells involved in the response to inhaled protein and, in particular, to inhaled allergens, triggering an allergic response (Robinson et al., 1992).

CYTOKINES OF T CELL SUBPOPULATIONS

There has been much research in the past few years to further subdivide the αβ T cell receptor-positive CD2/CD3/CD4+ population into two groups. One group, known as the TH1 subpopulation of CD4+ T helper cells, is involved in the mediation of delayed-type hypersensitivity; these cells have a unique protein profile that is secreted at activation. Known as cytokines, these proteins are one of the effector molecules generated by T cells that contribute to host defense. Therefore, for example, infection with an intracellular parasite with a preference for infecting macrophages leads (in both murine and human systems) to generation of a T cell response characterized by the production of gamma interferon and lymphotoxin (Mossmann and Coffman, 1989). IL-4 and IL-5, the T cell cytokines involved in antibody production, and the immunoglobulin E (IgE) and IgA isotype switch are not utilized.

By contrast, the other group of αβ T cells, the TH2 subpopulation of CD4+ T helper cells, is quite prominent in responses to allergens in the airway, both in vitro and in vivo (Wierenga et al., 1990). These CD4+ TH2 cells have an enhanced capability for producing IL-4, IL-5, IL-9, and IL-10, cytokines that are involved in the generation of allergic antibody responses and allergic inflammation. The next section discusses the potential role of these TH1 and TH2 CD4+ T cells in greater detail, describing the role of cell-mediated immunity in T cells and macrophages in the regulation of IgE-mediated disease and in hypersensitivity responses to inhaled proteins.

THE CD8 SUBPOPULATION

In addition to CD4+ populations in the airway, researchers have identified CD8+ T cell subsets. Known as cytolytic T lymphocytes (CTLs), these cells perform a cytotoxic function and are restricted to a set of targets different from those of CD4+ T cells. These CTL targets are protein antigens expressed on the surface of antigen-presenting cells in the context of class I MHC molecules. (Class I MHC molecules are ubiquitously expressed and play a prominent role in recognizing viral antigens after viral infection of all sorts of target cells, including B lymphocytes, macrophages, and epithelial cells.) The potential role of CD8+ T cells in regulating the response in the airway to inhaled proteins and allergens is not yet clear, but it is reasonable to assume that it involves limiting viral infection. In particular, it may involve limiting the potential damage that might result from inhaled allergenic proteins mediated by CD4+ TH2-type cells. CD8+ cells may also contribute to other regulatory phenomena, especially given that, early on, these cells were thought to act as suppressor cells in the regulation of antibody synthesis. Although antigen-specific suppressor cells have been quite difficult to identify in terms of cloning and structure, it is entirely possible that cells with a cytotoxic function bearing the CD8 marker may play some role in the regulating of CD4+ responses to inhaled proteins and allergens.

One potential role function of CD8+ T cells may be in the regulation and mediation of airway responses in patients with hypersensitivity pneumonitis (HP). While precipitating IgG to the offending antigen in HP is often an important diagnostic and possibly pathogenetic marker, recent studies have identified an enhanced utilization of CD8+ T cells in the bronchoalveolar lavage (BAL) of patients with HP. The function of these CD8+ BAL T cells is unknown, but their potential role as regulators or cytotoxic effectors for HP pathogenesis is intriguing.

OTHER T CELL SUBPOPULATIONS

Subpopulations of CD4+ and CD8+ T cells can also be differentiated according to whether they express the CD45 RA or CD45 RO molecule on their surfaces. CD45 RA and CD45 RO are protein molecules that are expressed as the cells undergo differentiation. One gene codes for each of CD45 RA and CD45 RO; and alternative gene splicing of messenger RNA yields either product as the T cells differentiate. For example, CD45 RA is expressed primarily on naive cells that have not yet been selected for their antigen specificity. The expression of CD45 RO on CD4+ or CD8+ cells is associated with a memory phenotype; it indicates that the cell has been stimulated previously through its antigen receptor. CD45 RA and CD45 RO both participate in the activation of T cells through enzymatic activities related to intracellular signaling.

B LYMPHOCYTES

B lymphocytes, the other major population of lymphocytes, are also involved in the immune and inflammatory reaction that occurs in the airways. B lymphocytes are derived from bone marrow precursors that mature through development in the bone marrow and by exposure to microenvironmental milieus that include a number of growth- and development-related cytokines. The process begins with the selection of B cells for their antigen-specific binding capabilities; these are reflected in membrane IgM on antigen-specific but immature cells. Following selection, the sequential effect of various T cell-derived cytokines, including IL-4, IL-5, and IL-6, drives the undifferentiated but antigen-specific B cells to differentiate further into antibody-secreting plasma cells. This process is accompanied by the specific selection of immunoglobulin isotypes different from the original surface IgM.

The products of B cells (i.e., secreted immunoglobulins) can be divided into five major isotypes depending on the heavy-chain structure of the immunoglobulins. These isotypes have different functions; of particular interest for inhaled indoor allergens are the isotypes associated with IgE, the major allergenic (or reaginic or atopic) antibody marker, and IgA, a mucosa-derived isotype involved in host defense. Therefore, in the airway, the generation of both IgE and IgA is extremely important, and the ability of CD4+ TH2 cells to produce the cytokines IL-4 and IL-5, factors involved in the specific selection of these reaginic and respiratory antibody isotypes, is critical (Wierenga et al., 1990).

In addition to their antibody-synthesizing capabilities, B lymphocytes are also of great interest because of several other immunological functions. These include the processing of antigens and proteins and the expression of both MHC class II- and class I-associated protein antigens for recognition by T lymphocytes.

Mast Cells

Mast cells are important immune effector and inflammatory cells in the airway. Like the T and B lymphocytes and macrophages, mast cells are derived from bone marrow precursors that generate mature airway tissue cells with significant host defense capability. Mast cell development is influenced by various cytokines including granulocyte macrophage colony-stimulating factor (GM CSF), IL-3, IL-4, IL-9, and IL-10; these substances seem to induce and drive the ability of precursors of mast cells to be selected for mature mast cell growth. Mast cells are of major importance in host defense because, other than basophils, these cells are the only cells that bear high-affinity Fc receptors for IgE. Bridging of IgE antibody and IgE Fc receptors by allergen on the surfaces of mast cells activates the cells to release the contents of their granules and other newly synthesized mediators. This cross-linking event and biochemical activation, utilizing subcellular and suborganelle activation molecules within the structure of the mast cell, lead to the release of both preformed mediators from the mast cell and granule-associated materials. These materials include proteoglycans, enzymes such as tryptase and chymase, and other kinds of mediators that can have proinflammatory effects. In addition to these preformed, granule-associated mediators, mast cells can generate newly synthesized mediators, including bioactive lipids such as PAF, prostaglandins (PG), and leukotrienes (LT) of the C, D, and E categories (Rankin et al., 1982). These mediators are also generated by stimulation of the allergen cross-linking of IgE Fc-receptor type I molecules on the surface of the mast cell. Recent studies have proposed an even greater role for the mast cell in allergic inflammation and long-term inflammation in response to allergens (Plaut et al., 1989).

Moreover, findings show that mast cells also have the capacity to synthesize cytokines in a manner akin to that of TH2 type CD4+ T helper cells (Robinson et al., 1992). Thus, the role of mast cells in the response to inhaled protein allergens is not a trivial one.

EOSINOPHILS

The eosinophil is another effector cell in the airway that can play a significant role in reducing the inflammation associated with responses to inhaled protein antigens. Eosinophils are derived from bone marrow precursors; production of mature eosinophils is stimulated by the action of cytokines, such as GM CSF, IL-3, and IL-5. GM CSF and IL-3 are also involved in ensuring eosinophil survival, both in vitro and in vivo; IL-5 is involved in activation, another aspect of eosinophil biology. The eosinophil can express an activated profile, the so-called hypodense profile, which is generally associated with eosinophils that have released their granules (Clutterbuck et al., 1989).

The eosinophils have a number of toxic capabilities—for example, they generate toxic oxygen-derived species and enzymes that can digest ground substances such as proteases and elastases. Another toxic capability is their expression of a family of proteins that are highly toxic to epithelia; these proteins include highly toxic basic proteins, such as major basic protein, the eosinophil-derived cationic protein, and eosinophil-derived neurotoxin (Sher et al., 1990). Through their unique structure and highly basic nature, these proteins are positively charged at a physiologic pH, permitting their generation of several toxic responses, including epithelial damage and antiproliferative effects. The fact that eosinophils are associated with reversible airway disease and airway damage clearly demonstrates their importance in the effector phase of airway responses to inhaled allergen; nevertheless, they depend on specific mechanisms for their recruitment. In particular, the activation by an allergen of T lymphocytes, and possibly mast cells, will generate the cytokines required to evoke a strong eosinophil response in airway inflammation.

NEUTROPHILS

At one time, neutrophils were thought to be extremely important in stimulating the asthmatic, destructive airway response that is found in patients with asthma. Now, however, the role of neutrophils appears to be less essential than was once thought, primarily because research has shown that eosinophils, mast cells, macrophages, and T cells are more central to the type of inflammation seen in this response. A number of factors, including cytokines, leukotrienes, and lipids, act to recruit neutrophils into the airways in an inflammatory and immune response. Once there and activated, the neutrophils have the same capabilities as eosinophils to generate toxic oxygen-derived species as well as to mediate tissue damage by generating elastases and other enzymes. Therefore, the contribution of neutrophils to airway inflammation and response to inhaled protein allergens should not be discounted (Boey et al., 1989).

MEDIATORS

Various kinds of mediators are derived from the immune and inflammatory cells discussed in the previous section and are important in the inflammatory response in the airways. These mediators include chemicals and autocoids such as histamine, serotonin, and adenosine; proteins such as secreted cytokines including tumor necrosis factor, IL-1, IL-8, and IL-6; enzymes and other toxic proteins such as collagenase, elastase, cathepsins, and the major basic protein; related cationic proteins; and neuropeptides. Table 4-3 shows the cytokine families involved in inflammation, and Table 4-4 lists other nonspecific mediators of host defense.

Under appropriate conditions, all of these mediators can stimulate enhanced mast cell and eosinophil-dependent inflammatory reactions in the airway. For example, substance P, neurotensin, and other factors made by stimulating C-type fibers in the airways can be associated with enhanced airway reactivity and inflammation. The actual interaction between the generation of substance P in the form of noncholinergic neurotransmitters and the hyperresponsiveness to cholinergic and α-adrenergic stimuli observed in patients with asthma appears to be well established.

Lipid mediators and bioactive lipids are also important in the inflamma-

TABLE 4-3 Cytokine Families

Family	Classification
Interferons (IFN)	IFNα, IFNβ, IFNγ
Interleukins (IL)	IL-1 through IL-12
Colony-stimulating factors (CSF)	Granulocyte CSF, granulocyte macrophage CSF, macrophage CSF, multi-CSF erythropoietin, IL-3, IL-5
Growth differentiative, reparative, and proliferative factors	Platelet derived growth factor, fibroblast growth factor, nerve growth factor, epidermal growth factor, insulin-like growth factor
Tumor necrosis factors (TNF)	TNFα, TNFβ
Histamine-releasing factors	Connective tissue activating peptide III, macrophage chemotactic activating factor

TABLE 4-4   Nonspecific Mediators of Inflammation and Host Defense

Mediator	Source	Function
Complement proteins $C_{5\alpha}$, $C_{3\alpha}$, membrane attack complex	Serum, liver, reticulo-endothelial system	Amplify immunoglobulin-mediated tissue damage
Enzymes Cathepsin, trypsin elastase, chymase neutrol proteases	Polymorphonuclear leukocytes, macro-phages	Tissue damage
Cationic proteins Major basic protein, eosino-phil derived neurotoxin, eosinophil cationic protein	Eosinophils	Tissue damage, anti-proliferative activity
Leukotrienes Leukotrienes, hydroxyeicosa-tetraenoic acids, etc.	Mast cells, neutrophils, eosinophils, macrophages	Smooth muscle contraction, edema and tissue swelling
Prostaglandins Prostaglandin F, prosta-glandin D, prostaglan-din E	Same as for leukotrienes	Immunosuppression, edema
Platelet-activating factor	Same as for leukotrienes, endothelium	Cellular influx and edema

tory response.   Products of arachidonate metabolism, including both lip-oxygenase and cyclooxygenase products, can be proinflammatory; such products may be mediators in activating other inflammatory cells at a site of inhaled protein antigen deposition.   Reactive oxygen metabolites, which are prod-ucts made by activated macrophages, neutrophils, and eosinophils, are an-other significant type of mediator, as are superoxide, peroxide, and hy-pochlorous acid—all of these substances may contribute to an inflammatory response to inhaled allergens.   There is also evidence indicating that the production of nitrites, nitrates, and nitric acid by the endothelium and mac-rophages may have a profound effect on both T and B cells as well as on smooth muscle function in blood vessels and in the airway.   These nitrogen-based mediators and/or their abrogation may have an effect on muscle tone in the airways during an inflammatory response.

Recent evidence has identified important new mechanisms for the re-cruitment of inflammatory cells in the airways.   An integral aspect of these mechanisms is the ability of inflammatory cells (such as those discussed earlier in this chapter) to express on their surface families of proteins known as adhesion molecules.   These molecules and their ligands allow polymor-phonuclear leukocytes (PMNs), for example, to bind to endothelial cells and to initiate migration of these cells into inflamed tissues. The adhesion molecules are upregulated by cytokines and can be inhibited by anti-inflammatory drugs such as steroids. The identification of IL-8 and other cytokines as chemotactic factors combined with our new understanding of cell adhesion and movement provides a useful framework for potential models of airway inflammatory cell recruitment in response to inhaled allergens.

GENERATION OF IMMUNE RESPONSE AND HOST DEFENSE TO INDOOR ALLERGENS

Nonspecific Mechanisms of Airway Host Defense

Several significant nonspecific mechanisms of airway host defense may play a role in the response of the upper and lower airways to inhaled protein antigens, allergens, and/or particles (see Tables 4-3 and 4-4). These mechanisms include cough and glottic functions in the upper airway and mucociliary clearance (i.e., through the production of mucus and the function of cilia); this second mechanism works to clear larger particles in the upper airway and in the upper respiratory tree of the lower respiratory tract. As mentioned earlier, the effectiveness of epithelial clearing mechanisms is tied to the ability of type III epithelial cells to generate proper ciliary motion; this serves to eliminate inhaled particles by beating particulate matter, caught in the bed of mucus, out of the airway.

In addition to this mechanism, goblet cells and submucosal cells contribute to airway host defense. These cells include many of the immune, inflammatory, and respiratory structure cells highlighted previously; they produce multiple soluble proteins in addition to mucus. A number of these proteins include enzymes such as collagenase and elastase, complement fragments, cytokines, fibroblasts, and other ground substance regulatory mediators such as platelet-derived growth factors. Antiproteases are also present in both the upper and lower respiratory tract and are derived either from serum or from respiratory cells themselves. These antiproteases include alpha-1-antitrypsin, alpha-2-macroglobulin, alpha-1-antichymotrypsin, and bronchial mucus inhibitors derived from airway cells. Liver cells are also involved in limiting the damage mediated by proteases such as elastase, collagenase, and cathepsins.

A number of cells whose functions were outlined earlier are involved in clearance in a nonspecific way; these cells act on particles and proteins that reach the upper and lower airways. Cells in this category include alveolar macrophages, bronchial macrophages, neutrophils, dendritic cells, and natural killer cells, a form of lymphocyte involved in nonspecific host defense against defined targets.

Allergen-Specific Mechanisms of Airway Responses

The generation of specific IgE and IgG against inhaled indoor allergens has been well established in both experimental animals and humans with allergic disease. Obviously, biological inhalant proteins carried into the airways on particles that range in size from 0.1 to 50 μm in diameter are most likely to generate an immune and allergic response in both the upper and lower airways. Little is known about the exact kinetics and concentration of inhaled allergen necessary to generate a response of this sort either in animals or in human subjects. It is enough to say that the airways and lungs can be an important portal of sensitization to foreign proteins and in particular to certain allergens found indoors. The kinetics of antibody formation and recognition are not well established because standardized methods of identifying allergen exposure on inhalation in the natural environment or under experimental conditions have not been developed. Nonetheless, one may be sure that the cells of the airways are interacting in such a way as to generate the appropriate phenotype of an allergic response in subjects who inhale protein allergens. Table 4-5 lists potential immunologic hypersensitivity reaction mechanisms in the allergic diseases associated with inhaled allergen exposure.

The potential role of T cells and cytokines in the regulation of IgE-mediated and hypersensitivity pneumonitis responses to inhaled proteins is of great interest. Ongoing research has found that inhaled protein allergens are processed by bronchial macrophages and other antigen-presenting cells in such a manner that allergenic peptides are then recognized by CD4+ T helper cells in the airway. These allergen epitope-specific T cells in the airways of atopic asthmatic individuals are critically involved in the generation of proinflammatory mediators; these mediators, which include the whole panoply of cells outlined in the previous section, subsequently recruit the various kinds of factors and cells related to and associated with airway inflammation and hyperreactivity. In particular, the CD4+ TH2 subset of T lymphocytes, when activated by allergen presented in the context of class II MHC molecules on an antigen-presenting cell, stimulates the generation of cytokines such as IL-3, IL-4, IL-5, and GM CSF. Cytokine production leads to the activation and recruitment of other allergy accessory cells in the airway (e.g., mast cells, eosinophils, basophils, and neutrophils). In fact, recent data suggest that the TH2 population of cells in humans is preferentially activated in response to the unique structures associated with allergens, and it has been shown that these kinds of cells produce these cytokines in response to allergen in the airways. An allergic proinflammatory process thus can be set up on the basis of recognition of an inhaled protein allergen. Other trigger factors may also play a role in this inflammatory milieu. An example is the potential action of a viral infection. When activated in a

TABLE 4-5 Immunologic Hypersensitivity Reaction, Types I to IV

Type of Reaction*	Onset	Antibody	Principal Cell/Effector	Airway Syndrome	Effector Molecule
Antibody dependent: anaphylactic (atopic reaginic)—type 1	Rapid (seconds to minutes)	IgE	Mast cells/basophils	Allergic rhinitis, asthma	Histamine, leukotrienes, platelet activating factor, prostaglandins, cytokines
Complement-mediated immune adherence (phagocytic reaction)—type II	Intermediate (minutes to hours)	IgM/IgG	Neutrophils, reticulo-endothelial cells	Immunologic pulmonary hemorrhagic syndrome	Imunoglobulins, complement fragments, enzymes
Immunocomplex (or tissue reactive antibody)—type III	Intermediate (30 minutes to 2 hours)	IgM/IgG	Neutrophils, reticulo-endothelial cells	Hypersensitivity pneumonitis	Complement fragments, enzymes
T-cell reactions: cell-mediated immune response—type IV	Prolonged (delayed 18–24 hours)	None	Responder T cell and monocyte/macrophage/antigen-presenting cells	Asthma, hypersensitivity pneumonitis	Cytokines, mediators

*Gell and Coombs (Gell et al., 1975) classification of Types I to IV.

SOURCE: Adapted from NRC, 1992a.

virus-specific manner, other populations of T cells, such as the CD8+ T cell, may spur the generation of GM CSF, IL-3, and gamma interferon (IFNα, all of which can lead to a proinflammatory milieu and the generation of factors related to allergic inflammation. Tables 4-6 and 4-7 summarize the mechanisms by which immune cells and cytokines induce airway reactions.

In addition to these mechanisms for pulmonary reactions to inhaled allergens, there are similar mechanisms for reactions at other tissue sites including the nose, the conjunctivae, and the skin. There are even data to suggest that indoor allergens may play some role in generating allergic inflammation of the skin resembling atopic dermatitis. Presumably, the skin has the same immunological mechanisms for allergy reactivity, except that there is also the possibility that epidermal and dermal cells, including Langerhans, cells, keratinocytes, and fibroblasts, will be involved in allergen-induced reactivity.

It is worth noting that the allergen response at all of these tissue sites adheres to a specialized pattern of kinetics after challenge. In the airways and nose, an initial immediate response is seen within the first 20 minutes to 1 hour after challenge. Beginning at 6–8 hours after challenge and continuing for 24–28 hours, a delayed response is seen; this response has been attributed to recruitment of inflammatory cells to the site of allergic reaction. This so-called late-phase response has become an important model for, or window into, the potential mechanisms of allergic inflammation.

TABLE 4-6 Sources of Cytokines in Allergen Induced-Airway Inflammation

Cell	Specificity	Cytokines
Macrophage	IgE—CD23	IL-1, IL-6, IL-8
CD4+ T cell (TH1)	Antigen presenting cell-allergen fragment complex T-cell receptor $\alpha\beta$	IL-2, IFNγ, IL-3, GM-CSF, leukotriene
CD4+ T cell (TH2)	Antigen presenting cell-allergen fragment complex T-cell receptor $\alpha\beta$	IL-4, IL-5, IL-9, IL-10, IL-3, GM-CSF
Mast cell	Fc receptor for immunoglobulin E—type I	IL-3, IL-4, IL-5, IL-9, IL-10, IL-1, tumor necrosis factor, GM-CSF

NOTE: IL, interleukin; IFN, interferon; and GM-CSF, granulocyte macrophage colony-stimulating factor.

TABLE 4-7 Cytokine-Induced Airways and Allergic Inflammation

Function	Cytokine	Activity
IgE regulation	IL-4	IgE isotope switch
	IL-2, IL-5, IL-6	B cell stimulation with IL-4
	IFNγ	Inhibits IL-4
	IL-10	Inhibits IFNγ (enhances IgE production)
Eosinophilia	IL-3, IL-5, GM-CSF IL-1, tumor necrosis factor (TNF)	Eosinophilopoietins and activators
Mast cell development and activation	IL-3, IL-4, IL-9, IL-10 hematopoietic stem cell factor	Mast cell growth factors
	Connective tissue activating peptide III, Neutrophil activating peptide 2, Macrophage chemotactic activating factor	Histamine releasing factors
Inflammation	IFNγ, GM-CSF, granulocyte-CSF, TNFs, IL-1, IL-4, IL-6, IL-8, platelet factor 4, macrophage inflammatory protein-1, macrophage inflamatory protein-2,	Neutrophil-endothelial activating factors
	GM-CSF, TNFs, IL-1, IL-3, IL-5	Eosinophil activating factors
	IFNγ, GM-CSF macrophage-CSF, TNFs, IL-1, IL-2, IL-3, IL-4	Macrophage activating factors

NOTE: IL, interleukin; IFN, interferon; and GM-CSF, granulocyte macrophage colony-stimulating factor.

Adjuvant Effects

Inhaled contaminants within the airways can serve as adjuvants and may have a costimulatory effect on airway function. This includes the ability of endotoxin and endotoxin-related proteins derived from bacteria to act as macrophage-activating factors in association with inhaled protein antigens. Considerable evidence has accrued, both from animal models and human subjects, to suggest that the adjuvant effects associated with macrophage activation are related to lipid A of endotoxin. Other relevant substances such as synthetic and water-soluble adjuvants are primarily macrophage activators involved in generating, from macrophages, proinflammatory cytokines such as IL-1, IL-6, and tumor necrosis factor. When administered with protein antigens, these proinflammatory cytokines can have an effect similar to that of standard endotoxin—or of any of the other water-soluble adjuvants and other materials that have adjuvant effects. It is entirely possible that inhaled contaminants, endotoxins, and endotoxin-like proteins, as well as fungal proteins and other irritants such as tobacco smoke, may activate macrophages and generate a proinflammatory milieu. Such a milieu will enhance the basic immunogenicity and allergenicity of inhaled proteins, which act as stimulants of the CD4+ T cell populations in a more standard immunological model.

CONCLUSIONS AND RECOMMENDATIONS

A great deal of knowledge has accumulated in the past 15 years concerning the role of inhaled proteins (including allergens) in airway inflammation. The prospects for the future are bright. Newer techniques of immunological science include the use of sophisticated methods for the detection of proteins. Indeed, such methods as monoclonal antibody-based radioimmunoassays and enzyme-linked immunosorbent assays as well as the potential use and recognition of the actual protein sequences of the allergens involved in inducing allergic responses have led us to consider a whole new approach to detection. Use of these reagents to enhance and better control environmental stimuli is a logical goal, as is the potential use of molecule-based products, including allergen fragments, cytokines, and cytokine inhibitors, to produce immune modulation in the host. Such promising approaches permit considerable optimism regarding the potential interruption of the immune system- and inflammatory-based circuits triggered by inhalation of proteins and allergens and long-term exposure to inhaled indoor allergens.

Evidence suggests that the adjuvant effects associated with macrophage activation are related to endotoxin. Additional research is necessary, however, to clarify this phenomenon.

Research Agenda Item: Determine whether bacterial products (such as endotoxin) or fungal products may act as adjuvants in the immune responses to indoor allergens.

As discussed in earlier chapters of this report, the magnitude of allergen exposure appears to be related to the risk of sensitization. Allergen exposure is also related to the risk of developing asthma and the age at which asthma develops. Genetic and other local host factors are also important in atopy and asthma. Understanding the relative importance of each of these individual factors and their interaction with each other is essential to understanding the mechanisms of immune function.

Research Agenda Item: Conduct research to identify risk factors other than exposure, and clarify their potential significance relative to indoor allergy. This effort should include an evaluation of the role of genetic and local host factors in allergen sensitization.

5

Medical Testing Methods

Methods to determine the effects of indoor allergens can be divided into two general categories: patient testing and environmental testing. Patient testing evaluates the health status of an individual (clinical) or population (epidemiological). Environmental testing characterizes environments with respect to sources of allergens, dissemination factors, ambient concentrations, and human exposure. Data from both kinds of testing can be used to direct treatment, control, and prevention of allergic disease. This chapter discusses common approaches to patient testing including the medical history, skin tests, in vitro serum tests, and evaluations of pulmonary function. Chapter 6 discusses environmental testing and the assessment of exposure and risk.

MEDICAL HISTORY AND DIAGNOSIS

History

A common dictum in allergy practice is that the patient's medical history is the primary diagnostic test. Laboratory studies, including skin and in vitro tests for specific immunoglobulin E (IgE) antibodies, have relevance only when correlated with the patient's medical history. Furthermore, treatment should always be directed toward current symptomatology and not merely toward the results of specific allergy tests. Several authors of current allergy textbooks reiterate these points:

• "The selection of appropriate diagnostic tests is fully dependent on the clinical history presented by the patient in question. It therefore follows that diagnostic tests should be ordered only after a careful history and physical examination have been obtained" (Kaplan, 1985).

• "The degree of success that will be achieved in the treatment of a patient's allergies will be proportional to the exactness of the history obtained" (Weiss and Rubin, 1980).

• "It is of utmost importance to begin with a thorough, perceptive general medical history. . . . If the general history suggests an allergic disease, one must ascertain what factors are important in producing the difficulty of the individual patient. The history is the major approach in making this assessment, and the most important clinical skill to be learned in evaluating allergy patients is to acquire facility in asking discerning questions so that logical deductions can be made about the cause of the patient's difficulty" (Korenblat and Wedner, 1984).

• "The purposes of the medical evaluation are to establish the diagnosis, to estimate the severity of the illness, to determine responses to previous treatment, to identify possible complications, and thus to guide appropriate further management. A thorough medical history is the most helpful tool in achieving these objectives in the field of allergy" (Bierman and Pearlman, 1988).

In spite of universal agreement about the primary importance of a patient's allergy history, the same textbooks from which these quotations were taken (and others) devote little or no space to this topic (Tables 5-1 and 5-2). Furthermore, review of the allergy literature reveals no discernible research on the subject.

Allergists use a variety of methods to obtain a history, including (1) an open-ended, nondirected question-and-answer session, (2) a series of questions ordered according to a formal protocol to ensure completeness, (3) a structured questionnaire history completed by the physician, or (4) a structured questionnaire history completed by the patient. Many allergists use a combination of these methods.

The use of particular history formats or questionnaires depends on the purpose of the examination—for example, whether it is the clinical evaluation of an individual patient or an epidemiological study of a general or selected population (e.g., that of a particular building, factory, or industry). A commonly used format for evaluating a patient's medical history contains the following eight components:

1. Chief complaint—This includes (a) the reason for the patient's visit, such as referral from a primary physician, need for treatment of a current problem, potential need to avoid an allergen (e.g., penicillin, cat), or disability evaluation, and (b) a concise definition of the symptom or complaint

TABLE 5-1 Textbooks of Clinical Immunology

Textbook	Publisher	Total Pages	Allergy Pages	Clinical Evaluation Pages	History Pages
Parker, C.W., *Clinical Immunology* (2 vols.), 1980	Saunders	1,438	248	0	0
Graziano, F.M., and Lemanske, R.F., *Clinical Immunology*, 1989	Williams & Wilkins	310	103	0	0
Freedman S.O., and Gold, P., *Clinical Immunology* (2nd ed.), 1976	Harper & Row	620	133	0	0
Lockey, R.F., and Bukantz, S.C., *Principles of Immunology and Allergy*, 1987	Saunders	335	214	26	1½
Lockey, R.F., *Allergy and Clinical Immunology*, 1979	Medical Examination Publishing Co.	1,175	588	9	3
Altman, L.C., *Clinical Allergy and Immunology*, 1984	G.K. Hall	473	268	0	0
Samter, M., *Immunologic Diseases* (2 vols., 4th ed.), 1988	Little, Brown	2,044	267	0	0
Stites, D.P., and Terr, A.I., *Basic and Clinical Immunology* (7th ed.), 1991	Appleton & Lange	794	64	8½	½

for which the patient is seeking treatment, preferably in the patient's own words.

2. Present illness—A complete description of current symptoms, including severity and duration. Questions such as the following should be answered:

— Do symptoms vary seasonally, monthly, weekly, or diurnally? Or are they randomly intermittent?

— What is the relationship of symptoms to location, such as in the home, at work, and while traveling?

TABLE 5-2 Textbooks of Allergy

Textbook	Publisher	Total Pages	Clinical Evaluation Pages	History Pages
Kaplan, A.P., *Allergy,* 1985	Churchill Livingstone	692	0	0
Lessof, M.H., *Allergy: Immunological and Clinical Aspects,* 1984	Wiley	464	0	0
Beall, G.N., *Allergy and Clinical Immunology,* 1983	Wiley	326	0	0
Korenblat, P.E., and Wedner, H.J., *Allergy: Theory and Practice,* 1984	Grune & Stratton	512	13	5½ (plus 5-page questionnaire)
Lawlor, G., and Fischer, T.J., *Manual of Allergy and Immunology: Diagnosis and Therapy,* 1981	Little, Brown	409 (plus 13 appendixes)	10	1½ (plus ½-page questionnaire)
Klaustermeyer, W.B., *Practical Allergy and Immunology,* 1983	Wiley	209	25	1½
Weiss, N.S., and Rubin, J.M., *Practical Points in Allergy* (2nd ed.), 1980	Medical Examination Publishing Co.	211	4½ (appendix)	1½
Middleton, E., et al., *Allergy: Principles and Practice* (2 vols., 3rd ed.), 1988	C. V. Mosby	1,597	0	0
Bierman, C.W., and Pearlman, D.S., *Allergic Diseases from Infancy to Adulthood* (2nd ed.), 1988	Saunders	787	6	3
Patterson, R., *Allergic Diseases* (3rd ed.), 1985	Lippincott	825	20	9 (plus 2-page questionnaire)

Patients should be asked about known precipitants (e.g., dust, animals, weather). If symptoms occur in discrete attacks, the allergist should ascertain the frequency of attack, and ask the patient to describe in detail a typical or recent episode. The description of the present illness should also include a list of all current medications and the duration of their use, as well as the efficacy of symptomatic medications. The practitioner should address specific efforts made by the patient to avoid certain allergens and the efficacy of such avoidance.

3. Past allergy history—In assessing this component, the practitioner reiterates questions about the patient's present illness but directs them toward other allergic manifestations that are either no longer occurring or not related to the current evaluation. In addition, he or she should specifically ask the patient about known allergies to foods, drugs, vaccines, and insect bites or stings. Results of past allergy testing and immunotherapy are also informative. Because of regional differences in both indoor and outdoor aeroallergens, a chronological listing of the patient's places of residence should be compared with a history of symptoms.

4. Current and past medical history, including review of systems—This is necessary because of the possible effect of other diseases on allergy, and vice versa. In addition, pregnancy may alter certain allergic manifestations.

5. Family history of allergy—The allergy history and general health of immediate family members should be determined.

6. Occupational history—This portion of the evaluation seeks clues to work-related sources of allergens that may explain a patient's illness. In cases evaluated for work disability, this step must include a complete list of all occupations in which the patient has engaged, including employer, location, and job description.

7. Social history—This set of questions may reveal a symptomatic role for psychosocial factors and "substance" use (tobacco, drugs, alcohol).

8. Environmental history—This is a unique feature of the allergy history. It describes specific features of the patient's indoor and outdoor environments and the effects of specific environmental agents on symptoms. It also serves as the basis of recommendations for allergen avoidance. Table 5-3 details appropriate information regarding the patient's environmental history.

Conclusion and Recommendation

In spite of universal agreement on the primary importance of a patient's allergy history, very little space in medical textbooks is devoted to the topic, and no standard exists for collecting appropriate information. A standardized, validated allergy-history questionnaire would be useful in both clinical and research settings.

TABLE 5-3 Environmental History for Indoor Aeroallergen Exposure

1. Home: Location _____
      Age _____ Years of Residence _____ Construction _____
      Heating/cooling system _____
         Filter type _____ Frequency of change _____
      Indoor plants _____
      Bedroom: Carpeting type _____ Age _____
         Carpet pad _____ Furniture _____
         Mattress type _____ Age _____ Dust cover _____
         Pillow(s) type _____ Age _____ Dust cover _____
         Quilt or comforter _____ Age _____
      Living room: Carpeting _____ Pad _____
      Family room: Carpeting _____ Pad _____
      Basement: Flooring _____ Dampness _____

2. Animals and Birds
      Dog(s):   No. _____   Years _____   In or out _____
      Cat(s):   No. _____   Years _____   In or out _____
      Others: No. _____   Years _____   In or out _____
      Bird(s)   No. _____   Years _____   In or out _____

3. Work: Employer _____ Years _____
      Occupation _____
      Effect on symptoms _____

4. Hobbies: _____
      Effect on symptoms _____

5. Exposure

Allergens	Irritants
House dust	Auto exhaust
Other dust	Diesel fumes
Mold, mildew	Gasoline
Feathers	Pesticides
Cats	New carpet
Dogs	Perfumes
Horses	Paints
Chickens	Cleaners
Birds	Solvents
Rats	Smoke
Mice	Newsprint
Guinea pigs	Others
Rabbits	

Research Agenda Item: Develop, test, and validate a standardized allergy-history questionnaire for use in multi-center studies.

Physical Examination

The physical examination should be thorough enough to rule out other causes for the patient's symptoms. Because allergic diseases may be episodic, the physical examination should be performed during a period of allergen exposure when objective signs of allergy can be seen. Negative results from a physical examination performed during a period of allergen avoidance does not mean that the patient does not have an allergy.

Daily Diaries

Daily diaries are sometimes necessary in cases in which the diagnosis is not clear from the history. They allow ongoing symptoms to be recorded and correlated with observed environmental exposures, ingested food, medication use, and other factors. Diaries can also be used in conjunction with such functional measures as the pulmonary peak flow rate (see the discussion later in this chapter). Clinical trials of medications or immunotherapy and studies of occupational asthma have used such diaries extensively and to good advantage. The literature indicates several specific uses that have been made of them. For example, daily diaries have been used to study the acute effects of environmental factors (Cohen et al., 1972; Finklea et al., 1974a,b; Lawther et al., 1970; Lebowitz et al., 1985; McCarroll et al., 1966; Quackenboss et al., 1989a; Schoettlin and Landau, 1961; WHO, 1982; Zagraninski et al., 1979). There have also been some efforts at partial standardization of diaries (Finklea et al., 1974a; WHO, 1982; Zagraninski et al., 1979) and validation for symptom and medication usage (Lebowitz et al., 1985; Zagraninski et al., 1979).

Daily diaries for use in surveys of morbidity (e.g., to estimate the prevalence of specific diseases) were developed, tested, and used in the early 1950s. They have higher reporting levels than standard health history questionnaires and may provide better information about minor health events. Diaries are especially useful for accurate reporting of acute episodes and disability (Allen et al., 1954; Laurent et al., 1972; Mooney, 1962; Muller et al., 1952; Peart, 1952; Verbrugge and Depner, 1981). Because symptom recording occurs soon after symptom development, recall is at a maximum, resulting in more comprehensive health information about the individual. Diaries compare favorably with interview surveys in response rate, segment completion rate, and low attrition. Physician visits reported in diaries are generally quite accurate, except for unusual events (e.g., x-rays only) or as part of general nonresponse (Marquis, 1978). Thus, diaries may provide important information with great accuracy for certain clinical research studies.

The format for diaries must be established on a case-by-case basis to obtain the information needed for the specific purpose of the study. Population studies that require statistical analyses may mandate simplified reporting at specified times of particular symptoms on a yes/no basis or by using a severity scale; clinical evaluation of an individual patient may be better served by open-ended narrative descriptions. Within a household, a daily diary can be completed by each adult and by one adult for all children. The data should include (1) symptoms selected from a list, (2) medication usage, (3) activity, (4) disability, and (5) physician visits.

For analysis, symptoms can be aggregated into sets (e.g., irritant, allergic, asthmatic, acute respiratory illness, nonspecific). Other information can also be aggregated—for example, (1) days of restricted activity ("unable to do usual activities"), (2) days of disability ("unable to work, go to school . . ."), and (3) days requiring physician or emergency room visits (as well as days requiring increased medication usage). In addition, acute respiratory symptoms such as irritative, infectious, or allergic responses can be tracked for periods of 2 or more weeks and then repeated in a different season of the year or at selected regular intervals to measure the reliability of findings and to study the effect of seasonal or other environmental exogenous stimuli.

SKIN TESTS

Allergen skin testing has been a primary diagnostic tool in allergy since the 1860s. Skin tests, in and of themselves, are not diagnostic of allergic disease but provide evidence of immunologic sensitization. These tests are of particular value in deciding on, and undertaking strategies to avoid exposure to indoor allergens that are causing allergic symptoms (NHLBI, 1992).

A positive skin test is the culmination of a number of events that begin with the interaction of the allergen with IgE on the surface of cutaneous mast cells, as described previously in this report. This interaction is followed by the release of chemical mediators such as histamine, which then exert their effects on the skin by causing the blood vessels to dilate and the plasma to leak into the tissue. Neuronal axon reflexes are also involved.

The characteristic skin reaction consists of a "wheal" produced by the leakage of fluid into the skin and a surrounding area of redness ("flare") produced by the dilated blood vessels—the "wheal and flare" response. Thus, the skin test is not simply a test for the presence of specific IgE antibodies; it also involves the sensitivity of the mast cell and the biological effect of the chemical mediators on tissue. The size of the reaction is related to several factors including the amount of allergen injected, the degree of sensitization of the cutaneous mast cells and their ability to release histamine, and the reactivity of the skin to the mediators released (H. S. Nelson, 1983). The advantages of skin testing are its simplicity, rapidity of results, low cost, and high sensitivity; however, skin tests can be subject to abuse through inappropriate use or overuse. Table 5-4 presents a comparison of some in vivo and in vitro methods for the diagnosis of IgE-mediated allergic disease.

Major Methods

Two major skin testing methods are commonly used: the skin prick or puncture test and the intradermal test. In the prick method, a drop of allergenic extract is applied to the skin, and the skin is pricked with a needle through the drop. A number of devices are commercially available for testing using this method (Demoly et al., 1991). The most commonly used skin prick method is to introduce the tip of a stylette or a 27 (or smaller) gauge needle into the epidermis at a 15- to 20-degree angle through the drop of test allergen and then to lift up until the tip of the needle pops loose. Alternatively, lancets or solid needles are often used.

One concern about the skin prick method is the possible transfer of allergens from one test site to another if the needle is not properly wiped off

TABLE 5-4 Comparison of Two In Vivo Skin Tests (Prick and Intradermal) with In Vitro Tests (RAST or Equivalent) for Diagnosing IgE-Mediated Allergic Disease[a]

| Parameter | In Vivo Skin Tests | | In Vitro (Specific IgE)— RAST or Equivalent |
	Skin Prick	Intradermal	
Sensitivity	Good	Excellent	Fair-good
Specificity	Good	Poor-fair	Excellent
Safety	Excellent	Good	Excellent
Reproducibility	Good	Good	Excellent
Convenience[b]	Excellent	Good	Fair
Cost	Excellent	Good	Fair

NOTE: RAST, radioallergosorbent test.

[a]Ratings are based on the assumption that the procedures are done by well-qualified personnel using properly standardized reagents and adequate quality control.

[b]Convenience is a composite of efficiency and ease of testing, lack of discomfort, and rapid availability of test results.

SOURCE: Adapted from Van Arsdel and Larson, 1989.

between each test. This problem can be avoided by using a fresh needle for each test site. Another potential area of difficulty involves placement of the allergens: if they are too close, overlapping reactions cannot be separated. It is therefore recommended that the extracts be placed at least 2 cm apart.

Infection and bleeding can lead to false-positive results. Likewise, insufficient penetration of the skin by the puncture instrument may lead to false-negative results. (This problem is more likely to occur with plastic devices [Bousquet, 1988].) Skin prick tests are applicable to children as young as 1 month of age if clinical indications are present. Use of the test usually requires allergen concentrations of 1:10 to 1:20, weight/volume. Skin prick tests pose an extremely low risk for the development of systemic anaphylactic reactions or fatalities.

In intradermal testing, allergenic extracts are diluted in a buffered saline solution containing 0.3 percent human serum albumin as a stabilizer. Volumes of 0.01 to 0.05 ml are injected intradermally. As noted above, individual, unitized syringes should be employed for each skin test to avoid any risk of contamination from syringes with removable needles that have been used repeatedly (Shulan et al., 1985). Solutions containing less than 1 percent glycerine may be used for skin testing; concentrations greater than this amount may induce nonspecific reactions.

Because a small but definite risk of anaphylaxis and fatalities exists with intradermal testing (Lockey et al., 1987), skin prick tests should be conducted prior to any such testing. In addition, patients who are suspected of having allergic disease but who have a negative skin prick test may be candidates for intradermal testing because intradermal testing is more sensitive. Properly conducted negative intradermal skin tests virtually exclude the presence of specific IgE antibody.

Scratch or abrasion methods should probably be abandoned (Van Arsdel and Larson, 1989). These tests have poor reproducibility because of the variable amount of allergen introduced into the skin. False-positive reactions may occur if bleeding is induced; there is also a risk of systemic allergic reactions (Guerin and Watson, 1988).

Variables and Controls

Several variables can affect the size of the cutaneous reaction. The magnitudes of both allergen and histamine reactions vary over different parts of the body. The upper and midback are more reactive than the lower back. The back is more reactive than the forearm. The ulnar side of the arm is more reactive than the radial. The wrist area is less reactive than the space in front of the elbow (H. S. Nelson, 1983). There is also some variability in skin test responsiveness at different times of the day. At 7 a.m., the reaction to allergen and histamine is less than that in the late afternoon and early evening. The age of the individual undergoing testing also influences the size of the skin test response. Young infants have wheals and flares that are smaller in diameter than those in adults. In addition, skin test reactivity declines after age 60.

The reproducibility of epicutaneous skin testing, particularly that using a skin prick test, depends on the degree of pressure applied to the skin. Medications can significantly affect the skin test response. For example, antihistamines and tricyclic antidepressants must be discontinued for at least 4 to 5 days before testing occurs (I. L. Bernstein, 1988). Some long-acting antihistamines such as astemizole (Hismanal®) may interfere with skin test results for up to 6 weeks. Long-term use of high-potency topical steroids also decreases skin test reactivity. Oral corticosteroids at doses of up to 30 mg a day for 1 week do not suppress the allergic skin test response.

Because skin tests may be affected by a number of factors, positive and negative controls should be applied. Some general recommendations for skin testing procedures are summarized in Box 5-1.

Shortcomings and Precautions

One of the major difficulties with reproducibility of skin test responsiveness is the lack of standardized reagents. Certain indoor allergenic extracts (e.g., dust mite and cat) now contain standardized amounts of major allergens, and further efforts at standardization are under way for other allergens. Because extracts often become less potent over time when stored under warm conditions, refrigeration is important. The addition of glycerine to skin prick test reagents enhances their stability, as does the use of

---

BOX 5-1 Allergy Skin Test Procedures

1. Avoid antihistamines and tricyclic antidepressants for 4 to 5 days prior to testing. Long-acting antihistamines e.g., astemizole (Hismanal®) should be avoided for 4 to 6 weeks prior to testing. Topical corticosteroids applied to the test site can reduce the response when used for several weeks. Topical, nasal, inhaled, or systemic corticosteroids have no effect on the skin test response.

2. Appropriate positive (histamine or codeine phosphatase) and negative (diluent) controls should be used.

3. The forearm or upper back skin are appropriate test sites.

4. Appropriately trained personnel and equipment to treat systemic anaphylaxis should be available.

human serum albumin for the more dilute solutions used for intradermal testing.

Because of safety issues involved in allergy skin testing, tests should never be performed unless properly trained personnel and emergency equipment to treat systemic reactions are available. In particular, patients receiving beta-adrenergic-blocking agents for treatment of hypertension or heart disease should be advised of this risk, because they are less likely to respond to epinephrine in the case of anaphylaxis. Skin testing should be avoided in these patients until a substitute treatment can be found (Executive Committee, American Academy of Allergy, 1989).

Interpretation of Results

The methods for interpreting skin tests are not well defined. It is known that an immediate skin test response to histamine reaches a peak at about 8–10 minutes; for agents such as codeine that act directly to cause histamine release from mast cells, peak response comes at 10–15 minutes, whereas allergens elicit responses in 15–20 minutes. At the time of peak response, the diameter of the wheal and flare is measured; sometimes a permanent record is obtained by outlining the size of the reaction with a pen and then blotting the marks onto cellophane tape, which is stored on paper.

Generally, a wheal 3 mm or greater in diameter is considered a positive test using the skin prick method. However, a grading system has been established (Bousquet, 1988) in which erythema (redness of the skin) and a wheal of less than 5 mm in diameter is a negative test. A 2+ reaction is a 5- to 10-mm wheal with 21–30 mm of erythema; this size response is often elicited by histamine control. A 3+ reaction is a 5- to 10-mm wheal with pseudopods and erythema of 31–40 mm. A 4+ reaction is a wheal of 15 mm or greater within any pseudopods and erythema of 40 mm or more. Other investigators have found that if a control site is completely negative, wheals of 1–2 mm with flare and itching are likely to represent a positive response. Although these reactions indicate immunologic sensitization, they do not necessarily indicate the presence of clinically relevant allergic symptoms.

From 2 to 8 percent of individuals with no personal history of allergy or respiratory disease exhibit positive skin test responses with intradermal testing. These positive tests may indicate the presence of specific IgE antibodies but not the presence of clinical allergy. A true false-positive test occurs as a result of irritant reactions; in the case of intradermal testing, it may be produced by the so-called splash response when air is injected into the skin. Occasionally, a false-positive reaction is the result of nonspecific enhancement, through the axon reflex, from a nearby strong allergic reaction. For this reason, skin tests should be placed at least 2–5 cm apart, and the positive histamine control should be at some distance from the allergen test sites. False-negative skin test results are usually due to poor quality, low potency, or lack of stability of the extracts used for testing. Most commercially available extracts exhibit wide variability in the concentration of allergens. Unrefrigerated extracts lose potency rapidly. Stabilizing agents such as glycerine or human serum albumin can improve this situation.

Correlation with Other Tests

Skin tests have been correlated with other tests for diagnosing allergic disease. A reasonably good correlation seems to exist between a strongly positive skin prick test and specific IgE assay by radioallergosorbent test (RAST) (see the discussion in the section below, "In Vitro Diagnostic Tests"). Similarly, a negative skin prick test and negative RAST also correlate well (H. S. Nelson, 1983). Small reactions to skin prick tests are less frequently associated with a positive RAST. Intradermal tests that are positive with high concentrations of allergens are only occasionally associated with a positive RAST (H. S. Nelson, 1983).

Patients may have positive skin prick tests before they develop allergic symptoms, and the tests may be predictive of future symptoms. In patients with negative histories and negative skin prick tests, the diagnosis of allergic disease can usually be excluded if the extract is of high quality.

Conclusions and Recommendations

A major unknown in skin testing is the identity of the more prevalent allergens involved in many indoor exposures. Studies to characterize these allergens are important for the development of reliable diagnostic reagents. Additional research is needed to identify, characterize, and standardize indoor allergenic extracts used for diagnostic testing.

Research Agenda Item: Develop standardized, well-defined indoor-allergen reagents for skin tests that can be used in clinical diagnosis and research studies.

Despite some relatively minor shortcomings, the value of skin testing has been well established over the past century. When correlated with an appropriate clinical history, skin prick tests often are a useful way of screening for the presence of allergic disease. Using appropriate positive and negative controls, intradermal testing can be used to demonstrate low levels of sensitization when allergy is clinically suspected. However, studies are necessary to determine the optimal concentrations and methods for skin testing and to address the relationship between defined skin test reactions and disease.

With respect to the specificity and sensitivity of skin tests within a population, more research is needed to determine the predictive values of both skin prick and intradermal tests. Moreover, the doses and criteria for positive skin tests used for such studies need further definition, together with criteria to define the relationship of a positive skin prick test to the wheal area or erythema that appears.

Recommendation: Encourage the development and use of improved standardized methods for performing and interpreting skin tests.

For safety reasons, appropriately trained personnel and adequate equipment need to be available to treat possible adverse systemic reactions.

IN VITRO DIAGNOSTIC TESTS

Confirmation of the diagnosis of an allergic reaction resulting from a given agent generally requires some immunologic test that demonstrates a specific antibody response to the agent. For allergic rhinoconjunctivitis and most allergic asthma, demonstration in vivo or in vitro of specific IgE would be appropriate. For hypersensitivity pneumonitis, in vitro demonstration of either specific immunoglobulin or specific cell-mediated immunity, or both, would be acceptable. Although allergic asthma is IgE mediated, the immunopathogenesis of non-IgE-mediated asthma has not been elucidated; thus, the appropriate corroborative immunologic tests are unknown. The exact role of cell-mediated immunity and T cell activation in asthma also has not been defined. Therefore, the role of assays to detect specific cell-mediated immunity or T cell activation in asthma is unclear.

An immunologic response to an agent is not sufficient to diagnose an allergic disease caused by that agent. Such a response means only that a prior sensitizing exposure to the agent or a cross-reacting agent has occurred. Diagnosis of a hypersensitivity disease such as allergic rhinitis requires a compatible clinical syndrome in addition to the appropriate specific immunologic response.

Several in vitro tests are available for use in clinical practice and research. These include a variety of serum antibody tests and tests of cell-mediated immunity. These tests are discussed below, along with the importance of reagent quality, the diagnostic value of the tests, and the importance of quality control.

Serum Antibody Tests

Several investigators have reported on in vitro tests for estimating IgE directed against an allergen. A widely used immunoassay is the RAST, in which the allergen is bound to a solid phase and then incubated first with patient serum and then with radiolabeled antihuman IgE (Adkinson, 1986).

The number of radiolabeled antihuman IgE molecules that bind to the allergen is indicative of the amount of specific IgE in the patient's serum. If antibody is not present in excess in the serum, however, or if there is competing allergen-specific antibody of another isotype, interference can occur.

The enzyme-linked immunosorbent assay (ELISA) is analogous to RAST except for the detection system (Voller and Bidwell, 1986). In the case of ELISA, the antihuman IgE is conjugated to alkaline phosphatase, which catalyzes a colorimetric change proportional to the level of serum-specific IgE. The same interferences described above for RAST can occur with ELISA. In addition, false positives may occur in both assays if the serum has a high total level of IgE resulting in nonspecific binding of IgE with the allergen being tested. Several other immunoassays for specific IgE have been reported but are not widely used.

By using antihuman antibody directed to other immunoglobulin classes, such as antihuman IgG, ELISA can be used to demonstrate specific isotypes other than IgE. The clinical significance of such antibodies in asthma, however, is not clearly apparent. In cross-sectional studies of asthma and hypersensitivity pneumonitis, the presence of specific IgG has been interpreted as a biological marker of exposure (Biagini et al., 1990; Lushniak et al., 1990). There is also evidence, especially following immunotherapy with allergens, that antibody of a non-IgE isotype can be protective and prevent exposed individuals from manifesting IgE-mediated clinical sensitivity (Cooke et al., 1935).

In patients with hypersensitivity pneumonitis, precipitin assays can be used to demonstrate high titers of specific antibody. The most commonly employed assay of this type is the double-immunodiffusion method of Ouchterlony (A. M. Johnson, 1986). In this assay, center and circumferential wells are cut in agar; sera are then placed in the circumferential wells, and the allergen is placed in the center. A precipitin band will form between the serum and the allergen wells if precipitating antibody, usually class IgG, is present in the serum. If a test serum is next to a positive control serum, researchers can determine whether the antibodies in the two sera recognize the same allergens, different allergens, or a combination; this judgment is based on the pattern of the precipitin lines, which show complete identity, nonidentity, or partial identity, respectively. Because commercial preparations of most hypersensitivity pneumonitis allergens are poorly standardized, the use of positive and negative control sera is critical. False-negative test results can also occur if a zone of equivalence (where precipitin lines form) is not obtained because of improper allergen concentration. Other techniques, including ELISA and the ammonium sulfate precipitation method of Lidd and Farr, have also been used as immunoassays for the evaluation of hypersensitivity pneumonitis (Lidd and Farr, 1962; Zeiss et al., 1977).

Tests of Cell-Mediated Immunity

Specific cell-mediated immunity can be evaluated by several methods (Fink et al., 1975; Stankus et al., 1982). In the lymphocyte transformation assay, the patient's lymphocytes are incubated with an allergen, and DNA replication is measured by tritiated thymidine incorporation. Other assays of cell-mediated immunity include production of mediators or cytokines, or messenger RNA for cytokines by sensitized lymphocytes that have been incubated with the putative agent. Although assays for detecting specific cell-mediated immunity may be more specific for hypersensitivity pneumonitis, they are generally regarded as research assays and are not usually performed in evaluations of patients or populations with hypersensitivity diseases.

Test Reagents

ALLERGENS

Ideally, the test reagents used in immunodiagnostic assays of allergic disease should be standardized extracts, the allergen contents of which are well characterized. Efforts to produce such reagents are in early stages of development. There are several requisites for allergen standardization and characterization that differ depending on whether the allergen is a complete protein allergen or a low-molecular-weight (LMW) chemical that requires conjugation to an autologous protein (Bush and Kagen, 1989; Butcher et al., 1989). Most protein allergens have not been standardized or characterized. To standardize an extract that contains protein allergens, the source of the allergen should be identified, as should the extraction procedures used, including time, temperature, extraction fluid, and method of filtration (Bush and Kagen, 1989). For many extracts the source is known, but the biochemical composition is not, although numerous procedures exist for such characterization. Properties that are useful for evaluation include total protein content, molecular weight distribution, isoelectric points of individual components, and allergenic composition. Protein content can be measured by techniques such as the biuret method (Kabat and Mayer, 1967) and the ninhydrin assay (Richman and Cissell, 1984). Using a variety of gels, column chromatography can determine molecular weight distribution. Isoelectric focusing can be used to ascertain the isoelectric points of individual components (Marcus and Alper, 1986). Numerous in vitro and in vivo assays are available for assessing such properties as immunologic or allergenic composition of an extract, including quantitative cutaneous endpoint titration (Norman, 1986), immunoblotting (Thorpe et al., 1988), RAST inhibition (Helm et al., 1988), leukocyte histamine release (Siraganian and Houk, 1986), and immunoelectrophoresis (Price and Longbottom, 1986).

The availability of recognized reference preparations, particularly in lyophilized form, would be of obvious value (Platts-Mills and Chapman, 1991). The International Union of Immunologic Societies (IUIS) in the World Health Organization has developed several reference preparations. The Center for Biologics Evaluation and Research of the U.S. Food and Drug Administration has also developed or sanctioned some preparations. They differ from the IUIS preparations, however, in character and unitage. In short, there are no universally agreed upon reference standards, methods of characterization, or assignment of unitage.

Because LMW chemicals are not complete allergens, they generally require conjugation to a carrier protein before they can be used in immunoassays (D. I. Bernstein and Zeiss, 1989). It is, of course, important to determine whether chemical protein linkage has occurred and the range of epitope density that has been defined, since the latter is known to determine the allergenicity of a chemical-protein conjugate. Any of a variety of techniques may be appropriate: free amino group analysis (Snyder and Sobocinski, 1975), spectrophotometric analysis (Zeiss et al., 1980), or gas chromatography (D. I. Bernstein et al., 1984). Thus, a reagent can be optimally prepared by selection of epitope density. The same uncertainties of characterization and standardization outlined above for complete protein allergens also exist for chemical-protein conjugates.

ANTIBODIES

Currently, there is no central serum bank or sharing of serologic reagents, and controls tend to be obtained by individual laboratories. In addition to reference allergen preparations, reference sera, reference serum pools, or reference monoclonal antibodies of known specific antibody content would be useful as positive controls in immunoassays.

Diagnostic Value of In Vitro Tests

The value of a laboratory test for diagnosing a given condition in a given population is judged by its sensitivity, specificity, and positive predictive value (Galen, 1986). These qualities of in vitro tests for the presence of IgE specific to common inhalant allergens have not been accurately determined. Evaluations of these tests would be useful before embarking on any large epidemiological studies of putative indoor allergens that cause allergic reactions. Among certain indoor allergens, such as various species of dust mite, cross-reactivity[1] of allergens is known to occur. To what

---

[1] The interaction of an antigen with an antibody formed against a different antigen with which the first antigen shares closely related or common antigenic determinants. The effect is to reduce the specificity and sensitivity of the test method.

extent this occurs in general with indoor allergens is not well established. In reports comparing in vivo and in vitro testing, however, in vivo testing (e.g., skin tests) has been found to be more specific (Johansson and Foucard, 1978; Wide and Bennich, 1967).

Almost all symptomatic hypersensitivity pneumonitis patients have positive precipitins against the inciting agent; however, a significant proportion of exposed, asymptomatic individuals (15–40 percent) also have positive precipitins (Pepys and Jenkins, 1965; Wilson et al., 1981). The sensitivity of the precipitin assay thus approaches 90 to 100 percent; the specificity is approximately 60 to 85 percent. These figures apply only when the inciting agent is identified and an antigen is available for in vitro testing. Cellular assays such as lymphocyte stimulation studies have been reported to result in fewer clinical false-positive results with consequent specificity of approximately 95 percent (Hanson and Penny, 1974). Virtually all symptomatic individuals also have positive cellular assay results; thus, the sensitivity of such tests approaches 100 percent. In addition, there is an epiphenomenon of patients with inflammatory lung diseases such as sarcoid or idiopathic pulmonary fibrosis who may have precipitins against common inhalant allergens.

Quality Control

Quality control of immunoserologic testing is similar but not identical to quality control for other diagnostic tests (Taylor, 1986). Moreover, it is unlikely that these criteria are universally followed; therefore, comparability of results from various laboratories should not be assumed. Periodically, a laboratory should assess inter- and intra-assay variability. Positive sera of known value and negative sera should also be assessed periodically. A standard curve is generally employed for quality control; the results of the assessments should fall within a certain range on the curves.

As new immunoassays are developed and reported, certain properties of the tests must be evaluated. These include specificity, parallelism, interassay variability, intra-assay variability, and comparability of tests to those results obtained from a recognized immunoassay.

Conclusions and Recommendations

The accuracy of any immunodiagnostic test is highly dependent on the characteristics of the test reagents, in particular, the allergen reagent. Standardization and characterization of allergen reagents used for immunodiagnostic tests are imperative. There are a variety of characterization methods and unitages that could be used. Similarly, the existence and characterization of control antibody, whether polyclonal or monoclonal, would be valuable for standardization and quality control of immunodiagnostic tests. Ideally, minimal standards for quality control should be devised for labs reporting results of tests to detect specific immunologic responses to indoor allergens.

Once the above are instituted, the specificity, sensitivity, and positive predictive value of immunodiagnostic tests could be determined for use in major epidemiological studies or to determine specific immunologic responses of individuals to indoor allergens. In addition, the degree of cross-reactivity of antibody developed in response to a given allergen could be determined by cross-inhibition with other allergens. It would be important to assess these factors prior to embarking on any large epidemiological studies. There are also several unclear aspects of the immunopathogenesis of allergic disease that need elucidation in order to define the role of various tests (e.g., tests of specific cell-mediated immunity in asthma). Further immunopathogenic studies of non-IgE-mediated asthma and cellular studies in all types of immunologic asthma will be required to clarify these issues.

Future studies in the development of in vitro diagnostic tests should include the following:

Research Agenda Item: Identify selected allergens of potential research usefulness, and prepare pure reference standards for the development of immunoassays, including those that can be used in large scale epidemiological studies.

Research Agenda Item: Develop and assess immunoassays for new allergens, including low molecular weight allergenic chemicals, that can be used for research and for the diagnosis of allergic disease.

PULMONARY FUNCTION TESTS

Applications and Methodological Challenges

Pulmonary function tests are well-established, practical methods that are widely used in the evaluation and monitoring of diseases due to indoor allergens. One of the major shifts in clinical practice recommendations is the suggestion that peak flow meters be widely used for outpatient self-monitoring of asthma (NHLBI, 1991). However, objective measurements of respiratory status have been available for decades.

The clinician must choose among an array of possible pulmonary function tests. Table 5-5 shows the types of practical questions that may be answered using pulmonary function tests, and which tests will address these questions. All diagnoses are made using pulmonary function tests as one part of a complete clinical evaluation. Patient cooperation and maximal effort are essential for valid test results. Proper training of pulmonary function technicians, supervision of the testing facility by a knowledgeable

TABLE 5-5 Choosing Lung Function Tests for Use in Clinical Medicine
and Research

Question	Test Method
Does my patient meet the objective criteria for a diagnosis of asthma?	
(1) Variable airflow obstruction	• An obstructive pattern on spirometry with an immediate positive bronchodilator response (15 percent improvement in FEV_1) *or* • Peak flow variability with serial measurements. *or* • Variable airflow obstruction with repeated spirometry (e.g., methacholine or histamine)
or (2) Bronchial hyperreactivity	• Positive response to nonspecific challenge
Is my patient improving with asthma treatment?	• Reduced bronchial hyperreactivity • >10 percent improvement in FEV_1 or peak flow • Reduced diurnal variation in peak flow
Does my patient have hypersensitivity pneumonitis?	• Reduced spirometry • Reduced diffusing capacity • Reduced lung volumes
Is my patient becoming ill in his/her home?	• Peak flow measurements with prolonged time in and out of the home
Could the reduction in lung function in my patient be due to a problem (disease) other than asthma?	• Spirometry • Lung volumes • Diffusing capacity
Why is my patient so short of breath with exercise?	• Spirometry and diffusing capacity *and* • If symptoms are unexplained by these tests, then perform exercise testing
What is the clinical value of histamine or methacholine challenge testing? (See Cockroft and Horgreave, 1990.)	• Excluding or confirming a diagnosis of asthma *or* • Diagnosis and management of occupational asthma *or* • Assessing the severity of asthma and monitoring asthma treatment

NOTE: Lung function tests are to be used in conjunction with a clinical evaluation.

professional, and attention to quality assurance generally result in high-quality tests (Enright, 1992).

Diseases such as asthma and hypersensitivity pneumonitis, which are caused by indoor allergens, are characterized by decrements in lung function that vary over time (Fink et al., 1971; Hodgson et al., 1987; Jacobs et al., 1989; Kawai et al., 1984; Lopez and Salvaggio, 1988). The variability occurs in recognizable patterns, yet is highly diverse among individuals (Perrin et al., 1991). This feature poses important challenges for clinical and epidemiological studies. Chief among these challenges is the requirement for repeated measures of pulmonary function and the need to assess the magnitude and pattern of variability in the test result. Clinicians must also develop practical schemes to monitor variations in lung function.

Pulmonary function tests have many applications in clinical medicine and research related to indoor allergens (Table 5-6). The appropriate choice of pulmonary function test depends on the requirements of the specific application.

Spirometry

Spirometry, the most reliable and commonly used pulmonary function test, measures the characteristics of a forced expiratory maneuver (ATS, 1991; Gold and Boushey, 1988). To perform this test, subjects are instructed to take in as deep a breath as possible, to seal their lips around the

TABLE 5-6 Applications of Pulmonary Function Tests

Application	Clinical Practice	Clinical Studies	Epidemiological Studies
Diagnosis			
Evaluate the effect of sensitization	X		
Identify specific diseases	X	X	X
Assess disease severity	X	X	X
Suggest causal relationships*	X	X	X
Demonstrate causal relationships*	X	X	
Treatment			
Evaluate the response to therapy			
Medication	X	X	X
Environmental modification/ allergen modification	X	X	X
Impairment/disability evaluation	X		
Disease incidence or prevalence			X

*Suggestion or demonstration of causal relationships usually requires serial measurements of pulmonary function coupled with challenge testing or a change in environments.

TABLE 5-7 Coefficients of Variation (percentage) for Spirometric Measurements of Different Subject Groups

Subject	Slow VC	FVC	FEV_1	$FEF_{25-75\%}$
Normal	3	5	7	13
Obstructed		11–14	14	18

NOTE: VC, vital capacity (maximal volume of air exhaled from the point of maximal inhalation); FVC, forced vital capacity (total volume of air exhaled); FEV_1, volume of air exhaled in 1 second; and $FEF_{25-75\%}$, forced expiratory flow between 25 and 75 percent of forced vital capacity.

SOURCE: ATS (American Thoracic Society), 1987.

spirometer mouthpiece, and then to blow out as hard and as fast as they can for at least 6 seconds or until instructed to stop. The spirometer measures changes in volume as a function of time (a volume-based system) or changes in airflow as a function of time (a flow-based system) and plots the tracing on an x–y axis. The results of each tracing are described as volume exhaled in 1 second (FEV_1), the total volume exhaled (FVC, forced vital capacity), and the percentage of the total volume exhaled in 1 second ([FEV_1/FVC] × 100). Subjects repeat the maneuver until three acceptable tracings are obtained. Criteria for acceptable tracings have been established by the American Thoracic Society. One important criterion is that the technician must judge that the patient has exerted a maximal effort. The best value of three acceptable efforts is taken as the actual measurement.

To be a valid measure of lung function, reproducibility criteria must be met. Reproducibility criteria state that the best and the second-best FEV_1 and FVC should be within 5 percent or 100 cc, whichever is greater (ATS, 1987). However, the use of these standards may tend to produce underestimates of pulmonary function changes in worker populations, because those with the most disease are the least likely to produce acceptable curves. Diseased workers producing nonreproducible or unacceptable spirometry results would be deleted from the study and thus would not be identified as having disease. This example of "ascertainment bias" has been discussed more extensively by Eisen (1987). Studies summarized in Table 5-7 have quantitated reproducibility in spirometric measurements across days and weeks (ATS, 1991; Lebowitz et al., 1987). Less precise measurement could lead to more variability over time; conversely, stringent quality assurance can result in coefficients of variation below 6 percent for repeated spirometry, even for subjects with obstructive lung disease (Enright et al., 1991).

Spirometry results are compared to published reference values, and presented as percent predicted derived from cross-sectional population-based studies (Crapo et al., 1981; Knudson et al., 1983; Morris et al., 1971). Predicted values depend on an individual's height, age, and sex. Tall, young men have the greatest predicted lung function. When an individual's actual function is below predicted values, the magnitude of the abnormality may be described according to various schemes such as that presented in Table 5-8 (Engelberg, 1988).

Spirometry interpretation characterizes the type of abnormality that is present. Typically, asthma is characterized by airflow obstruction, meaning that a disproportionate decrease in FEV_1 relative to FVC exists and peak flow rates are reduced. Hypersensitivity pneumonitis is characterized by a reversible restrictive pattern; that is, FEV_1 and FVC are reduced in parallel, and peak flow rates may be unchanged despite significant drops in other measures of lung function. Exceptions to these generalizations, however, are well recognized. In asymptomatic or mild asthma, lung function may fall within the normal range. In more severe asthma, symmetric reductions of FEV_1 and FVC occasionally occur and may be misinterpreted as restrictive lung disease. Measurement of lung volumes (see below) will show an increased total lung capacity and air trapping. Clinical evaluation using bronchodilators may also result in improvement. The use of flow rates is helpful in limited circumstances and requires more careful interpretation (ATS, 1991). Spirometry may be repeated after administration of bronchodilators; improvement in pulmonary function indicates the presence of reversible airway obstruction, a characteristic feature of asthma.

Spirometry interpretation includes determining whether lung function has changed over time. Comparison with previous tests by the same subject

TABLE 5-8 American Medical Association/American Thoracic Society: Description of Respiratory Impairment

Parameter	None	Mild	Moderate	Severe
FVC[a]	≥80	60–70	51–59	≤50
FEV_1	≥80	60–79	41–59	<40
FEV_1/FVC	≥70	50–69	41–59	<40
DLCO[b]	≥80	60–79	41–59	<40
or				
VO_{2max}[c]				
ml(kg · min)	>25	20–25	15–20	<15

[a]Predicted values of FVC are from Crapo and Morris, 1981; Crapo et al., 1981.
[b]DLCO, diffusing capacity of the lungs for carbon monoxide.
[c]VO_{2max}, maximum volume of oxygen exhaled.

SOURCES: Engelberg, 1988; Renzetti et al., 1986.

adds greatly to the validity of spirometry because it narrows the range of expected values. With this approach, normal is defined with reference to the subject's original value instead of defining "normal" as any value between 80 percent and 120 percent of standard, predicted values which are derived from population studies. In principle, there is a 95 percent certainty of a decline in lung function if the measured fall is 1.65 times the coefficient of variation for repeated studies (Pennock et al., 1981). However, these guidelines are not uniformly applied or accepted. For example, the cotton dust standard of the U.S. Department of Labor defines a 5 percent, or 200 ml, change across a work shift as significant, yet some studies have shown this magnitude of change to be within expected cross-shift variability for some workers (Glindmeyer et al., 1981; Sheppard, 1988a). Enright and others (1991) achieved a confidence interval of 5.9 percent for duplicate spirometry performed (on average) 25 days apart among subjects with mild to moderate airflow obstruction. Intrasubject variability from year to year, as reported by Nathan and colleagues (1979) for more than 1,000 individuals, indicates that most subjects have greater yearly variability than short-term changes (Lebowitz et al., 1982). Roughly speaking, the expected annual decline in FEV_1 is 1 percent.

In the clinical setting, a source of potential variability is the use by patients of different laboratories with different equipment and technicians. In theory, of course, these differences should be minimal if proper calibration procedures are used.

For the most effective evaluation of individual pulmonary function, epidemiologists must also designate proper criteria by which to define exposure-response relationships. Lebowitz and colleagues (1987) propose that such criteria represent clinically meaningful—as distinct from statistically significant—responses.

Spirometry equipment is relatively inexpensive, portable, and accessible. It may be transported to work sites and is available in most hospitals and some physician offices. The equipment is sturdy and usually retains precision. Spirometers are now frequently coupled with computers for automated data calculation. Results appear on a printout or on a computer screen, and sequential tracings are superimposed for ease in assessing reproducibility. "Hard-copy" tracings can also be made and may be required for medicolegal reports.

Extensive efforts have resulted in improved precision for spirometry testing. Table 5-9 shows minimum standards for spirometry equipment that have been established by the American Thoracic Society. Nevertheless, S. B. Nelson and coworkers (1990) found that some commercially available spirometers produced FVC errors as large as 1.5 liters (a 25 percent error) primarily as a result of problems with computer software. Technicians and subjects can be trained to perform the test with precision, accuracy, and

TABLE 5-9 Technical Standards for Peak Flow Meters and Spirometers

Device	Range of Accuracy	Interdevice Variability	Precision	Reproducibility
Peak flow meters				
Children	100–400 liters/min	±5%	±10%	5% or 10 liters/min
Adults	100–700 liters/min = 10%			
Spirometers	720 liters/min ≥7 liters 15 seconds		±3% or 0.05 liter	5% or 100 cc

NOTE: The test signal for spirometers is the 24 standard waveforms; for peak flow meters, only waveform no. 24 should be used. The waveforms, developed by Hankinson and Gardner, can be used to drive a computer-controlled mechanical syringe for testing actual hardware and software.

SOURCES: ATS (American Thoracic Society), 1987; NHLBI, 1991.

reproducibility. Technician training courses, certified by the National Institute for Occupational Safety and Health, are available throughout the United States. Efforts to standardize spirometry methods (ATS, 1987) and improve technician training have greatly increased precision, which is now adequate for such applications as clinical diagnosis, disease management, evaluating severity of impairment, and for clinical and epidemiological studies. However, this precision is inadequate to allow a demonstration of subtle longitudinal declines in lung function, because the expected rate of decline for an individual is only about 1 percent per year. Conversely, spirometry is a sensitive measure of disease, since reductions can be detected before severe impairment occurs.

Determination of the reversibility of airflow obstruction is made by performing spirometry before and after administration of a bronchodilator. A 15 percent improvement in FEV_1 after using a bronchodilator is considered evidence of reversible airflow obstruction.

Peak Flow Measurements

Peak flow measurements are especially useful to patients in asthma self-management. Patients produce peak flow measurements by inhaling as deeply as possible, sealing their lips around the mouthpiece, and briefly exhaling at maximum velocity. The tube is attached to a measuring device that records the maximum flow rate achieved during exhalation. Patients then read the result and manually record it on a paper record. Patients perform 3 efforts, and the best value is taken as the actual measurement.

Peak expiratory flow occurs within the first second of a forced expiratory maneuver. Maximum flow occurs within the first few hundred milliliters of volume expired from total lung capacity and is volume and effort dependent. To measure peak flow accurately, the exhalation must begin at maximum inspiration and must be performed with maximum effort. In contrast to spirometry, there is no requirement for a prolonged smooth exhalation, which is an advantage for asthmatics. (Asthmatics often cough immediately following a forced expiratory effort, which can interfere with attempts to obtain acceptable spirometry.) Reduced peak expiratory flow is considered a valid measure of airflow obstruction, and it correlates well with FEV_1.

Peak flow meters are lightweight, compact, inexpensive ($15–$50) instruments that are small enough to be carried in a purse or coat pocket. Hospitals use peak flow meters for bedside monitoring of asthma severity and response to bronchodilators. Current asthma management guidelines recommend daily home monitoring of peak flow using a peak flow meter (NHLBI, 1991). Peak flow measurements are used extensively in clinical trials of asthma therapy.

Efforts are proceeding to improve peak flow meters. The European Respiratory Society and the American Thoracic Society are developing standards for their construction and use. In addition, peak flow meters with a computer chip to allow recording of the time of peak flow effort should be on the market soon. This innovation will increase the cost of peak flow meters but will be a significant advantage for clinical studies.

The accuracy of peak flow meters varies among models. Table 5-9 shows technical standards suggested by the National Asthma Education Panel of the National Heart, Lung, and Blood Institute (NHLBI, 1991). Deficiencies in commonly used meters are outweighed by their benefits (Lebowitz, 1991; NHLBI, 1992). Although a flow range of 0–720 liters per minute has been recommended, some units are highly inaccurate within this range. Nevertheless, peak flow meters are invaluable for correlating changes in respiratory obstruction with a variety of activities and events experienced by the patient.

Current guidelines suggest that peak flows should be reproducible within 5 percent or 10 liters per minute (NHLBI, 1991). Reproducibility of peak flow measurements using mini-Wright peak flow meters was determined by having 10 subjects perform 30 forced expiratory maneuvers. Coefficients of variation ranged from 2 to 14 percent (Lebowitz et al., 1982). Significant training effects were seen during the first 2 days of a fortnight's study, and Quackenboss and colleagues (1991b) excluded these data from consideration in an epidemiological study. Measurement by standard and mini-Wright peak flow meters has been found to be stable over 6 months (Morrill et al., 1981; Van As, 1982). This stability over time may be more important than absolute accuracy, given that the usual application for peak flow involves an assessment of serial changes in peak flow in the same subject over time (Van As, 1982).

Interpretation of peak flow variability is best performed by visual inspection of a graph that plots peak flow over time. Computer-based algorithms for interpreting peak flow measures have been shown to be no better than the "eyeball" method for diagnosing work-related asthma (Perrin et al., 1992).

Detection of Airway Hyperreactivity

Bronchial hyperreactivity, a cardinal feature of asthma, is measured as follows. Patients perform initial spirometry, and the resultant FEV_1 is defined as the baseline value. The patient then inhales saline, followed by increasing concentrations of an agonist such as methacholine or histamine, or a stimulus such as cold air or a distilled water aerosol. After each provocation, spirometry is repeated and FEV_1 is determined for that dose. The test is stopped when a predetermined reduction in lung function is achieved or the maximum dose is administered. Exercise testing is also used occasionally to measure bronchial reactivity. Typically the test ends when a 20% reduction in FEV_1 or a 35 percent reduction in specific airway conductance has occurred, and a provocative dose is calculated.

The challenge method and type of data recorded differ among laboratories (Chai et al., 1975; Chatham et al., 1982a). Some record the concentration of agonist at the endpoint, whereas others record the cumulative dose. The percent decrease in FEV_1 (targeted as a positive end point) ranges from 10 to 40 percent, with a 20 percent decrease being most common. Popa and Singleton (1988) attempted to determine which of the current provocative doses for histamine optimally separates normal from asthmatic subjects. They concluded that new normative data for diagnostic provocation were needed because of a high misclassification rate using current methods. Inhalation challenge testing with methacholine, histamine, distilled water, exercise, etc., is commonly termed non-specific challenge testing to distinguish these from specific allergen challenge testing. However, each of these chemical and physical agents act on different bronchial receptors and therefore reflect different aspects of non-specific bronchial hyperreactivity. Furthermore, animal studies indicate that inherited hyperreactivity to these agents is transmitted at distinct genetic loci (Levitt et al., 1990).

Despite recognized limitations, testing for airways reactivity is widely used because it is a practical test which has great utility in clinical medicine and research. Asthmatics as a group develop reductions in lung function with provocation at much lower doses than do non-asthmatics. The considerable within-subject overlap in hyperreactivity to methacholine, histamine and distilled water challenges means that each of these agents is widely used. Increased reactivity has been described in non-asthmatics, particularly in first degree relatives of asthmatics, cigarette smokers, people with allergic rhinitis, and some apparently normal individuals. In general, asthmatics are the most reactive (i.e., require the smallest concentration of agonists to effect a reduction in lung function). The risk of bronchial reactivity increases with increasing skin test reactivity (Burrows and Lebowitz, 1992; Lofdahl and Svedmyr, 1991). In asthmatics, the degree of reactivity correlates with other measures of disease severity. People with allergic rhinitis demonstrate an intermediate level of reactivity. A high proportion of cigarette smokers with airflow obstruction demonstrate increased airways reactivity, but the population distribution of airways reactivity among smokers without airflow obstruction is unknown at present.

Data from Pattemore and colleagues (1990) show a 24 percent prevalence of frequent wheezing in children with normal airway responsiveness and a lack of current asthma symptoms in 41 percent of people with hyperresponsiveness. This has led some to criticize the utility of measurement of methacholine or histamine responsiveness in clinical practice. Cockroft and Horgreave (1990) disagree, and point out that when spirometry is normal, methacholine and histamine hyperresponsiveness is a sensitive measure of abnormal airway function that correlates closely with the presence and degree of variable airway obstruction.

Cockroft and Horgreave (1990) specifically identify three areas of clinical utility for histamine and methacholine inhalation tests:

1. Exclusion or confirmation of a diagnosis of asthma, especially if the presentation is atypical.
2. Diagnosis and follow-up of occupational asthma.
3. Assessment of the severity of asthma and monitoring of asthma treatment.

Bronchoprovocation Testing with Specific Allergens

Allergen-specific bronchoprovocation testing is a research tool used to diagnose specific immunologic diseases, identify new etiologic agents, and study the pathogenesis of asthma and hypersensitivity pneumonitis (Chan-Yeung and Lam, 1986; Pepys and Hutchcroft, 1975). Guidelines for allergen challenge have been proposed by the American Academy of Allergy and Immunology (Chai et al., 1975).

The limits of allergen inhalation challenge testing outside of a research setting are several. It is time-consuming and staff intensive, and may not be reimbursable. Patients must be monitored for up to 24 hours to detect and treat late reactions. Potentially toxic reactions must be avoided. When an individual has been removed from exposure to an allergen, several days' reexposure may be required for a measurable response to occur. Allergen extract in aerosol form is deposited differently in the bronchial system compared with allergen in its natural state.

Allergen challenge has been used in connection with bronchoalveolar lavage, in which a bronchoscope is used to sample the fluid in the alveoli. Bronchial biopsies can also be performed. Demonstration of lymphocytosis in bronchoalveolar fluid can suggest the diagnosis of hypersensitivity pneumonitis (Reynolds, 1988). Beasley and others (1989a) performed specific inhalation challenge with allergen followed by bronchial biopsy and lavage, and, later, histamine challenge. Their studies demonstrated an inverse correlation between the number of epithelial cells in lavage fluid and histamine reactivity.

Despite its limitations, specific inhalation challenge testing will continue to have a unique place in the study of the health effects of indoor allergens.

Lung Volumes

Measurement of lung volumes is useful to help evaluate reductions in forced vital capacity. The usual method (gas dilution) involves breathing a known concentration of an inert gas in a closed-circuit system, followed by measurement of the functional residual capacity (volume in the lungs at the end of a normal breath). Total lung capacity is then determined by summing the functional residual capacity and the inspiratory capacity (determined by spirometry).

Diffusing Capacity Testing

Measurement of diffusing capacity is indicated when the clinician suspects that gas exchange is impaired by the disease process. Diffusion capacity measurements are not necessary in most cases of suspected asthma, but they may be indicated occasionally to exclude interstitial lung disease.

In diffusing capacity testing, the patient inhales a known mixture of an inert (nondiffusible) gas and a readily diffusible gas such as carbon monoxide (ATS, 1987). The exhaled gas is collected and analyzed, and the uptake of carbon monoxide is expressed in ml/min/mmHg. Similarly to spirometry, reference values have been determined by cross-sectional studies; interpretation consists of comparing actual values with reference values and with any previously measured values for the patient. Possible confounding factors are also considered. For example, diffusing capacity may be falsely elevated by conditions that raise the metabolic rate, lung blood volume, or red blood cell count; it may be lowered by the presence of carboxyhemoglobin in the blood or by anemia (Gold and Boushey, 1988). Reference standards exist for the performance of diffusing capacity testing. As of 1987, however, interlaboratory variability was too great to recommend one set of equations for all laboratories. Rather, laboratories are advised to use internal controls, such as repeated measurements on a normal individual.

Exercise Studies

Exercise pulmonary function studies can be used as part of a clinical evaluation to quantitate exercise-induced bronchoconstriction in asthmatics (Chatham et al., 1982b), and to exclude the diagnosis of asthma in other cases. Often, however, serial peak flow monitoring during normal activities will be sufficient to document exercise-induced asthma. Exercise testing is also used to evaluate an individual with suspected interstitial lung disease. The occurrence of desaturation or a widening alveolar-arterial oxygen gradient suggests significant interstitial disease (Whipp and Wasserman, 1988). Finally, exercise testing can be used as part of the impairment evaluation when symptoms are disproportionate to static lung function due to deconditioning or unrecognized cardiovascular or pulmonary pathology (Engelberg, 1988).

Measures of Upper Airway Function

Measures of upper airway function are primarily used in research settings at present. The one exception is the obtaining of a nasal swab with cytologic examination for the presence and characteristics of inflammatory cells.

For physiologic measurement of nasal airflow (rhinomanometry), the pressure-flow characteristics during panting maneuvers are used to derive nasal airway resistance (Cole, 1982). Ten to 15 percent of patients are unable to perform the maneuver successfully; normal values are problematic because of normal fluctuations in nasal airway caliber. Nasal resistance measures correlate to a variable degree with congestive symptoms. Peak flow measurements of nasal airflow have been made by adapting a standard peak flow meter with a pediatric anesthetic mask (Ahman, 1992). This method is not without its problems or constraints; for example, technical questions remain about whether to record a nasal inspiratory or expiratory maneuver.

Acoustic rhinometry is a new technique that describes the cross-sectional area of the interior of the nose as a function of the distance from the nares. An acoustic pulse is generated, and reflections are recorded by a microphone and analyzed with a computer program using Fourier analysis. Subjects must stop breathing only for approximately half a second; the test is therefore easier to perform than physiological measures of nasal airway resistance. The technique has been validated by comparison with cadavers and computerized tomography (Hilberg et al., 1989), but much more extensive clinical trials are needed to evaluate this technique for clinical use.

Nasal lavage can be readily performed following nasal inhalation challenge. Analysis of cells, mediators, and proteins can then be conducted to study the pathogenesis of allergic and nonallergic rhinitis (Bascom et al., 1986). However, normal values have not been determined for most measures.

Conclusions and Recommendations

There are simple, reliable measures of lung function that may be used for studying diseases caused by indoor allergens. Indeed, objective measures of respiratory function should be a part of protocols to determine the efficacy of therapeutic strategies for these diseases. Predicted values for pulmonary function fall along a normal distribution curve with 95 percent confidence intervals for FEV_1 and FVC of approximately 80–120 percent. The lower limit of variation in population studies for the midlevel expiratory flow rate (FEF_{25-75}) is approximately 60 percent.

Spirometry is limited in its ability to detect impairment of ventilatory function in asymptomatic individuals (Morris et al., 1971) because of the wide range of normal values, even with predicted levels that control for age, sex, and height. Significant inaccuracy can result from errors in spirometry performance, almost all of which lead to underestimation of the true respiratory function. These tests can, however, help to evaluate the effects on an individual of sensitization to specific allergens (J. M. Smith, 1988). They can also help to diagnose respiratory diseases that may be caused or worsened by indoor allergens (Lopez and Salvaggio, 1988; NHLBI, 1991; Woolcock, 1988) and to assess disease severity, which is often critically important in clinical decisionmaking (NHLBI, 1991). Serial pulmonary function testing in the home or workplace can demonstrate causal relationships between the indoor environment and respiratory illness. Serial pulmonary function testing coupled with bronchoprovocation can demonstrate the causal relationship between specific allergens and respiratory responses (Chan-Yeung and Lam, 1986).

In epidemiological studies, measures of environmental factors and of pulmonary function can be evaluated for associations that suggest causal relationships (S. Weiss et al., 1983). Pulmonary function tests may also be used to assess the efficacy of therapy, determine response to treatment, or determine the effect of environmental modification (Ehnert et al., 1991; Platts-Mills et al., 1982). Such tests are required when physicians are asked to determine impairment resulting from a respiratory disease for insurance or benefit systems such as workers' compensation and social security disability (Engelberg, 1988). Finally, estimates of disease incidence or prevalence often result from epidemiological studies in which pulmonary function tests are used to ascertain disease.

Recommendation: Include pulmonary function tests in epidemiological studies to help improve estimates of disease incidence and prevalence. Because they are portable and can be self-administered, tests that utilize peak-flow measurements are most desirable for this purpose.

One drawback of many pulmonary function tests is that they must be administered by technicians. Peak flow measurements are less reliable but are highly portable, can be self-administered, and are therefore often more sensitive in the diagnosis of asthma.

Recommendation: Include objective measures of respiratory function in experimental protocols designed to determine the efficacy of therapeutic strategies (e.g., pharmacotherapy, environmental modification, avoidance) used to treat respiratory diseases caused by indoor allergens.

Bronchial hyperreactivity is a feature of asthma that correlates with clinical severity and does not require repeated measurement (Boushey et al., 1980). It is unclear, however, whether bronchial hyperreactivity can be correlated with exposure to indoor allergens.

Research Agenda Item: Determine whether changes in bronchial hyperreactivity can be correlated with exposure to indoor allergens. If such a correlation exists, determine how reducing the level of allergens affects bronchial hyperreactivity.

6

Assessing Exposure and Risk

This chapter has two major parts: exposure assessment and risk assessment. The exposure assessment section addresses issues that include environmental monitoring for indoor allergens, exposure measures, and internal dose. The risk assessment section describes the general nature of the process, including exposure assessment, and presents an example that uses dust mite exposure data in assessing the risk of sensitization.

EXPOSURE ASSESSMENT

Assessing exposure involves numerous techniques to identify contaminants, contaminant sources, environmental exposure media, transport through each medium, chemical and physical transformations of the contaminant, routes of entry to the body, intensity and frequency of contact, and spatial and temporal concentration patterns of the contaminant (NRC, 1991). Once an estimate of personal exposure is constructed for the time period of interest, it may be adjusted to estimate the internal dose; it may also be further refined to estimate the biologically effective dose for a critical target tissue. In practice, personal exposure measurements may be qualitative or quantitative. The use of quantitative estimates to reflect biologically active concentrations requires a detailed understanding of deposition and the kinetics of transport, metabolism, and clearance.

Assessment of exposure is a multistep process (Figure 6-1). The specific elements in each step of the process are usually tailored to the types of information that are available or that can be collected, and to the agent of interest.

Aeroallergens are common both outdoors and indoors. For example, exposure to pollens and spores from plant material may occur during time spent outdoors; the pollens and spores may be transported inside offices or other buildings in which additional exposure may occur. Indoors, dust mites, mammals, and fungal growth are examples of sources for aeroallergens. For both indoor and outdoor environments, sources must be recognized,

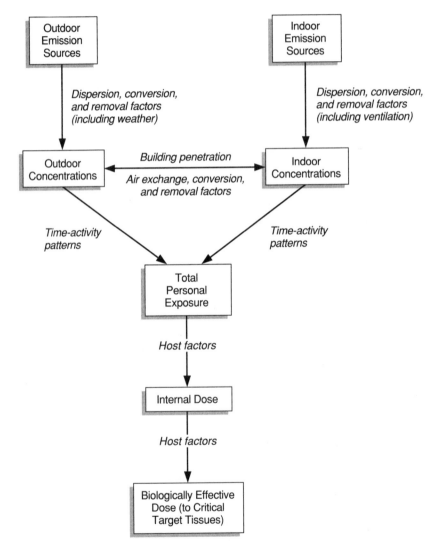

FIGURE 6-1 Framework for exposure assessment. Source: NRC, 1985.

methods of dispersion within the space determined, and factors that may aid in removal evaluated. Once sources have been identified, sampling strategies that allow some estimation of exposure can be devised.

Monitoring for Indoor Allergens

GENERAL PRINCIPLES

Monitoring for allergens can help determine the cause of one or more cases of allergic disease (in which instance, sampling and analysis modes that yield information on the range of possible organisms must be chosen) or characterize environments with respect to specific allergens. Measurements can be semiquantitative (e.g., "presence or absence" or "low, medium, or high"); or, alternatively, modes can be chosen for sampling and analysis that will give accurate, precise documentation for a specific population regarding a more or less well-defined allergen.

Although monoclonal antibody immunoassays measure specific allergens, most allergen monitoring includes sampling for indicators rather than the actual aerosolized allergen. Table 6-1 is a list of aeroallergens and commonly used indicators. In developing a sampling strategy to collect the data for testing a particular hypothesis, important decisions must be made on where, when, and how to sample.

Where to Sample

The majority of allergen samples are collected either from ambient air in one or more environments or from potential allergen reservoirs. Ambient

TABLE 6-1   Aeroallergens and Sampling/Analysis Modes

Common Name	Aeroallergen	Indicators	Sample	Analysis
Thermophilic actinomycetes	Glycoproteins	Living cells	Air, bulk	Culture
Molds	Glycoproteins	Living cells Spores Antigens	Air, bulk	Culture Microscopy Immunoassay
Arthropods	Glycoproteins	Antigens Whole animals Guanine	Bulk Bulk Bulk	Immunoassay Microscopy Chemistry
Mammals	Glycoproteins	Antigens	Air, bulk	Immunoassay
Pollen	Glycoproteins	Pollen grains Antigens	Air Bulk	Microscopy Immunoassay

air sampling sites should be chosen to represent the range of conditions that might occur. Often, practical considerations require that a single site in each community or in each interior be chosen for all air sampling. Outdoor monitoring is usually done at single sites on rooftops to avoid undue contributions of local sources. Indoors, the site at which a group of people spend a majority of their time (e.g., the living room of residences, open office spaces in large buildings) or at which one or more individuals experience symptoms is usually chosen for air sampling. Material is collected from reservoirs from which exposure can be assumed or extrapolated using mathematical models (Swanson et al., 1990).

When to Sample

Allergen levels in both air and reservoirs change over time as well as in space. Ideally, it should be possible to evaluate air samples continuously over time. As a substitute, multiple discrete samples can be taken over relatively short periods of time, or long-term samples can be taken and analyzed as a time-weighted average or in discrete units.

Sampling for allergens always occurs in a complex environment, and consideration must be given to factors that might modify source strengths, emission rates, accuracy and precision of sample collection, or sample analysis methods and health effects (O'Rourke et al., 1990). Controls are necessary for each step of the investigation.

How to Sample

Sample collection methods include observation, bulk or reservoir sampling, and air sampling. Observational sampling can include sensory perceptions of an indicator (e.g., odors or visible fungus growth) and observation of factors known to be related to specific kinds of sources. (For example, mites are likely found in carpeting on grade-level concrete floors.) Observations can be formalized or casual, the type used by most environmental investigators.

Source or reservoir samples have been used as indicators of exposure to most indoor aeroallergens. Dust samples have been collected for analysis of microorganisms (Gravesen, 1978; Saad and el-Gindy, 1990), arthropod allergens (Ishii et al., 1979; Swanson et al., 1989), and mammalian allergens (Ohman and Lorusso, 1987). Dust collection can be standardized by using a dust collector that maintains a constant flow, vacuuming a given surface for a standard length of time or vacuuming carefully measured sections of a surface, or by using a combination of these methods. A number of measured areas of each substrate can be sampled to estimate variability in space. Vacuum-collected dust samples are weighed before or after sieving to remove hair and other large irrelevant fibers and particles. The levels of allergen per gram of dust can then be used to calculate source strengths at the time of sampling. Similar sampling at different times allows an estimation of patterns and rates of change in source strengths. Similar principles apply to the collection of liquid samples. If liquid reservoirs are well-stirred, a single sample can be representative of the entire reservoir.

Surface samples are prepared by swabbing or pressing a plate of culture medium or a sticky tape against a surface; such samples are useful for identifying obvious microbial contamination. Many samples must be taken to allow characterization of the surfaces in a space, and such data may or may not be relevant to allergen exposures.

Ideally, human respiratory exposure is measured using air samples taken near the breathing zone of individuals (personal sampling; Macher and First, 1984). Most allergen sampling, however, is done to characterize ambient aerosols. Air sample collection involves drawing a representative sample of the aerosol into a collection device and removing particles in an unbiased way and in a form that allows appropriate analysis.

Volumetric air samples for indoor allergen analysis are usually collected by suction devices. To accurately represent the aerosol, suction samples should be collected under isokinetic conditions in which ambient air flows into the sampler parallel to and at the same rate and direction of suction. In still air and in cases in which the sampling orifice is at an angle to ambient air movement, suction samplers tend to oversample small particles, which are easy to divert from their original path. When air is moving into the sampler orifice faster than the suction rate, small spores will tend to follow streamlines around the orifice and be undercollected. In most sampling protocols, isokinetic conditions are not present. For the small filter cassette samplers and the low flow rate suction impactors (e.g., the Andersen type) used in indoor environments, the error introduced is small. However, high-volume filter samplers and, possibly, the portable suction impactors (SAS, RCS) pull air into the sampler at a rate much higher than ambient air speeds; consequently, small particles are preferentially collected. It is also important to collect a small enough sample that the aerosol in the space is not changed during sampling. High-volume filtration devices process large amounts of air and can actually act as air cleaners.

Once particles have entered the sampler, they must be removed from the airstream. The two most commonly used methods are inertial impaction and filtration. Inertial impaction allows the collection of particles that are able to cross the airstream lines inside a sampler and thus stop at the collection surface. In general, large particles are more efficiently collected by inertial impaction than are small particles. The impaction samplers are often rated by the 50 percent cutpoint (the particle diameter at which 50 percent of particles entering the sampler will be retrieved; ACGIH, 1989). Ratings are set for the commonly used aeroallergen samplers listed in Table 6-1.

Because particle capture is related to both particle diameter and to the speed at which the particle is moving, one can control the size of the particles collected by changing the speed with which particles approach the collection surface. The cascade impactors (e. g., Andersen, 1958) fractionate aerosols by accelerating air through smaller and smaller orifices.

The nature of the impaction surface will also influence the efficiency of particle collection. Unless a particle is traveling at exactly the right speed to stop at the collection surface, it may hit the surface and bounce back into the airstream. This bounce is a function both of particle inertia and of the stickiness of the collection surface. For culture plate impactors, bounce is probably minimal; particles tend to penetrate the agar surface rather than bounce off. For spore traps, however, the collection surface must be coated with adhesive, and it gradually becomes less sticky as more and more particles are trapped. Sampling times must therefore be short enough so that overloading does not occur. Spore traps not only collect spores but all kinds of other particles that might be in the air. In aerosols in which levels of nonbiological particulates far exceed spore concentrations, surfaces often overload before a significant number of biological particles have been collected.

The nature of the impaction surface often determines the kind of analysis that can be used. Adhesive (greased) surfaces are usually analyzed by light microscopy. Agar surfaces are most commonly used for cultural analysis, but they can also be homogenized and assayed using immunochemical or biochemical techniques (Tovey et al., 1981a; Yoshizawa et al., 1991).

Modifications of the Hirst spore trap (the Burkard, Kramer-Collins, and Lanzoni spore traps [Solomon et al., 1980b]) are increasingly used for evaluating outdoor allergen aerosols. These suction devices collect 10 liters of air per minute, impacting particles on a moving, greased tape (over seven days) or on to a microscope slide (over 24 hours). The orifice of the sampler is designed to collect particles as small as about 3 μm efficiently. The trap commonly used outdoors has a wind vane so that the orifice faces into the wind. Similar indoor traps have no wind vane, and the orifices face upward. The recording Burkard version collects particles continuously over 7 days or 24 hours. The 7-day and 24-hour slide samples can be analyzed as 24-hour averages or in time increments of as little as 1 hour. Other devices collect samples in discrete bands over 24 hours (Samplair) or on one spot over a few minutes (Burkard personal).

The rotorod has been used for many years to evaluate outdoor allergens, in spite of the fact that it grossly underestimates fungus spore levels. The particle collection efficiency of these rotating arm impactors depends on the rotational speed and the width of the collecting surfaces (the rods). The wide collection surfaces commonly used for allergen monitoring (1.59 mm) efficiently collect particles as large as 10 μm; however, small particles tend to follow the streamlines around the surfaces and are missed. The rotorod efficiently collects most pollen types but is less useful for allergen-carrying particles common in indoor air.

Impingers are suction devices that impact particles onto a surface submersed in a liquid. Impingers will not collect hydrophobic particles (like most fungus spores) efficiently unless surfactants are added to the impinger fluid. These devices are especially useful for collecting bacterial aerosols in highly contaminated environments because the resulting suspension of cells in a liquid can be diluted for analysis (Macher and First, 1984; Morey, 1990b).

Suction samplers that collect particles by filtration efficiently trap all particles above the rated pore size of the filter. In addition, particles smaller than the rated pore size are collected by diffusion or impaction. For example, at low flow rates, a 0.8-μm filter probably collects 50 percent or more of the particles in an aerosol larger than 0.1 μm (see Sakaguchi et al., 1989). As the flow rate increases, the filter will capture a higher and higher percentage of smaller particles. The filter's stickiness and kind of pore structure, as well as whether it can be adequately pulverized, all contribute to the ease with which particles can be removed from it for assay.

Extraneous chemicals that remain in some filters after manufacturing can directly change the nature of some aerosols or interfere in assays used to analyze the collected material. Endotoxin assays are particularly susceptible to this kind of interference (Milton et al., 1989), and carefully designed controls should be used each time a new filter medium is employed. Unless spore concentrations are very high, the low flow rate of standard personal cassette samplers (as are commonly used for asbestos) collects too few spores to allow precise estimates of concentration. However, in agricultural or some industrial environments, filter collections can produce valuable information on fungal spore and bacterial counts.

Gravity or settle sampling is the simplest method of collecting an air sample, but these samples are never representative of the aerosol. Large particles tend to be overrepresented, and even small changes in air speed and direction drastically change the fraction of the aerosol collected. Other methods can be used to collect particles from aerosols, including electrostatic precipitation, which is commonly used to clean the air. These devices are not commonly used for collection of aeroallergens.

TYPES OF SAMPLE ANALYSIS

Once an aerosol or a reservoir sample has been collected, it must be analyzed before meaningful data are produced. Assay types used for analyzing allergens include culture, microscopy, immunoassays, biochemical assays, and bioassays. Of these, the most commonly used are culture, microscopy, and immunoassay.

Cultural assays evaluate viable units as indicators of the presence of allergen-carrying particles. The content of the allergen itself may or may not be accurately represented. Cultural assays always quantitatively underestimate levels of allergen-carrying microbes (because only culturable cells are counted) and are biased with respect to the kinds of microbial allergen sources recovered. No single culture medium or set of environmental conditions will allow capture of all the different organisms in a mixed aerosol; however, investigators usually choose culture media that allow recovery of the broadest range of different organisms. Special-purpose media are likely to underestimate even the organisms they are designed to recover. Moreover, some organisms in air or in relatively hostile reservoirs have adapted to minimal conditions and will not grow if rich culture media are provided. Although minimal media are often used to recover bacteria from water samples, they have not been routinely used for fungi. Many of the species of *Penicillium* and *Aspergillus* that are common in indoor air grow well in media containing large amounts of salt, sugar, or glycerol (Verhoeff et al., 1990a,b). There are also media that inhibit fast-growing fungi such as *Rhizopus* or *Trichoderma*, which can quickly multiply and mask other organisms (Burge et al., 1977; Verhoeff et al., 1990a,b).

Because researchers have yet to discover a generic biomarker for fungal growth and because it appears that fungi differ allergenically at the species level, each fungus that releases allergens into the environment must be identified accurately and precisely so that specific diagnostic and treatment materials can be produced. For some kinds of fungi, generic identification can be completed on the isolation medium. For others (including most fungal species) and for all bacteria and yeasts, subculturing on diagnostic media is required. In addition, for bacteria and some fungi, physiological tests are necessary for species identification.

For particles that are morphologically distinct, direct microscopy is a straightforward approach to studying aerosols. Although very few fungi or pollens are identifiable to the species level microscopically, broader categories can be identified and useful information obtained. Categories of fungus spores that can be identified and counted range from spores of *Epicoccum nigrum* (a monospecific genus with very distinctive spores) to categories such as "colorless basidiospores," which include hundreds of different kinds of fungi. If a cultural sampler is used in conjunction with a spore trap, spore morphology comparisons can be made and the kinds of identifiable spores expanded. Although pollen is usually stained for microscopic examination, fungus spores often do not take up stains well, and important color characteristics can be masked.

Sample methods that allow microscopic evaluation include suction impaction (as used in the Burkard and Lanzoni spore traps) and rotating impactors (the rotorod). Filter samples can also be analyzed microscopically, although the clearing necessary to produce optically good samples requires the use of organic oils and solvents, which tend to cause fungus spores to collapse. Relatively smooth surfaced filters (e.g., polycarbonate, teflon) can be mounted in 1% acid fuchsin in lactic acid for counting. Filter samples can also be analyzed using fluorescent stains and epifluorescence microscopy. Water samples are commonly analyzed for bacterial content using the fluorescent dye acridine orange, which stains nucleic acids inside the bacterial cells (Palmgren et al., 1986). The melanin pigments in many fungus spores mask this fluorescence, and some fungal spore walls apparently prevent entry of the dye. However, the method is promising for indoor aerosols of bacteria (including actinomycetes) and colorless fungus spores.

Scanning electron microscopy has also been used to analyze filter samples, and bacteria and fungus spores can be readily seen and counted on smooth-surfaced filters (Pasanen et al., 1989). The level of identification is not as high as for light microscopy, however, and preparation methods are cumbersome for routine counts.

Immunoassays measure the actual allergen rather than an indicator (e.g., viable spores) and are essential for most amorphous allergens including those from mites, cats, and cockroaches. The most sensitive and specific environmental immunoassays for allergens are based on immobilized monoclonal antibodies that bind specific allergen in unknown samples.

Immunodiffusion is a relatively crude immunoassay that is used as a diagnostic test to detect allergen-specific immunoglobulin G (IgG) antibodies in patient sera. It has also been used to demonstrate exposure under the theory that specific IgG may represent exposure to allergens rather than actual disease. Extracts of air or reservoir samples have been used in place of defined allergens to indicate such exposures (Reed and Swanson, 1987).

Finally, measuring the amount of guanine in a sample (a biochemical assay) has been used to estimate the amount of mite allergen in dust samples. Other types of sample analyses have not been applied to allergens.

THE IDEAL AIR MONITOR

No single method is sufficient to detect and monitor all of the different allergens that can be present in indoor environments (Samson, 1985). Currently, the most adaptable collection devices are probably the filter collectors because filter samples can be examined directly by using either visible light, epifluorescence, or electron microscopy. In addition, they can be washed to provide allergens for immunoassay or viable cells for culture. For some allergens, methods of analyzing filter samples have been well defined. For others, especially viable cells, much work remains to be done to define the conditions under which recoveries are optimal. The advantages of filters (especially personal samplers) are many. They fulfill most of the characteristics of the ideal sampler:

- They are efficient for collection of small particles.
- They can be efficiently analyzed using several modes.
- They can be used for both personal and ambient sampling.
- They are inexpensive and portable.

Designing a sampling strategy requires consideration of the nature of the allergen source, the nature (including the size and expected concentrations) of the allergen-containing particles, and parameters that influence respiratory exposure to the allergen-containing particles. Each of these parameters influences the choice of sampling method (i.e., observation or reservoir or air sampling) and analytical approach, the sampling plan (amounts of sample to be collected, times and locations to be sampled), and approaches for analysis and interpretation of the data.

Exposure

As mentioned earlier in this chapter, many of the methods used for estimating environmental concentrations of aeroallergens are not truly representative of an individual's exposure to actual allergens. For example, ambient air monitoring is often used to represent personal exposure, and indicators rather than specific inhalable allergens are usually monitored.

AMBIENT DATA AND TIME-ACTIVITY LOGS

One of the ways in which ambient data can be used to estimate personal exposure is to use questionnaires or time-activity logs to catalogue the amount of time people spend in the various areas for which concentration data are available. Examples of the use of area monitoring combined with activity logs have been reported in several studies (Lebowitz et al., 1989; O'Rourke et al., 1989, 1990; Quackenboss et al., 1989a,b, 1991a,b). In these studies, the contribution of indoor concentrations to total personal exposure exceeded that of outdoor concentrations because of the difference in time spent in the two environments. For example, indoor exposure to total pollen, *Gramineae*, total mold, *Cladosporium,* and *Aspergillus/Penicillium*-type spores was two to four times greater than outdoor exposure.

Questionnaires may also be used to determine occupancy rates or activity in the indoor environment under study (Lebowitz et al., 1989; Quinlan et al., 1989). Results from the questionnaires or logs can then be linked with the results of observations to develop categorical estimates of exposure potential in areas for which no concentration information is available.

INDICATOR DATA TO REPRESENT EXPOSURE

An example of the use of indicator data to represent exposure to an allergen is the use of counts of whole house dust mites in dust samples. The presence and quantity of mites in a sample viewed through the microscope indicates only the potential for allergen exposure, and is not a direct measure of exposure itself. House dust mite allergen is associated with fecal particles rather than with the mite itself. Even the counting of fecal particles must be considered a surrogate measure since the concentration of allergen on any one particle cannot be predicted. An accurate measure of exposure to dust mite allergen, as well as other allergens, requires the use of specific immunoassays.

NATURE OF ALLERGEN-CARRYING PARTICLES

Little or no *Der p* I has been detected in the air in undisturbed rooms (de Blay et al., 1991b; Platts-Mills et al., 1986; Swanson et al., 1989; Yasueda et al., 1989), and after a disturbance in a room, concentrations of airborne mite allergen fall rapidly. It is likely, therefore, that the majority of allergen is carried on particles with large aerodynamic diameter (e.g., greater than or equal to 10 μm). The levels that become airborne depend critically on the form of the disturbance and vary from 5 to 200 ng of *Der p* I/m^3. Because the mean allergen content of mite fecal material and the size of the particles are known, it is possible to estimate the actual amounts of allergen that could enter the lung at a given ambient concentration. For particles of 10-μm aerodynamic diameter, only 5–10 percent would be expected to enter the lung during gentle mouth breathing (Svartengren et al., 1987; Task Group on Lung Dynamics, 1966). Given that each particle contains 0.2 ng of *Der p* I (Tovey et al., 1981b), an airborne level of 20 ng/m^3 means that about 100 fecal particles could be expected to enter the mouth per hour and that from 5 to 10 of these would reach the lung. Use of this technique requires the collection of sufficient mass to measure concentration reliably, a difficult task for aerosols as transient as those of mite fecal particles.

Inhaled allergens from other sources can also differ from generally measured arthropods, particles, grains or spores. For instance, far more allergen is present on amorphous pollen (grain fragments) than assumed from whole grain measurements (Schumacher et al., 1988).

Internal/Actual Dose

Exposure measurements, within the constraints of the methods used, characterize the ambient environment. For example, the number of pollen grains counted may include some that are outside the respirable range and that cannot be inhaled. Similarly, reported concentrations of allergen are not "adjusted" to reflect the percentage that may be exhaled. Measured concentrations are thus estimates of exposure but not doses to the body. To estimate dose, the following must be known for particulates that enter the respiratory tract:

- Relationship between the exposure measure and allergen concentration
- Variability of allergen content in the respirable fraction
- Percentage of deposition
- Breathing rate(s)
- Duration of exposure(s).

No examples have been published in which all of these variables are known, thus allowing a calculation of the internal dose of an aeroallergen.

BIOLOGICALLY EFFECTIVE DOSE

Once contact has occurred, the kinetics of uptake for exposure measures, distribution to the target tissue, and clearance/detoxification must be calculated to estimate a biologically effective dose. These data are not currently available for aeroallergens.

RISK ASSESSMENT

In general, risk assessment is a process designed to evaluate the potential relationship that may exist between exposure to a particular agent, e.g., aeroallergen, and a particular effect, e.g., sensitization or allergic disease. This section describes a number of issues that may be considered in conducting a risk assessment for aeroallergens and an example using data on environmental levels of dust mite allergen and the occurrence of sensitization. The discussion is not exhaustive; rather, it shows the range of issues that can be included for study when data are available or that can be included in a research design to provide data for performing a risk assessment.

Steps in a Risk Assessment

Risk assessment is a process that is composed of four basic components, or steps: (1) hazard identification, (2) exposure assessment, (3) dose-response assessment, and (4) risk characterization (NRC, 1983b). Risk assessment for noncoarcinogens is currently in the process of being formalized in the literature (Pierson et al., 1991; Shoaf, 1991). Figure 6-2 shows the elements of research and exposure assessment and their relationship to the processes of risk assessment and risk management. The four basic steps of risk assessment are described briefly below, followed by an example.

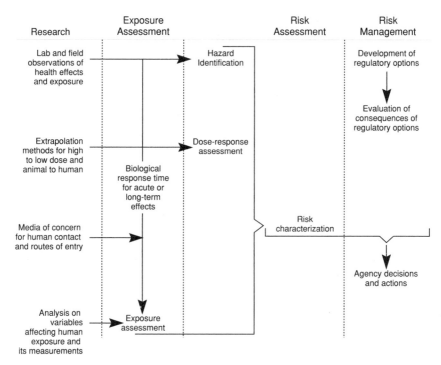

FIGURE 6-2   Elements of research and exposure assessment and their relationship to the process of risk assessment and risk management.   Source: NRC, 1991.

STEP 1: HAZARD IDENTIFICATION

Important aspects of hazard identification include a complete description of the aeroallergen and its potential to affect human health adversely at some level of exposure or dose, an understanding of the number of people who are potentially exposed, the background levels of exposure in the environment, the nature of the health effect of interest, and the level of exposure thought to be associated with an increased risk of the health effect. Other exposures that potentially contribute to the development of the health outcome should also be identified.

STEP 2: EXPOSURE ASSESSMENT

The measurement of exposure to allergens is a rapidly advancing field. Instrumentation and approaches were described earlier in this chapter; this discussion summarizes some of the issues that arise in using aeroallergen exposure data for a risk assessment.

Generally, hazard identification (step 1 above) comprises the review of a number of studies from the literature. Often in those reports, more than one sampling protocol will be described, and different instrumentation or interview methods are likely to have been used. It may be difficult to synthesize data presented in different formats (e.g., range only, graphically only, mean values only). In addition, the material collected for environmental samples may be analyzed for different components. For example, in the literature on dust mites, several different exposure metrics are reported, including mite counts, immunochemical assays of allergen, and guanine determinations (Platts-Mills and de Weck, 1989). Within this framework, individual assays may differ among laboratories (Lau et al., 1989; Sporik et al., 1990).

An understanding of how the exposure and thus the health effect of interest may be influenced by other factors is essential in interpreting the data. For example, seasonal variation strongly influences exposure to some allergens and resultant reporting of symptoms (O'Rourke, 1992; O'Rourke et al., 1989; Platts-Mills et al., 1987). Characteristics of the exposure site, including type of flooring, kind of ventilation system, and other such features, may affect the nature of exposures as well. The possible role of altitude in allergen exposure has also been investigated (D. Charpin et al., 1991).

Available exposure data must be summarized in spite of all of these issues and problems. In cases in which differences in technique or analysis are thought to provide incomparable results, appropriate conversion factors are often estimated.

Step 3: Dose-Response Assessment

The information developed in steps 1 and 2 above is used to estimate the dose from the exposure, and to model the dose-response relationship. The exposure-response relationship may be estimated when too little information is available to allow extrapolation from exposure to dose. For noncarcinogens, this effort has traditionally focused on identification of a "threshold exposure" below which no health effects are observed (Pierson et al., 1991); this threshold is now known as the No Observed Adverse Effect Level (NOAEL; Shoaf, 1991).

The value of the NOAEL, as determined from modeling, depends on the quality of the data and the model selected. Examples of various models include linear, probability (Rose and Gerba, 1991), and exponential and beta-Poisson forms (Regli et al., 1991). Shoaf (1991) reviews the statistical considerations in their use.

Once the NOAEL has been determined, it is divided by an uncertainty factor (UF). Traditionally, a factor of 10 has been used to account for variability in responses among a human population. When a Lowest Observed Adverse Effect Level (LOAEL) is determined, a UF of 10 is applied. Additional uncertainty in the data is accounted for by use of a modifying factor, varying from 1 to 10, to further reduce the NOAEL (Shoaf, 1991).

STEP 4: RISK CHARACTERIZATION

The relationship between dose (or exposure) and response is used to characterize the risk within a population. For aeroallergens, this relationship might be expressed as the number of new cases expected at a given exposure level or the distribution of expected change in the severity of symptoms with changing exposure levels.

Example of a Risk Assessment:
Exposure to Dust Mite Allergen and Sensitization for Asthma

STEP 1: HAZARD IDENTIFICATION

When dust mites were first reported in the literature in the 1970s, it was suggested that levels of more than 500 mites/g of dust in a house were likely to produce symptoms of asthma in allergic individuals. During the 1980s, data accumulated demonstrating a dose-response relationship between reservoir concentrations of dust mite allergens at home and both sensitization and asthma (Table 6-2). Results of this kind led to the proposal of threshold levels for reservoir concentrations. Two separate thresholds have been considered: more than 2 μg of *Der p* I/g of dust has been associated with increased prevalence of sensitization, whereas 10 μg increases the risk of symptomatic or acute asthma. In an 11-year prospective study of 68 children, Sporik and colleagues (1990) found that exposure[1] to high levels of allergen at age 1 increased the risk of asthma and was inversely related to the age of onset of asthma in atopic children. In subsequent studies of children hospitalized in the south of England, it was found that about 80 percent were both sensitized and exposed to high levels of relevant allergen at home (Sporik et al., in press). These results suggest that in areas in which all houses contain high levels of mite allergen, sensitivity to mites is a major risk factor, not only for wheezing but for hospitalization of children with asthma.

Evidence to support a quantitative relationship between exposure to mites and asthma has also been found in studies from Denmark, Australia, Germany, and France (Table 6-2). Two studies in particular strongly imply that high levels of domestic exposure can increase the prevalence of asthma.

---

[1]For this discussion, reservoir concentration is used as an indicator/surrogate for exposure.

TABLE 6-2 Evidence for Regarding Specific Levels of Dust Mite
Exposure as Risk Factors for Asthma

Study Topics	Location of Study	Levels Associated with Increased Risk of Disease
Mites in houses of patients with asthma		
Voorhorst et al., 1967	The Netherlands	500 mites/g
Korsgaard, 1983b	Denmark	100 mites/g
Peat et al., 1987	Australia	100 mites/g
Dowse et al., 1985	Papau New Guinea	1,000 mites/g
Mite allergen levels in houses of patients with asthma		
Platts-Mills et al., 1987	United States	2–10 µg of group I/g
Lau et al., 1989	Berlin	2 µg of Der p I/g
D. Charpin et al., 1991	Marseilles	2 µg of group I/g
Sporik et al., 1990	England	10 µg of Der p I/g
Arruda et al., 1991*	São Paulo, Brazil	10 µg of Der p I/g
Level of exposure associated with change in bronchial reactivity		
Platts-Mills et al., 1992	London, U.K.	13.5 µg/g reduced to 0.2 µg/g
Provisional standards for dust mite exposure levels		
Platts-Mills et al., 1992		2 µg of group I allergen/g of dust (equivalent to 10 mites/g)
		10 µg of group I allergen/g of dust (equivalent to 500 mites/g)

*In São Paulo, all houses have high levels of mite allergen.

In Papua New Guinea, the disease appeared in a group of highland villages and increased to a prevalence of about 7 percent; the previous rate had been less than or equal to 0.5 percent. This increase coincided with the introduction of blankets that the men wrapped around their heads at night. The men with asthma all tested positive on skin tests for dust mites, and dust from the blankets contained an average of 1,140 mites/g of dust (standard deviation = 868; Dowse et al., 1985).

D. Charpin and others (1991) compared children growing up in a suburb of Marseilles, which is a seaport, with those growing up in Briançon at an elevation of 6,000 feet. The average concentration of mite allergen in mattress dust from Marseilles was 15.8 µg/g, compared with 0.36 µg/g in Briançon. In parallel, the prevalence of skin sensitivity to dust mites in Briançon was 4.1 percent compared with 16.7 percent in Marseilles. Finally, the prevalence of asthma in the mountains was lower (4 percent) than that at sea level (6.7 percent).

Because there is a general misconception that levels of allergen will (or should) be higher in the houses of patients with allergic disease, it is important to understand the actual situation. Within a defined area, most houses have similar levels of allergen. For example, on the Gulf Coast, in São Paulo, Brazil, or in southern England, "all" houses have high levels of dust mite allergen; similarly, in inner-city Atlanta, Georgia, or Wilmington, Delaware, most houses have high levels of cockroach allergen. In suburban areas, about 50 percent of all houses have a cat and consequent high levels of cat allergen. In each of these areas, it is the individuals who develop IgE antibodies and have continued exposure to the *relevant* allergen who are at risk of developing allergy and asthma.

Two hazards exist: sensitization and allergic disease. From the data, it appears that the exposure initiating each outcome may be different. For the purposes of this example, sensitization was chosen as the endpoint.

STEP 2: EXPOSURE ASSESSMENT

Because the common allergens are thought to cause or exacerbate asthma by the inhalation route, measuring inhaled allergen might seem to be the best method for determining exposure (Price et al., 1990; Swanson et al., 1985; Tovey et al., 1981b). However, there are two reasons for using values from reservoir samples instead in this example. First, the majority of exposure data available from the literature are based on measurements of allergen in dust collected from carpets, mattresses, sofas, and other such items. Second, the quantities that become airborne apparently are quite small, commonly 5–50 ng/m^3, and appear to depend on the level of physical activity in the home during the time of sampling.

This example uses data from four studies that contain both exposure information and data on the prevalence or incidence of sensitization (D. Charpin et al., 1991; Lau et al., 1989; Price et al., 1990; Sporik et al., 1990). The underlying assumption of this risk assessment is that the cumulative exposure (i.e., duration multiplied by intensity) is proportional to the sensitization concentration; this approach differs from that of previous reports where the exposure value at the time of the survey was related to the health outcome. In using the literature data presented in Table 6-3, three operational guidelines were developed a priori:

1. The exposure concentration is expressed as micrograms of *Der p* I per gram of collected dust. In instances in which the range is reported, the midpoint is used as the average; if the range is not given, the maximum value is estimated from the graphical display of the data and the midpoint is estimated.

2. Exposure that is reported as the total of allergens *Der p* I and *Der f* I is assumed to consist of 50 percent *Der p* I.

TABLE 6-3 Data for Dust Mite Risk Assessment: Example

Exposure concentration (µg *Der p* I/g dust)		Years of Exposure	Cumulative Exposure (µg of *Der p* I/ g dust/year)	Percent Sensitized	Reference
Reported	(Estimated)				
<2	(1.0)	6[a]	6	0[b]	Sporik et al.,
2–10	(6.0)	6	36	38	1990
11–50	(31.0)	6	186	50	
>50	(75.0)[c]	6	450	78	
<0.2[d]	(0.1)	10[e]	1	17[f]	Lau et al.,
>5	(100.0)	10	1,000	86	1989
0.2–0.5	(0.35)[f]	10	3.5	36	
0.5–5	(2.75)[f]	10	27.5	67	
8[d]	8.0	10[e]	80	17[g]	D. Charpin et al., 1991
0.2	0.2	10	2	4	
<0.5	(0.35)	8[e]	2	19[h]	Price et al.,
<1	(0.5)	8	4	32	1990
<2	(1.0)	8	8	43	

[a]Taken as average age at sensitization; 40 percent are sensitized by age 5, and 60 percent are sensitized by ages 5–11 on average, 10 percent per year. Therefore, 50 percent are sensitized by age 6.
[b]See Table 2 in Sporik et al., 1990.
[c]See Figure 1 of Sporik et al., 1990.
[d]Concentration of antigen is estimated.
[e]Years taken as average age.
[f]See Figure 3 of Lau et al., 1989.
[g]See Table 4 of D. Charpin et al., 1991.
[h]See Table II of Price et al., 1990.

3. Time is expressed in years. For a population that is surveyed cross-sectionally, exposure is taken to be equal to age; the average years of exposure thus are equal to the average age. For populations that are surveyed prospectively, the age at sensitization is equal to the years of exposure; if an age is not given, years of exposure are assumed to equal the average follow-up period.

STEP 3: DOSE-RESPONSE ASSESSMENT

Exposure is estimated in this example from several different studies that use slightly different protocols for reporting the results from reservoir sampling. For example, Sporik and colleagues (1990) report the highest concentrations from samples in different parts of the house. Price and coworkers (1990) report the value from a single, composite sample collected from several locations in the house.

Appropriate models to relate exposure information to dose are not known for these data sets, and two assumptions were made:

- log normal distribution of the exposure variable, and
- a linear relationship in exposure-response.

The base-10 logarithm of the estimated cumulative exposure was used in the analysis.

Figure 6-3 shows the results of the committee's analysis. The data fit an equation given by:

$$\text{percent sensitized} = 21.5 \log_{10} (\text{cumulative exposure}) + 10.6. \qquad (1)$$

This can be simplified (by rounding off) to:

$$\text{percent sensitized} = 22 \log_{10} (\text{cumulative exposure}) + 10. \qquad (2)$$

Using the linear assumption, neither a NOAEL nor a LOAEL is identified, because the line intercepts the y-axis at a positive value (approximately 10%). This indicates the lack of a threshold concentration below which sensitization would not occur, i.e., it is not possible to achieve zero risk of sensitation. The R^2 value is 0.6, indicating that 60 percent of the variability in linear trend is explained by the model.

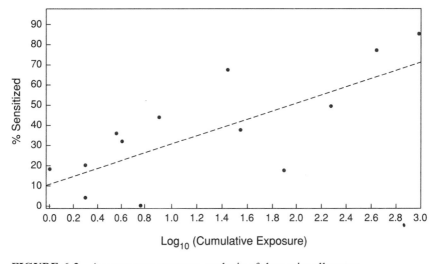

FIGURE 6-3 An exposure-response analysis of dust mite allergens.

STEP 4: RISK CHARACTERIZATION

The equation developed from the exposure-response model can be re-written as

$$\text{percent sensitized} - 10 = 22 \log_{10} (\text{years} \times \text{concentration}), \quad (3)$$

or

$$\frac{\text{percent sensitized} - 10}{22} = \log_{10} (\text{years}) + \log_{10} (\text{concentration}). \quad (4)$$

Once the risk (percent sensitized) is set, the relationship can be solved for concentration in the reservoir, since the value of *years* will be known for a given population. Alternatively, if the duration and concentration are known, the risk (percent sensitized) can be estimated. For example, assuming 5 years of exposure at a concentration of 2 μg/g, the risk of sensitization would be 32 percent.

These calculations are based on data that were abstracted from reports by researchers using various methods, and must therefore be considered "soft"; the results are presented only as an example of how a risk assessment for sensitization could be performed.

CONCLUSIONS AND RECOMMENDATIONS

Some of the issues to be considered when undertaking a risk assessment for aeroallergens have been discussed and an example has been given by using data from the literature on dust mite exposure and the development of sensitization. The data can be described by a linear model and indicates that there is a positive relationship between cumulative exposure to dust mite allergen and the risk of sensitization.

Some residual sensitization (i.e., approximately 10 percent in this example) will occur irrespective of exposure to dust mite allergen according to these estimates. This finding is consistent with the knowledge that other factors may also result in sensitization. For this reason, information on cross-reactivity of allergenic agents in study subjects is desirable, and important to the analysis of potential mechanisms of sensitization.

Research Agenda Item: Determine whether a practical method could be developed to measure concentrations of dust mite allergens that are capable of sensitizing humans.

Variability in the methods used in the multiple study protocols reported to date is high. More uniformity in the collection of exposure data would be useful in the risk assessment process. Although reservoir sampling has yielded meaningful results that assist in remediating exposure, further development of standardized air sampling collection and analytical methods is needed.

Research Agenda Item: Standardize methods of collecting and analyzing indoor allergen samples to facilitate comparative and collaborative studies.

Improved, standardized methods of collecting and analyzing indoor allergen samples would be particularly valuable in establishing the relationship between reservoir samples, personal exposure measures, levels of activity, and the potential for airborne exposure of sufficient magnitude to induce negative health outcomes.

Research Agenda Item: Quantitate the relationship of allergens in reservoirs (and on surfaces) to aerosols and develop monitoring methods for quantitating airborne-allergen concentrations in personal breathing zones.

Assessment of exposure is a rapidly advancing, complex, and multistep process that entails numerous variables and estimations. Most monitoring, for example, is often based on sampling for indicators rather than the actual allergen. There is a need for developing improved methods for estimating environmental concentrations of aeroallergens and the resultant individual exposures.

Research Agenda Item: Develop appropriate exposure metrics for specific indoor allergens that are analogous to time-weighted averages and permissible-exposure limits for industrial chemicals.

Methods for determining the effects of indoor allergens can be divided into two general categories: patient testing and environmental testing. Data from both kinds of testing can be useful to the physician in directing the treatment, control, and prevention of allergic disease. There are, however, no effective means currently available to physicians or other medical professionals for obtaining quantitative information on environmental exposures.

Recommendation: Establish effective mechanisms for medical professionals to acquire assessments of potential exposure to indoor allergens in residential environments.

7

Engineering Control Strategies

The fundamental objectives of environmental control are to prevent or minimize occupant exposures that can be deleterious and to provide for the comfort and well-being of the occupants. For acceptable control both objectives must be achieved simultaneously. In many cases, the most effective control strategy is removal of the source of contamination. However, source removal is not always feasible or sufficiently effective, and other methods of reducing exposure to allergens and other contaminants must be employed. This chapter characterizes the existing building stock in the United States and describes relevant engineering principles and practices that can be employed to prevent or minimize occupant exposures to indoor allergens.

There are approximately 4 million commercial (i.e., nonindustrial and nonresidential) buildings and 84 million detached single- and two-family residential buildings in the United States (U.S. Department of Energy, 1985, 1986). Current estimates are that the existing building stock is being replaced at a rate of about 1–2 percent per year, while another 1–2 percent per year is being added to the stock. Thus, 75–85 percent of the buildings that will exist in the year 2000 have already been built.

The percentage of existing commercial buildings in which occupants are exposed to environmental conditions that result in complaints of symptoms or illness (i.e., problem buildings) has been estimated to be 20 to 30 percent (Akimenko et al., 1986; Woods et al., 1987). This percentage has not been determined for residential buildings.

Problem buildings can cause two general types of problems: Sick Building Syndrome (SBS) and Building Related Illness (BRI) (Cone and Hodgson,

1989; Molhave, 1987; Morey and Singh, 1991; NRC, 1987b; Stolwijk, 1984). Sick building syndrome is suspected when occupants complain of symptoms associated with acute discomfort (e.g., eye, nose or throat irritation, sore throat, headache, fatigue, skin irritation, mild neurotoxic symptoms, nausea, odors) that persist for more than 2 weeks at frequencies significantly greater than 20 percent; a substantial percentage of the complainants report almost immediate relief upon exiting the building. Building-related illness is suspected when exposure to indoor pollutants results in clinical signs of a recognized disease that is clearly associated with building occupancy (e.g., some kinds of infections, building-related asthma, humidifier fever, hypersensitivity pneumonitis).

Accurate estimates of the relative occurrences of SBS and BRI in problem buildings have not yet been established. However, an initial analysis of more than 30 problem building investigations indicated that approximately two-thirds of the cases involved complaints and symptoms associated with SBS, while about one-third involved symptoms and signs associated with BRI and SBS—no cases of BRI without concomitant SBS were observed (Woods, 1988). The percentage of existing buildings that are causally associated with occupant exposure to allergens in either the SBS or BRI categories is unknown.

The most frequently reported characteristics in problem building investigations include inadequate quantity or quality of outdoor air provided by heating, ventilation, and air-conditioning (HVAC) systems for ventilation (incidence rates of 64–75 percent), and inadequate distribution of air supplied to and returned or exhausted from occupied spaces for thermal and air quality control (46–75 percent incidence) (Woods, 1989a, 1991). These characteristics are described as ventilation efficiency or ventilation effectiveness (ASHRAE, 1989b).

Equipment problems that have been most frequently reported in problem building investigations include inadequate specification and installation of air filters for removal of inert particulates and bioaerosols (incidence rates of 57–65 percent), inadequate specification and installation of drain pans and drain lines for removal of water condensed from cooling coils and humidifiers·(60–63 percent incidence), inadequate specification and installation of duct work to prevent microbial contamination (38–45 percent incidence), and inadequate specification of humidifiers to prevent microbial or chemical contamination in the humidifier and subsequently in the air (16–20 percent incidence) (Woods, 1989a; 1991).

PRINCIPLES AND STRATEGIES

To achieve the twofold objectives of indoor environmental control (i.e., prevent or minimize deleterious exposures and provide for occupant comfort and well-being), building design and management principles must be applied throughout each of the four stages of a building's life cycle: (1) preconstruction period, (2) construction period, (3) long-term occupancy, and (4) adaptive reuse and eventual demolition (NRC, 1987b). These principles apply to both commercial and residential buildings.

Two basic strategies for controlling occupant exposure to allergens can be identified: *source control*, which can eliminate occupant exposure, and *exposure control,* which can minimize but not eliminate occupant exposures by methods of dilution or air cleaning. A simple, one-compartment model of a control system for a uniformly mixed occupied space, shown in Figure 7-1, illustrates the interrelationship that exists among the variables that affect air quality (Woods, 1991; Woods and Rask, 1988). In this model, C_i is the indoor concentration, and V_r is the recirculation airflow rate. The dilution rate, V_o, represents infiltration, natural ventilation, or mechanical ventilation with outdoor air. In terms of controlling airborne allergens, the removal rate, E, represents the capacity of air conditioners, filters, and other such removal devices. The efficiency of the removal device, e, is defined in terms of the contaminant being removed. Exposure to allergens can be controlled by removing or minimizing the sources, i.e., by reducing the net

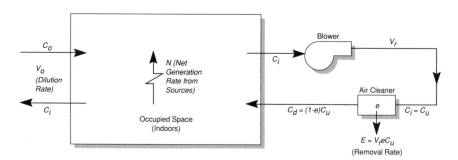

FIGURE 7-1 One-compartment, uniformly mixed, steady state model for indoor air quality control. In steady state, an energy or mass balance for the model can be expressed as $C_i - C_o = N - E / V_o$ where:

C_o = concentration of contaminant in outside makeup air,
C_i = concentration of contaminant indoors (occupied space),
C_u = concentration of the contaminant upstream of the air cleaner,
C_d = concentration of the contaminant downstream of the air cleaner,
V_o = dilution rate (i.e., infiltration, natural or mechanical ventilation rates),
V_r = air flow rate through air cleaner,
e = efficiency of the air cleaner,
N = net generation rate, and
E = removal rate.

generation rate (N), by increasing the rate of removal of allergens from air (E), or by diluting airborne allergens (V_o). For some allergens, which are airborne only sporadically and in association with specific human activities (e.g., house dust mite allergens), source control is the best option. For others (e.g., outdoor allergens such as pollen) removal by filtration (E) is the usual method.

The physical causes usually associated with problem buildings include two basic inadequacies: (1) design inadequacies, which include system and equipment problems, and (2) operational inadequacies, which consist of inappropriate control strategies, lack of maintenance and housekeeping, and ineffective load management of building systems (Cone and Hodgson, 1989; EPA, 1991a; Molhave, 1987; Morey, 1984, 1988; Morey and Singh, 1991; NRC, 1987b; Robertson, 1988; Stolwijk, 1984; Woods, 1988, 1989a, 1989b, 1991). Table 7-1 summarizes some of the data on the physical causes of problem buildings.

The remainder of this section on principles and strategies discusses various source and exposure control strategies as they relate to building design, HVAC system design, building operation, and remediation. This is followed by a section that summarizes some practical control strategies in general, and as they apply to residential and commercial buildings. The chapter ends with conclusions and recommendations.

TABLE 7-1 Frequencies of Occurrence of Physical Causes of Problem Buildings Reported by Two Independent Investigative Teams

Problem Category	Physical Cause	Frequencies (%) of Occurrence	
		Woods	Robertson
Design	System problems		
	Inadequate outdoor air	75	64
	Inadequate air distribution to occupied spaces (supply and return device)	75	46
	Equipment problems		
	Inadequate filtration of supply air	65	57
	Inadequate drain lines and drain pans	60	63
	Contaminated duct work or duct linings	45	38
	Malfunctioning humidifiers	20	16
Operations	Inappropriate control strategies	90	NA*
	Inadequate maintenance	75	NA
	Thermal and contaminant load charges	60	NA

*NA, not available.

SOURCES: Woods, 1989a; Robertson, 1988.

Building Design for Source Control

Building design as it relates to controlling indoor allergens is discussed below in terms of moisture, carpeting, and other substrates and reservoirs.

MOISTURE

As described in previous chapters, control of relative humidity or water vapor pressure in the occupied space and in the HVAC systems is important for allergen control in both residences and commercial buildings. It is therefore important to determine and define the indoor air, temperature, and relative humidity conditions that will provide for occupant comfort and also suppress the growth of allergen-producing microorganisms and mites.

The development of allergen-containing reservoirs depends on available water in the microenvironment of the allergen-producing organism. The amount of water in these environments depends on the relationship between the amount of airborne water vapor and the temperature in the environment (which controls condensation), the ability of substrates in the environment to absorb water, and the presence of liquid water sources (e.g., flooding, water reservoirs). The primary sources of water vapor are outdoor air that is used for ventilation or that infiltrates into the building, bathing, washing, and cooking processes, and evaporation of perspiration (latent heat dissipation) from building occupants. Secondary indoor sources of water (vapor or liquid) include condensation on surfaces that are colder than the dew point temperature (e.g., on walls and floors) and materials that can adsorb moisture (e.g., carpets, wood products, mattresses, clothes).

Ambient relative humidity is often considered the major controlling factor for indoor allergens. Performance criteria for acceptable temperature and humidity ranges are specified in ASHRAE (American Society of Heating, Refrigerating and Air-Conditioning Engineers) Standard 55-1992, primarily for human comfort (ASHRAE, 1993). This 1992 standard considers an acceptable range of relative humidity to be from less than 30 percent (depending on operative temperature) to an upper limit of 60 percent which was selected for control of microbial and arthropod growth and prevention of conditions that lead to condensation. The lower limit of dew point temperature (the temperature at which moist air becomes saturated—100 percent relative humidity—with water vapor when cooled at constant pressure) for comfort is 35° F (ASHRAE, 1993). The operative temperatures (i.e., simple averages of dry-bulb and mean radiant temperatures) that define the comfort zone (ASHRAE defines acceptable conditions as those that satisfy 80 percent or more occupants) range from 68° F to 78.5° F and vary according to seasonal considerations. This temperature range is influenced by factors such as the motion (velocity) of air in the room, the insulation value of clothing, and the metabolic rates of the occupants.

The operative temperature and relative humidity/dew point ranges in ASHRAE Standard 55-1992 are specified for the occupied zone, which is defined as the space approximately 18 to 72 inches above the floor. For thermal acceptability, this standard allows no more than a 5° F vertical difference in air temperature within the occupied zone, a 9° F vertical radiant asymmetry, and an 18° F horizontal radiant asymmetry. As such, the ASHRAE thermal environmental criteria for comfort in occupied spaces have limited influence on the thermal environmental conditions in microenvironments, such as on the surfaces of walls or windows, or within porous furnishings such as carpets, mattresses, and upholstered furniture.

Thermal environmental conditions that are conducive to the growth of some arthropods are well defined, although the actual relationships between ambient and microenvironmental conditions have not been examined. Mites reproduce over an ambient relative humidity range of 45 to 80 percent, with 75 to 80 percent being optimal (Andersen and Korsgaard, 1986). The temperature range of 65° to 80° F is optimal for mite growth. This range just brackets the range of thermal acceptability defined in ASHRAE Standard 55-1992 (ASHRAE, 1993). Relationships between fungal and bacterial growth and relative humidity have been less clearly defined, and probably vary with specific organisms.

Condensation, which will usually lead to fungus growth, will occur on walls, ceilings, and floors when their temperatures are below the dew point temperature of the surrounding air. In cold climates, condensation is often seen on the surfaces of windows, ceilings, and walls. Condensation is especially evident in residences at the inside surfaces of corners of walls, the inside surface junction of the ceiling and the external wall, and at the inside corners of the building that have minimum or no protective insulation (White, 1990). In cold climates, condensation will also occur within the building envelope itself when moist indoor air exfiltrates through leaky construction and encounters cold surfaces on the "weather side" of the building (Lstiburek, 1989).

In hot and humid climates, condensation can occur on the inside surfaces of exterior walls in air-conditioned buildings when warm moist outdoor air infiltrates through exterior facades and encounters a surface at a temperature below that of the dew point of the infiltrating outdoor air (Morey, 1992). Vinyl and other wall coverings often have low water vapor permeabilities. When these coverings are used in hot and humid climates, they can result in condensation at gypsum board wall covering interfaces at or near room temperatures.

Condensation may also occur on the uninsulated surfaces of chilled water or cold water pipes that may be present in building locations such as ceilings and walls. The surface temperature of chilled water pipes may be as low as 45° F, which means that condensation will occur on pipes when the dew point temperature of the air in the wall of the ceiling cavity exceeds 45° F.

To reduce condensation, vapor retarders (a layer of material with low moisture and air permeance) are installed in the envelopes of buildings (ASHRAE, 1989a). The vapor retarder is usually located near or at the surface exposed to the higher water vapor pressure. An effective vapor retarder keeps interior moisture out of the low-temperature space between the inner and outer surfaces of the wall. Since moisture is transported by both diffusion and air movement, a retarder can be made ineffective by the presence of seams, gaps, and tears which allow moisture to migrate into the interwall space. In buildings (including residences) in cold climates, the higher water vapor pressure generally occurs in the occupied spaces and the vapor retarder is located on the side of the insulation facing the occupied spaces. In hot and humid climates, vapor retarders may be located on the exterior surface of air-conditioned buildings because the high vapor pressure occurs in outdoor air (ASHRAE, 1989b). In slab, on-grade construction in hot and humid climates, the vapor retarder is placed under the slab to reduce water wicking through the slab (Lotz, 1989).

Primary sources of liquid water are rain and groundwater that leak into buildings, indoor water reservoirs that leak or rupture (e.g., water piping systems and storage tanks), or appliances that are designed to add water to the air (e.g., humidifiers). Ultrasonic humidifiers which tend to release smaller-size droplets, often contain bacterial but not fungal contaminants. Cool-mist humidifiers, which characteristically emit large water droplets, are often contaminated by bacteria and hydrophobic fungal spores (Solomon, 1974). Portable humidifiers, used primarily in residences or offices, have been evaluated for microbiological emission potential by Tyndall et al., (1989). Residential humidifiers containing water sumps near furnace hot air plenums have been associated with emission of thermophilic microorganisms that cause hypersensitivity pneumonitis (Fink et al., 1971; Sweet et al., 1971). In order to prevent microbiological amplification, cleaning and maintenance of these humidifiers must be fastidious. Scale that accumulates on wet surfaces must be removed to prevent microbiological amplification.

In spaces where humidity is a problem, portable dehumidifiers are sometimes used. Portable dehumidifiers operate on a mechanical refrigeration cycle, with the room air first being blown over a set of cold coils, to be dehumidified and cooled, and then over a warm set of coils, to be reheated. The net result is substantial dehumidification and a modest temperature rise. The condensed water that collects in portable dehumidifiers is best removed from the building by ducting it to a drain. Alternatively, emptying and cleaning the water reservoir to minimize fungal growth should be done daily.

Moisture can also enter buildings for a number of primarily climatic reasons. Wind-driven rain, especially in coastal regions, can penetrate the building envelope and saturate construction materials, especially if roof and window flashings are inadequate. Moisture that enters walls and roofs must be removed by drainage to the outside or by indoor ventilation air. Wind-driven snow can also enter HVAC system outdoor air inlets, especially those that are located at grade level or flush with horizontal roof surfaces.

CARPETING

Carpeting, which is widely used in homes and offices, can provide niches for both the accumulation (e.g., *Fel d* I and nonviable spores) and production (e.g., viable mites and xerophilic fungi) of allergens. Carpeting typically consists of fibers made into tufted yarn and looped through a backing by means of an automated manufacturing process. The fibers are anchored to the backing with latex, which itself may be an allergen (see Chapter 3), and mixed with a filler such as crushed marble. A secondary backing is added to give the carpet body and to promote dimensional stability. Carpets are installed over a padding, traditionally felted jute, but urethane paddings are also widely used. The installed carpet system therefore consists of three distinct layers: the fiber, the backing, and the padding.

A carpet has some similarity to an air filter, in that dirt particles can become mechanically trapped and are not easily dislodged during cleaning. Particle adhesion is increased if the carpet is moist and if the particle and the carpet fiber are both wettable. Oil, greasy dirt, and sticky substances such as detergent residues have a strong tendency to adhere to the carpet fiber. Sticky substances such as detergent residues can also bind dust to fiber.

Carpets have been characterized when wetted as "cultivation media" for microorganisms (Gravesen et al., 1983). Indeed, the concentration of fungi and bacteria in the air above carpet has been found to be consistently higher than that over noncarpeted floors (Anderson et al., 1982; Gravesen et al., 1983). The log count of bacteria per square inch of new wool carpet reaches a steady state of about 5 within 2 weeks after installation (Anderson et al., 1982). By contrast, the log count of bacteria varied between 2 and 3 for bare floors in the same facility. In addition to microorganisms, carpeting can function as a reservoir for pollen and pollen fragments that are tracked into the indoor environment from outdoor sources.

Carpet moisture is a special problem when it penetrates into the carpet backing and the padding because carpet backing is essentially a porous barrier that permits downward flow by gravity, but essentially blocks the passage of water vapor upward. Therefore, if the padding is wetted by cleaning, flooding, or water vapor migration upward through a cracked concrete slab floor, it may not dry out as long as the carpet above the padding is left in place.

Chronic flooding of carpet will result in the amplification of indoor microorganisms. Carpet that has been chronically flooded can function as a significant fungal reservoir long after drying has occurred—higher concentrations of microorganisms are found in the occupied spaces of flooded versus nonflooded floors (Kozak et al., 1980b; Morey, 1984). The dust and debris obtained from the backing of chronically flooded carpet may be heavily contaminated by fungal spores, most of which are nonviable (Kozak et al., 1980b).

OTHER SUBSTRATES AND RESERVOIRS

Many substrates found in indoor environments can support microbiological growth. Materials found in buildings, such as wood, cardboard, and paper, all contain carbon sources adequate to support growth. The dirt and debris present in HVAC systems, including internal insulation, also provide adequate substrates for growth. However, water must be present in the substrate if growth is to occur.

Some unusual substrates in indoor environments have been shown to support the growth of indoor organisms. For example, the occurrence of *Aspergillus* infections in cancer patients in a new hospital was associated with the growth of fungi on cellulosic fireproofing materials present above the ceiling in patient rooms (Aisner et al., 1976). Also, casein-based self-leveling compounds used in concrete flooring in the late 1970s in Sweden provided an unusual substrate that supported the putrefactive fermentation of various microorganisms including *Clostridium* species (Bornehag, 1991; Karlsson et al., 1984).

In addition to providing nutrient materials in situ, many existing building or construction materials contain greatly elevated fungal and bacterial concentrations compared to new materials (i.e., as they left the factory). For example, old chipboard may contain up to 50,000 times more bacteria and fungi than new chipboard (Strom et al., 1990). Plastic flooring and sheeting contain variable degrees of microbiological contamination possibly associated with the presence of additives such as plasticizers, oils, and resins that can act as carbon and nitrogen sources. When some natural building materials such as cork are used, they may be heavily contaminated by microorganisms such as *Streptomyces* (Strom et al., 1990) and *Aureobasidium* species.

HVAC System Design for Source Control

The primary purpose of HVAC systems is to provide for the thermal and air quality requirements of the occupants. Well-designed and maintained HVAC systems will exclude most atmospheric aeroallergens such as pollen and fungi from interior spaces. By contrast, poorly designed HVAC systems may provide for amplification of fungi and actinomycetes in wet niches in the system. Pollen and fungi may enter indoor environments through the air conveyance system itself or through infiltration of the building envelope when the HVAC system is improperly operated or maintained. (EPA, 1991a; Morey, 1984, 1988; Robertson, 1988; Woods, 1988, 1989b). The following section discusses the various components of the HVAC system (e.g., outdoor air intakes, filters, heat exchanges, humidifiers) in the context of source control.

Outdoor Air Intakes

Some residential and most commercial HVAC systems are designed and installed to provide outdoor air for ventilation through "makeup air intakes" that are directly connected by duct work to the HVAC systems. Microbiological contaminants from sanitary vents, toilet or building exhaust air, cooling towers, evaporative condensers, swimming pools, and saunas may contaminate poorly located outdoor air intakes. Outdoor air intakes of some HVAC systems may be located at grade or below grade levels. These outdoor air intakes and the pathways (e.g., metal or concrete ducts) connecting them to the air-handling unit of the HVAC system can collect leaves and other debris that can plug bottom drains. In addition, water and debris that collect in poorly maintained and drained outdoor air pits and areaways provide amplifications sites for microorganisms.

The protection of outdoor air intakes is unfortunately given only minor and inadequate attention in ventilation codes and standards. Section 5.12 of ASHRAE Standard 62-1989 (ASHRAE, 1989b) states that special care should be taken to avoid entrainment of moisture drift from cooling towers into outdoor air intakes. Mechanical codes, such as the Southern Building Code Congress International (SBCCI), state that it is acceptable to locate outlets such as those of chimneys or sanitary sewers as close as 10 feet away horizontally and 2 feet above HVAC system outdoor air intakes (SBCCI, 1990, Section 513). Guidelines for construction and equipment of hospital and medical facilities state that outdoor air intakes shall be located at least 25 feet from exhaust outlets and vents (DHHS, 1987, Section 7.31). The protection of outdoor air intakes from cooling towers is not specifically mentioned in this guideline.

When outdoor air is provided to HVAC systems, it is usually mixed in a compartment or plenum with return air from the occupied spaces before it is filtered or thermally treated. This mixed-air plenum can collect debris such as leaves and feathers if bird or leaf screens on the upstream outdoor air inlet are defective. If rain or snow is carried over into the air-handling unit (AHU), water or rust in the mixed-air plenum may occur.

FILTERS

In air-conditioned residences without outdoor air inlets (the majority of the residential building stock), pollen and other atmospheric allergens are excluded by filtration in the envelope itself or by physical factors such as sedimentation. In large buildings, air from the mixed-air plenum is usually cleaned by one or more sets of particulate air cleaners before it is thermally treated (i.e., heated, humidified, cooled).

The most common type of filter provided for most residential and commercial HVAC systems is installed primarily to prevent dirt and debris from depositing on the heating and cooling coils (Morey, 1988; Woods, 1989b; Woods and Krafthefer, 1986). These filters, rated in terms of weight arrestance (ASHRAE, 1976), have little effectiveness in removing respirable particles from the air moving through the HVAC system (Morey and Shattuck, 1989). The capacities of filters of varying efficiencies to remove bioaerosols (size 1 to 5 μm) from air have been studied (H. Decker et al., 1963). Roughing or prefilters were shown to remove 10 to 60 percent of bacterial particles from the airstream. Medium- and high-efficiency filters (including bag filters) removed 60 to 99 percent of bacterial particles. Kuehn et al. (1991) published a recent review of the filtration of bioaerosols. Figure 7-2 shows the approximate particle sizes of various potential contaminants.

In HVAC systems that contain filters, the air to be used for room ventilation passes through a filter "dust cake." The dust cake often contains contaminants such as human skin scales, fungal spores, pollen, tobacco smoke components, and atmospheric dust and debris. Filters that become moist or wet can function as significant amplification sites for microorganisms, especially fungi (Morey, 1984; Schicht, 1972). Fungal populations in filters can amplify by 2 to 4 orders of magnitude when incubated at 96 percent relative humidity for 10 days (Pasanen et al., 1991). The protection of filters from moisture and careful, periodic replacement of the filters (i.e., without leaving residue from its dust cake in the system) are essential for controlling potential allergen emissions from this portion of the HVAC system.

Dirt, debris, and fungi can be expected to accumulate in AHUs and main and branch supply air ducts, especially in HVAC systems with inefficient filters, where filters do not fit properly in filter frames, or in poorly designed filter banks where significant volumes of air can bypass the filter bank. One study in Kuopio, Finland, found that pollen from outdoor air made up 9 percent of the weight of supply air duct dust (Laatikainen et al., 1991). Dust and debris in supply air duct systems can be expected to be

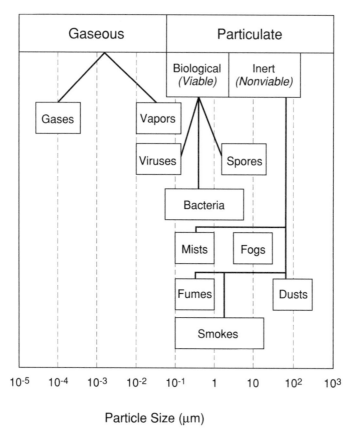

Particle Size (μm)

FIGURE 7-2 Dimensional continuum of potential contaminants. Source: Adapted from Woods, 1982.

most abundant near elbows and turning vanes, as well as zones where airflow restriction occurs (e.g., in reheating coils).

HEAT EXCHANGERS

The heat exchanger section of the air handling unit (AHU), is where heat is either added to or removed from the airstream. During the cooling and dehumidification process, moisture is condensed from the airstream as it passes over the cooling coils in the AHU heat exchanger. This occurs when the dew point temperature of the airstream is greater than the surface temperature of the cooling coil. Consequently, the relative humidity in the air supply plenum downstream of the coils will approach 100 percent. Water that condenses on the surface of the cooling coils then collects in drain pans and exits the AHU through drain lines. During this cooling and dehumidification process, microorganisms can amplify in the heat exchanger section because of the presence of stagnant water in drain pans. A biofilm or slime on pan or coil surfaces is an indicator of microbiological amplification. Moisture that can promote the growth of microorganisms in locations downstream from the cooling coils can originate from water droplets being blown off coil surfaces when the air velocity through the coils is too great. Organic dust in the airstream, especially in HVAC systems with inefficient filter banks (e.g., most residences), can impact on moist surfaces or settle in moist duct work and provide nutrients for the amplification of microorganisms, especially fungi.

Water spray systems (air washers) with recirculated chilled water (in place of cooling coils) represent another type of system that can be used to extract heat and moisture from the airstream. These systems, which are found in some office buildings (Hodgson et al., 1987) and in industrial operations (Reed et al., 1983), can become strong microbiological amplifiers because the dirt and debris that are also extracted or scrubbed from the airstream serve as nutrients for the microorganisms present in the sumps of these systems. Water spray systems used for air conditioning were originally designed to use sterile water or water that was disinfected by biocides (Yaglou and Wilson, 1942). The aerosolization of biocidal chemicals into the ventilation airstream is unacceptable because of their potential adverse health effects on occupants (CFR, 1987).

Another form of air conditioning, prevalent in hot, dry areas of the United States, is achieved with evaporative coolers. In these systems, outdoor air is drawn through mats or pads that are wetted by recirculating water from a sump. In these systems, the dry-bulb temperature of the air entering the unit is cooled adiabatically to approach the wet-bulb temperature, and the air exiting the unit is thus nearly saturated with moisture. Amplification of microorganisms can occur on the evaporative mats or in the sumps of these units (Macher and Girman, 1989). One case of hypersensitivity pneumonitis has been attributed to the presence of thermophilic actinomycetes found in the straw evaporative mat of the unit and in the house dust of the residence where illness occurred (Marinkovich and Hill, 1975).

HUMIDIFIERS

Humidifiers are used to add moisture to the air. These devices are usually installed in the supply air plenum or duct work, downstream from the heating coils. In some residential installations, however, they may be found in the return air. Injection nozzles of humidifiers should be located in areas of AHUs or duct work that are devoid of porous insulation (Morey,

1988). In addition, humidifier moisture should never wet nearby filters. Water spray humidifiers in AHUs must be fitted with downstream demisters or eliminator plates to remove carryover of unevaporated droplets (Ager and Tickner, 1983).

Water sumps in cold water humidifiers must be fastidiously maintained in order to prevent the amplification of microorganisms that can cause building-related illnesses such as hypersensitivity pneumonitis, humidifier fever, and asthma (Morey et al., 1986). Humidifiers that use steam require less maintenance than those that use cold water. However, steam emitted into the supply airstream should not contain corrosion inhibitors such as volatile amines, because they can be nitrosated and are potentially toxic to occupants in the humidified zones (NRC, 1983a). The temperature of the moisture emitted by steam humidifiers is biocidal. Humidifiers that function by evaporating or emitting water molecules (only) are also not direct emission sources of bioaerosols.

Humidifiers that emit water droplets may do so by discharging all water from a supply line or only a portion of the water from a recirculation system. The potential for bioaerosol emission from these humidifiers is directly related to the microbiological contamination in the water supply that is aerosolized. Emission of microorganisms from spray-type humidifiers that use recirculated water is greatly reduced by installation of highly efficient upstream filters that remove dusts that would otherwise enter the humidifier and serve as growth nutrients.

ASHRAE Standard 62-1989 (ASHRAE, 1989b, Section 5.12) recommends steam as the preferred moisture source for humidifiers. However, if cold water humidifiers are used, water should originate from a potable source and units that use recirculated water should be subject to frequent maintenance and blow-down. The specific protocol for maintenance of humidifiers that use recirculated water is not specified in ASHRAE Standard 62-1989.

The Nordic Committee on Building Regulations (NKB, 1990) provides advice with regard to humidification of the air. Section 4.6.6 of this standard recommends that where humidification is required, a type of device "which does not involve the risk of microorganisms being released into the air shall be chosen."

Air Supply Plenum and Duct Work

After passing through the heat exchanger and supply fan (the location of the fan may be upstream or downstream of the heat exchanger), conditioned air is distributed through a system of ducts to the occupied spaces. The AHU plenum housing the fan is usually thermally insulated on its internal surfaces with a fiberglass lining that also acts as a sound attenuator.

The main supply air duct work immediately downstream of the fan and the heat exchanger is usually constructed of galvanized sheet metal and can be either externally or internally insulated with fiberglass. However, in some residences and commercial installations, duct work itself is constructed of a rigid fiberboard that is intrinsically insulated. Internal fiberglass or fiberboard liner either in the fan or main air supply plenum should have sealed surfaces to prevent the erosion of fibers into the airstream. The internal insulation in these plenums should be protected from emissions from humidifiers or from water droplet carryover from heat exchangers to reduce the likelihood of microbiological amplification.

PERIPHERAL UNITS

In some commercial buildings, a separate system may be installed to heat and cool perimeter zones that are more affected by outdoor climatic conditions than interior zones that are not in contact with the building envelope. Fan coil and induction units often mounted in sheet metal enclosures along exterior walls, are two common types of peripheral units that are used to condition air in perimeter zones.

Fan coil units contain small fans, low-efficiency filters, and small heat exchangers with small drain pans. Fan coil units condition and recirculate room air (often without any outdoor air) in peripheral zones. Sheet metal enclosures are usually lined along interior surfaces with porous insulation. Large buildings may contain hundreds of these units, and consequently, maintenance is often neglected. Fan coil units often accummulate dirt and debris, becoming amplifiers and disseminators of microorganisms during the air-conditioning season, when the heat exchanger actively contributes moisture to each unit's enclosure (Morey et al., 1986).

Induction units are provided with primary air from a central AHU. This air, containing a percentage of outdoor air, exits each induction unit through a series of nozzles that induce a flow of room air, and with which it is mixed and supplied to the room. Induction units usually contain low-efficiency filters and a heat exchanger that removes some sensible heat but little latent heat or moisture. Under design operating conditions, condensate pans in induction units are expected to contain little if any water. Thus, they are less likely to be sites for microbiological amplification, but they are likely to be reservoirs for pollen and phylloplane fungi, especially if the primary air has been poorly filtered.

RETURN AIR SYSTEMS

In many commercial buildings, the cavities or plenums above the finished ceilings are used as unducted passageways for air returned from the occupied spaces to the central HVAC system (Morey and Shattuck, 1989). The return air plenum in the ceiling can become a source (or amplification site) of microorganisms when fire and acoustical insulation and ceiling tiles become wet. Thus, roof leaks resulting in high relative humidity above ceiling cavities can cause microbiological amplification on wood and gypsum board, as well as condensation of water (with subsequent microbial growth) on uninsulated air supply duct work surfaces (during the air-conditioning season), and on the uninsulated upper surfaces of diffusers.

Instead of cavities or plenums, some residential and commercial buildings use return air ducts to transport the air from occupied spaces to the central HVAC systems. These ducts are often insulated for sound attenuation. Because the air entering these ducts is usually not efficiently filtered, if at all, the likelihood of house dust accumulation over time is high. Moreover, if care is not taken to ensure that the relative humidity in the return air does not exceed 70 percent, amplification of xerophilic fungi and mites is possible.

POROUS INSULATION

As mentioned previously in this chapter, porous insulation is often installed on the inside surfaces of HVAC system components such as AHU plenums and the sheet metal of fan coil units, induction units, unit ventilators, and return air ducts. The fiberglass insulation often used in HVAC system components when new, can accumulate dirt and debris. Unlike bare sheet metal surfaces, which may be subjected to vigorous cleaning processes, it is difficult or impossible to remove the dirt and debris that becomes entrained in insulation pores. Hydrophilic dirt and debris in the fiberglass insulation can absorb moisture from the airstream up to a level of about 0.5 pounds of water per pound of dry fiberglass (West and Hansen, 1989). Moisture is readily available for absorption by hydrophilic dirt in insulation pores because the relative humidity of the air leaving the cooling coil section approaches 100 percent as a result of the air-conditioning process.

Porous HVAC insulation was found to be a source of fungi or bacteria in 9 of 18 buildings evaluated for microbiological contaminants (Morey, 1988). Porous insulation can become a secondary emission source (or amplification site) for microorganisms when the equilibrium moisture content and water activity in the entrained substrate (including binder resins) become adequate to support growth (Morey and Williams, 1991). The kinds of microorganisms that may grow in insulation are often different from those commonly found in the outdoor air or normally present in indoor air.

Porous insulation or fleecy surfaces in components of HVAC systems that never become wet, such as forced-air heating ducts, can become reservoirs for allergens from outdoor sources such as common atmospheric fungi and pollen. In buildings with inefficient filters, spores accumulate over the years in supply and return air ducts and perimeter unit insulation (Morey and Williams, 1991). The insulation then becomes a secondary emission source of fungal spores when the HVAC system component is disturbed, such as during maintenance activities.

Building Operation for Source and Exposure Control

Building operation as it relates to controlling indoor allergen sources and exposures is discussed below in terms of control strategies, and maintenance and housekeeping practices.

CONTROL STRATEGIES

Inappropriate control strategies have been associated with nearly all problem buildings (Woods, 1989a, 1991). The two most common problems are: (1) the complexity of the control system is not within the comprehension of the current building operators, and (2) overaggressive energy-saving strategies have compromised the ability of the control strategies to provide acceptable indoor environmental quality (NRC, 1987b; Woods, 1991).

Changes in thermal and contaminant loads during the operational lives of the facilities without commensurate changes in the system capacities have been reported as a significant finding in problem building investigations (e.g., 60 percent incidence; Woods, 1989a, 1991). The need for changes in capacities is seldom evaluated when occupancy loads are changed or when renovations are implemented. As a result, significant mismatches occur. Of particular significance to the control of allergens, these changes nearly always affect the ability of the systems to control humidity.

Humidity problems may also occur in buildings because of inappropriate HVAC system operation parameters. As an energy management strategy, building operators may be tempted to raise the temperature of chilled water entering the cooling coils from a design value (e.g., 45° F) to a higher value (e.g., 55° F). Although the dry-bulb or operative temperature in the occupied spaces may continue to be adequate for comfort, insufficient latent heat may be removed from the airstream and the indoor air in the building will become too humid.

Incipient problems can exist without detection in buildings for indefinite periods. During these periods, small changes in system performance can occur. These changes can result in small increases in discomfort complaints and symptoms. The frequency of occurrence of buildings in this category remains unknown, but an initial postulate has been made that 10–20 percent of the existing buildings are candidates (Woods, 1989a). It is important to recognize this category of buildings because it is the basis for a continuous source of problem buildings if diligence is not maintained in mitigating the incipient problems.

MAINTENANCE AND HOUSEKEEPING

Inadequate maintenance may be the second most frequently reported operational problem (i.e. approximately 75 percent incidence) (Woods, 1989a, 1991). These problems include obvious oversights such as missing or dirty filters, dirty makeup air intakes and duct work, and inoperative equipment. However, they also include less obvious problems such as degradation of component performance as the building and its systems approach failure at the end of their useful lives and lack of preventive maintenance programs.

Housekeeping is probably the most common means of removing potential allergens, and vacuum cleaners are the most commonly used housekeeping tool. Dry vacuum cleaning is traditionally used to remove dirt and debris from the fibrous pile of carpets. Little information, however, is available on the effectiveness of this cleaning method in removing the various types of particles, including specific allergens that may adhere to pile fibers and carpet backing. The work of Wassenaar (1988b) suggests that viable mites are less readily removed by vacuum cleaning than other kinds of entrained particles.

The actual physical cleaning process itself (the movement of a vacuum cleaner across carpet fibers) may be sufficient to disperse fine particles such as *Fel d* I and some fungal spores into indoor air. The presence of significant populations of some allergens less than 2.5 μm in size (for example, *Fel d I*) in carpet suggests that vacuum cleaners with high-efficiency particulate arresting (HEPA) air filters may be necessary to prevent reaerosolization of fine particles in occupied spaces. Vacuum cleaning of carpet with instruments without HEPA filters leads to about an order of magnitude increase in fungi such as *Cladosporium* and *Penicillium* species (Hunter et al., 1988). Airborne levels of *Fel d* I have been shown to rise approximately 1 order of magnitude when carpeted floor is disturbed by an operating vacuum cleaner without a filter (Luczynska et al., 1990). However, work by Dybendal and colleagues (1991), suggests that HEPA filtration is not necessary to collect most allergens such as those from pollen, fungi, and mite fecal pellets that may be present in carpets and other fleecy furnishings.

Little scientific information on the effectiveness of carpet cleaning shampooing methods for allergen removal is available. However, any wet cleaning method that provides a moist microenvironment in the backing and padding of carpets also provides conditions that promote potential amplification of fungi and mites.

The use of ultra-high-speed floor burnishing instruments to polish hard-surface floors in health care facilities is known to produce elevated concentrations of dust particles and microorganisms in indoor air. Special air restraint assemblies can be fitted onto ultra-high-speed burnishers, with the net effect that significant increases in the levels of dust particles and microorganisms are prevented (Schmidt et al., 1986). Similar devices are apparently not available for vacuum cleaning devices used for carpet maintenance.

Renovation and repair work, such as replacing windows or repairing furniture in residences, generally results in increased concentrations of fungi in occupied spaces (Hunter et al., 1988). The disturbance of walls visually contaminated by fungi in both residences (Hunter et al., 1988) and commercial buildings (Morey, 1990a) generally results in a 2 to 4 order of magnitude increase in airborne fungi, generally of a single type, at various distances from the disturbance site. In large buildings, the disturbance of fungus-contaminated insulation in the HVAC system can result in order of magnitude increases of fungi throughout the occupied spaces (Morey and Williams, 1991). As an example, renovation on a floor above a renal transplant ward was causally associated with an outbreak of nosocomial *Aspergillus* infection (Arnow et al., 1978). Movement of heavy equipment on the floor above the transplant ward probably caused sufficient vibration to allow dusts, including *Aspergillus* spores, to aerosolize in the ceiling space above the ward and to disseminate into patient areas.

Remediation

Once contamination of a building has occurred, exposure can be controlled by removing the contamination (source control), by cleaning the air (E in the model), or by dilution control (V_o). Air cleaning and ventilation control can be accomplished by using local and/or central ventilation system components.

SOURCE CONTROL

Source control in indoor environments is most effectively achieved by limiting moisture, which promotes the growth of some microorganisms (xerophilic fungi), and by limiting the use of fleecy finishing and furnishing materials (e.g., carpet), which by their porous nature permit the accumulation of allergens. Restricting the use of carpets, for example, to those that can be removed from the house for periodic cleaning is an effective means of helping to control house dust mite allergens.

Removing the source is the most effective means of remediating existing contamination. Furnishings and construction materials that are visibly contaminated by microorganisms should be discarded. Disinfection of contaminated surfaces is not sufficient because dead microbial particulates are antigenic and still capable of reacting with the immune system.

AIR CLEANING

Protocols for cleaning air conveyance systems are being developed by the National Air Duct Cleaners Association (NADCA, 1991). Most duct-cleaning procedures involve the physical removal of dust and debris using vacuum systems with HEPA filters. The mere presence of microorganisms in an air supply duct, however, is not an adequate basis for the initiation of duct cleaning. The presence of sufficient dust and debris to restrict airflow or to result in the dissemination of particulates through diffusers into occupied spaces is a valid reason for cleaning air supply ducts.

In occupied buildings, duct-cleaning procedures that require the use of biocides in place of physical removal of dust or that require encapsulation of dust and debris that may contain microorganisms are of questionable value. Biocides, (e.g., acaricides), that are used to directly lower viable mite populations have been variously effective (Lau-Schadendorf et al., 1991; Lundblad, 1991; Tovey et al., 1992). Polyphenolic materials added to acaricide formulations are thought to denature or modify the antigenic characteristics of mite allergens so that the IgE response is not elicited (Green et al., 1989). The beneficial effect of chemicals used to control mite and mite antigens must be balanced against the effect of these agents on nontarget populations (for example, the irritating effect of benzyl benzoate on humans). More information can be found in a recent forum on the duct-cleaning industry (IAQU, 1991).

As for portable air filtration devices, their beneficial effects over and above those associated with the central air conditioning are thought to be minimal (H. S. Nelson et al., 1988). Additionally, the beneficial effect of an air cleaner in removing aeroallergen is small when considered against the large amount of some allergens (such as those from mites and cats) that exist in surface reservoirs. Luczynska and colleagues (1990), for example, found that the beneficial effect of a portable air cleaner with a HEPA filter in lowering the air concentration of *Fel d* I occurred only when surface reservoirs were undisturbed and only when the air exchange rate was unusually high (20 air changes per hour).

The use of portable HEPA-filtered air cleaners does not lower mite allergen levels in settled dusts (Antonicelli et al., 1991). Since mite allergens are predominantly large particles (greater than 10 μm) that settle rapidly from the air after disturbance of surface reservoirs, it would not be expected that air cleaning itself will significantly lower allergen contents in surface reservoirs.

Reisman et al. (1990) found that use of a portable HEPA-filter air cleaner was without effect in reducing allergic respiratory symptoms, although patients subjectively believed that use of air cleaners was beneficial. Measurements made with a direct reading particle counter showed that the air cleaner used in these studies did lower the concentration of particles equal to or greater than 0.3 μm over placebo controls. However, no attempt was made to identify the kinds of particles excluded by the air cleaners, which could include small aeroallergens (e.g., some fungi) and non-specific irritants (e.g., environmental tobacco smoke).

A laminar flow air cleaner with HEPA filter when attached to beds of mite-allergic asthmatics was beneficial in reducing patient symptoms. This device is useful in producing a small zone of high-quality air in the breathing zone of the resting patient but, because of its small capacity will likely have little beneficial effect in the entire patient room or residence.

The inability of portable HEPA units to remove *Fel d* I emitted from carpets (de Blay et al., 1991a) is analogous to the failure of HEPA filter units in hospitals to remove small-diameter spores such as those of *Penicillium* and *Aspergillus* species emitted from strong fungal reservoirs in rooms housing immunocompromised patients. For example, Streifel and colleagues (1987), found that rotting wood in a medication room beneath a sink was the source of thermophilic *Penicillium* species found in a corridor in bone marrow transplant ward that was supplied with air that was HEPA filtered. *Penicillium* spores are approximately 2 to 4 μm in size. A baseline thermophilic fungi concentration of 812 colony-forming units/m^3 was present in the medication room (when doors to the rotted cabinet were open), even though the room contained a portable HEPA filter unit operating at 20 air changes per hour. This demonstrates that unexpected emissions of small-diameter (2- to 5-μm) allergens from indoor reservoirs can overwhelm even the best possible ventilation systems.

Important considerations in the potential use of air cleaners for removal of aeroallergens include the volume of room or building air that passes through the filter and the particle size of the air contaminant to be removed. If the airflow rate through the device is low and the emission rate of allergen is high, then the beneficial effect of the air cleaner is likely to be nonsignificant.

DILUTION

Dilution is seldom effective in controlling existing contamination, because strong emission sources of allergens (reservoirs) overwhelm the dilution capacity of highly filtered (high-quality) outdoor air (see example given by Streifel et al. [1987] above).

RECONTAMINATION

Recontamination of indoor spaces can occur if the fundamental reasons for the initial contamination were not adequately addressed. For example, cleaning of air conveyance systems and replacement of gypsum board in a building contaminated by xerophilic fungi will be ineffective if the elevated indoor moisture that led to the growth of fungi is not addressed. Replacement of carpet containing *Der p* I with new carpet will not prevent recontamination unless the carpet removal is associated with other actions such as lowering the relative humidity in the indoor air and in the new carpet.

SUMMARY OF PRACTICAL CONTROL STRATEGIES

The following summary of engineering principles and practices that can be employed to prevent or minimize occupant exposures to indoor allergens is organized into three categories: general, residential buildings, and commercial buildings.

General

As indicated by the list of general control strategies that follows, ambient relative humdity is often considered to be the major controlling factor for indoor allergens.

1. Control ambient relative humidity. For example:
   — Use air conditioning to remove moisture that has entered indoor air. Natural ventilation does not necessarily remove moisture from indoor air.
   — Use dehumidifiers to remove moisture from air in occupied spaces that are not adequately air conditioned.
   — Exhaust strong indoor moisture emissions (e.g., steam from simmering foods, bathroom moisture, clothes dryer emissions) directly outdoors.
2. Prevent condensation. For example:
   — Install vapor retarders in the building envelope.
   — Install insulation in the building envelope so as to prevent condensation in wall cavities and on wall or ceiling surfaces.
3. Use nonporous floors, walls, or ceilings whenever possible or cover existing fleecy surfaces with impervious sheeting.
4. Design buildings and systems to minimize the potential for flooding. If floods occur, then ensure comprehensive cleanup.
5. Avoid the creation of moist microenvironments. For example, carpet should not be installed on floors that are likely to be flooded (near bathtubs) or on concrete floors in basements.

Residential Buildings

A primary function of mechanical ventilation systems is to provide comfort to the occupants. Providing comfort means controlling the moisture content of the air. Controlling the moisture content of the air also controls allergen content as a corollary benefit. Appropriate filtration can provide additional allergen control as a part of cleaning the air. For residences with forced air heating and cooling systems, a number of principles and practices can be employed for controlling allergens as outlined below:

1. Use forced-air heating and cooling systems to maintain thermal environmental conditions at all times in the occupied spaces as recommended in ASHRAE Standard 55-1992 (ASHRAE, 1993). For allergen control it is especially important to use the air-conditioning system to keep the relative humidity below 70 percent (preferably below 60 percent) in occupied spaces, including basements and in parts of the building where that air mixes with the occupied space.

2. Include in the central forced air system filters with at least a moderate atmospheric dust spot efficiency (30 to 50 percent) to remove aeroallergens from the ventilation airstream. Care is needed to ensure that the fan capacity of the system is sufficient to overcome the additional airflow resistance (pressure drop) imposed by these higher-efficiency filters.

3. Ensure that outdoor air is provided at least at the minimum rates (0.35 air changes per hour, but not less than 15 cubic feet per minute per person) recommended by ASHRAE Standard 62-1989, Table 2.3. This minimum outdoor air ventilation rate is also recommended for naturally ventilated residences.

4. Consider design options where outdoor air ventilation requirements are met through provision of outdoor air directly into the HVAC system. This would allow outdoor aeroallergens to be removed by HVAC system filters. Additionally, overall pressurization in the conditioned space could be made slightly positive compared with that in the atmosphere so that aeroallergens from outdoor sources do not infiltrate through loose construction or through cracks in window or door frames.

5. Ensure that the forced-air ventilation system itself does not become a source of allergens. For example, accumulation of water in ventilation systems should be prevented. Ventilation systems should be kept clean by regular maintenance because dirt and debris can accumulate in poorly maintained systems, and the dirt and debris can contain allergens or function as substrates for microbial growth. Access panels into plenums and the air conveyance system is essential to allow for regular maintenance. Internal surfaces in forced-air systems should be smooth and not provide substrates for fungal growth.

In residences without forced-air mechanical ventilation, maintain the minimum outdoor air ventilation rate recommended in Table 2.3 of ASHRAE Standard 62-1989. Dehumidification or supplemental ventilation (air conditioning) may be required to maintain thermal environmental conditions specified in ASHRAE Standard 55-1992 (ASHRAE, 1993).

Commercial Buildings

A number of principles and practices can be employed for controlling indoor allergens in commercial buildings, as outlined below:

1. Outdoor air intakes should be located at a site (preferably on the roof) where the ambient air quality is the best. Grade-level sites should be avoided. Outdoor air inlets should be located at sites so that possible entrainment of contaminants from cooling towers, exhaust and relief vents, and other contaminant sources is avoided. Keep the outdoor air intake plenums clean.

2. HVAC systems should be accessible for cleaning. Ceiling AHUs, rooftop AHUs, and central system AHUs must be designed for easy access for cleaning. Access panels should have gaskets and smooth inner surfaces.

3. Filter banks should be changed frequently and kept dry.

4. Avoid stagnant water in the heat exchangers of HVAC systems; drain pans should self-drain. Biocides that can be aerosolized into indoor air should not be used in operating AHUs (drain pans), water spray systems, and humidifiers.

5. Keep the porous insulation in HVAC systems clean (by protecting with adequate filtration) and dry. If the insulation is contaminated, consider cleaning it with a HEPA filter vacuum, replacing the insulation, externalizing the insulation, or placing the insulation between metal surfaces.

6. Avoid the use of materials that cannot be cleaned in common return air plenums; avoid the use of cellulose; avoid the use of high-surface-area materials. Do not locate air-handling units in a common return air plenum.

7. Peripheral HVAC systems should be accessible for periodic maintenance. These units should not be used in buildings if maintenance is impossible.

CONCLUSIONS AND RECOMMENDATIONS

The fundamental objectives of environmental control are to prevent or minimize occupant exposures that can be deleterious and to provide for the comfort and well-being of the occupants. Well-designed and maintained HVAC systems will exclude most aeroallergens (e.g., pollen, fungal spores) from interior spaces. Poorly designed or maintained systems, however, can provide for amplification and/or infiltration and dissemination of allergens. Inappropriate control strategies have been associated with nearly all problem buildings.

Recommendation: Improve the design, installation, use, and maintenance of residential and commercial HVAC equipment, for both new and existing construction, in order to minimize allergen reservoirs and amplifiers. These improvements should be based on recommendations developed by the American Society of Heating Refrigerating and Air Conditioning Engineers (ASHRAE).

Carpeting can provide niches for both the accumulation and production of allergens, and has been characterized by some as a "cultivation medium" for microorganisms when wetted. Carpeting can also serve as a reservoir for pollen and pollen fragments. The magnitude of the potential significance of carpeting as a source and reservoir of indoor allergens indicates that it should be given consideration as a serious problem.

Recommendation: Expand the scope of the Carpet Policy Dialogue Group of the Environmental Protection Agency to consider the serious problem of carpets as a source and reservoir of indoor allergens.

Standards have been established by the American Society of Heating, Refrigerating and Air-Conditioning Engineers (ASHRAE) for acceptable temperature, humidity, and ventilation as they relate to human comfort. However, little attention is given in these standards to the protection of buildings, furnishings, and construction materials from water damage, and the potential for subsequent adverse health effects.

Recommendation: Develop consensus standard recommendations for controlling moisture in naturally and mechanically ventilated buildings. These recommendations, designed to help control microbial and arthropod aeroallergens and allergen reservoirs, should be developed by ASHRAE and be included in their Standard Series 55 (thermal environmental conditions for human occupancy) and Standard Series 62 (ventilation for acceptable indoor air quality).

Dry vacuum cleaning is traditionally used to remove dirt and debris from the fibrous pile of carpets. Little information is available, however, on the effectiveness of this cleaning method in removing the various types of particles, including specific allergens that may adhere to pile fibers, carpet backing, and other furnishings. In addition, the physical cleaning process itself may be sufficient to disperse fine allergenic particles.

Research Agenda Item: Develop standardized tests for rating the effectiveness of vacuum cleaners in removing allergen-containing particles of known size from carpets, upholstery, drapes, and other materials. The tests should take into account the possible dispersion of particles from carpet caused by the cleaning process itself.

The effectiveness of air cleaning devices and practices depends on variables such as the volume of air that passes through the filter, the particle size of the air contaminant to be removed, and the source emission rate. If the air flow rate through an air cleaning device is low, for example, and the emission rate of the allergen is high, then the beneficial effect of the air cleaner is likely to be nonsignificant.

Research Agenda Item: Develop standardized test procedures for rating the effectiveness of air cleaning devices and other methodologies for removal of known size classes of particles containing allergens. The tests should address the capability of the device or methodology in removing airborne particulates from entire rooms or zones of buildings.

Restricted airflow and dissemination of particulates into occupied spaces are valid reasons for cleaning air supply ducts. Protocols for cleaning air conveyance systems are currently in development by the National Air Duct Cleaners Association. However, the effectiveness of duct cleaning in controlling allergic disease is yet to be determined.

Research Agenda Item: Evaluate the role of duct cleaning in controlling allergic diseases.

As described throughout this report, ambient relative humidity is often considered to be a major controlling factor for indoor allergens. Control of relative humidity, or water vapor pressure in occupied space and in the HVAC system is an important part of allergen control in both residential and commercial buildings.

Research Agenda Item: Develop a public-use guideline on moisture and allergen control in buildings. The guideline should describe the proper use of vapor retarders and other techniques for moisture control in both naturally and mechanically ventilated buildings.

There are approximately 4 million commercial and 84 million detached residential buildings in the United States. About 75–85 percent of the buildings that will exist in the year 2000 have already been built. Maintenance, operation, renovation, and housekeeping practices affect the useful life span of a building and the quality of the indoor air. Cost effective strategies for source and exposure control are needed to address the problems associated with normal degradation of the HVAC performance that occurs as a building ages.

Research Agenda Item: Determine the relative efficacy of currently recommended environmental control strategies and develop cost-effective strategies for controlling aeroallergens throughout the lifetimes of residences and other buildings.

8

The Role of Education

Education is an important component of outreach for the prevention and control of allergen-induced diseases. It is fundamental to the knowledge base of those who either suffer from such diseases or treat them, and provides the foundation for behavior. Education that improves awareness of health risks is basic to disease prevention and health promotion; by disseminating information to health care providers and to patients, prevention of diseases associated with indoor allergens becomes not only realistic but may offer a cost-effective means of reducing morbidity. Many questions remain unanswered, however, as to the best way or ways to disseminate information and provide such education.

This chapter discusses the role of education in the control and prevention of allergen-induced diseases. It also reviews educational tools and materials for use in these efforts and makes recommendations for future research and for addressing the educational needs of patients, health care providers, and the general public.

EDUCATING PATIENTS AND THE GENERAL PUBLIC

Educational programs involving the control and prevention of asthma far outnumber those developed for other allergen-induced diseases. By necessity, therefore, much of what is known about the effects of education with these diseases has been derived from studies on asthma. In the late 1960s, approximately two dozen residential centers in the United States specialized in the treatment of childhood asthma. Publications from the

Children's Asthma Research Institute and Hospital (CARIH) in Denver have been credited with establishing the background for asthma patient education and self-management (Decker and Kaliner, 1988). Those publications show the emerging awareness that patients and their families first had to learn about asthma management skills and competencies; patients then had to demonstrate that they could perform these skills to help control the disorder. This approach reflects the learning-performance distinction inherent in social learning theory (Bandura, 1986). Effectively performing management skills achieved the aim of permitting patients to become partners with their physicians in controlling their asthma.

By the mid-1970s, it was apparent to many health professionals in the United States that greater emphasis had to be placed on developing and evaluating self-management programs for asthma. By the end of the decade, more than two dozen such "first generation" programs for childhood asthma had been developed and evaluated. A similar number, many of which were still undergoing testing, had been created for adults. Recently, Creer and coworkers (1990) and Wigal and colleagues (1990) reviewed 19 education and self-management programs for childhood asthma (Table 8-1).

A number of positive results were obtained from the earliest of these programs. Significant findings included decreases in the number of attacks, in hospital use (including visits to hospital emergency rooms), in school absenteeism, and in asthma-related costs. Investigators also found increases in peak flow rates; improved attitudes toward asthma, including the perceived ability of patients that they could help control it; and improved self-esteem. The investigators who conducted these studies included both medical and behavioral scientists who had worked together for a number of years on the problems of asthma. Other positive findings suggested by Wigal and colleagues (1990) included the following:

• In most instances, children and their parents became partners with their physicians in the management of the children's asthma.
• The programs were applied in a broad array of settings, including those serving children from different socioeconomic backgrounds.
• There was evidence that some dependent measures were valid and reliable, although flaws were found in many questionnaires.

Perhaps the greatest contribution of the programs was the development and availability of a variety of educational materials and techniques that could be used to teach self-management to patients and their families. Indeed, a veritable smorgasbord of programs is available to teach asthma management skills to patients (Creer et al., 1992).

The overall impact of these programs was positive, and virtually all of them were able to demonstrate the acquisition of knowledge about asthma and self-management skills. Yet most of these early programs did not as-

TABLE 8-1 Nineteen Self-Management
Programs for Childhood Asthma

A.C.T. for Kids
Air Power
Air Wise
Asthma Command
Asthma Summer Camp Program
Camp Wheeze
Children's Hospital of Pittsburgh
Community Program for Childhood Asthma
Educational and Exercise Program
Family Asthma Program
Living with Asthma
Open Airways
Self-Care Asthma Education
Self-Care Rehabilitation in Pediatric Asthma
Self-Help Education-Exercise Program (CASH-IN)
Sunair
Superstuff
Teaching My Parents/Myself about Asthma
You Can Control Asthma

SOURCE: Wigal et al., 1990

sess the degree to which patients applied the skills (Creer et al., 1990). The procedures and methods used by children and their families to bring an attack under control were not assessed; neither were the factors that led to an amelioration of the attacks. Also not examined was the relationship between (1) reduced hospitalizations and emergency room visits and (2) the behaviors recommended by physicians and in educational programs. It has yet to be determined whether children who learn these self-management skills can and will apply them to manage their asthma attacks more effectively. Although some studies show that people who learn management skills use them in the short term, it is less clearly documented that these behaviors persist over the long term.

Not all educational programs on asthma have been successful. For example, in an educational program consisting of classes conducted in group settings, only 38 percent of a group of Chicago inner-city children with asthma and their families attended one or more of four scheduled classes (Shields et al., 1990). The median household income of these families was 6 percent below the overall city median of Chicago. Among the participants who remained for the entire study period (29 months), there was no significant improvement (reduction) in postintervention health care utilization for respiratory-related illnesses, and the program did not achieve its goal of decreasing emergency room visits for children with asthma.

The failure of this program probably hinged on two elements: implementation and program effectiveness. Program implementation was difficult because it was offered to all children with asthma, regardless of severity; those with mild asthma may have been less motivated and less likely to benefit than those with more severe disease. Program effectiveness suffered primarily because the program addressed knowledge rather than behavior modification, and specific behavior modification techniques such as reinforcement and contingency contracting were not utilized. Moreover, the socioeconomic, educational, and ethnic profiles of the families may have contributed to problems associated with both program implementation and effectiveness. Thus, future programs designed to teach patients how to manage and control their allergic diseases should (1) include a strong behavioral component, (2) be targeted to key groups who will actively participate and benefit from them, (3) be monitored to ensure that immediate goals are reached, and (4) be validated for specific socioeconomic groups before implementation.

Self-monitoring of asthma by objective means appears to be effective for some patients. In a study of adults with asthma, the use of home monitoring of pulmonary function played a significant role in both decreased bronchial hyperresponsiveness and increased compliance with medication regimens (Beasley et al., 1989b; Woolcock et al., 1988). To explain such findings, the authors suggested that once a patient observed an improvement in daily readings, he or she was more likely to take medications as prescribed. In a more recent study of 39 adults with asthma, home peak flow monitoring in association with a treatment plan was used to assess individuals over the course of 6 months. Substantial improvement was seen in both subjective and objective measurements of severity; however, no attempt was made to identify which features of the management plan were responsible for improved control of symptoms. Thus, it was unclear if regular assessment of peak flow, adequate inhaled corticosteroids, education of the patient, prevention of acute episodes, regular clinic attendance, or improved adherence to prescribed regimens had separately or collectively contributed to the overall improvement that was noted.

Allergen Avoidance and Environmental Control

The role of allergen avoidance is a primary method of promoting good health and controlling diseases that are initiated or exacerbated by exposure to indoor allergens. For example, as decided previously in this report, avoidance is considered an important method of treatment for dust mite allergy (Buckley and Pearlman, 1988; Melan, 1972) and is also associated with improvement of asthma when rigorous methods of avoidance are employed (Murray and Ferguson, 1983; Platts-Mills et al., 1982). Box 8-1 presents a concise example of instructions for reducing exposure to house dust mites.

BOX 8-1 University of Virginia Allergy Clinic Instructions for Reducing Exposure to House Dust Mites

House dust mites require humidity (greater than 50 percent relative humidity) and warmth (above 70° F) to grow. Because mites avoid the light and because surfaces dry out rapidly, mites flourish in mattresses, bedding, upholstered furniture, carpets, pillows, and quilts. Under really humid conditions, mites will also grow in clothing, curtains (drapes) and any material.

Procedures to reduce dust mites should focus first on the bedroom because more time is spent there than any other room and it is generally easiest to change. However, in the long run, it is best to modify much of the house and this should certainly be considered when moving.

PRIORITY OBJECTIVES

• Mattresses and pillows should be enclosed in a zippered, plastic cover or a special vapor permeable, allergen proof fabric. Damp wipe the mattress cover every two weeks.

• Wash all bedding including mattress pad, pillow cases, and blankets in hot cycle (approximately 130° F) *weekly*. Comforters (or duvets) should be replaced with dacron or orlon which can be washed with the bedding or covered with vapor permeable covers.

• Small objects that accumulate dust such as knickknacks, books, stuffed animals, and records should be placed in drawers or closed cabinets. Clothing should be stored in closed cupboards or closets. Unused clothing should be stored away from the bedroom.

• Carpets should be vacuum cleaned weekly using a vacuum cleaner with an effective filter. The patient should either avoid vacuum cleaning or wear a mask during cleaning. In general, dust mite allergens will take about 20 minutes to fall after cleaning.

MEDIUM TERM OBJECTIVES

• Removing carpets from bedrooms makes it much easier to control mites. This is because carpets are very difficult to clean and will tend not only to grow mites (in humid seasons) but also to act as a source to reinfest bedding, clothing, etc.

• Replace curtains/drapes with washable cotton curtains or venetian/slat blinds.

• Control humidity in the house; this can be achieved by increasing ventilation if the outdoor conditions are cold and/or dry; alternatively, reducing humidity can be achieved with central air conditioning. Dehumidifiers are helpful in basements. The objective is to keep relative humidity below 50 percent.

continued

BOX 8-1 *Continued*

• Treat carpets with acaricides (e.g., acarosan) or tannic acid (e.g., allergy control solution).

CHOICE OF HOUSES/APARTMENTS

• Basements are not recommended for any allergic patients (in some cases moving out of a basement may be urgent because it is so difficult to control mite and/or fungal growth in a basement). Bedrooms should be upstairs.
• Carpets fitted to a concrete slab either in a basement or on the ground floor tend to become damp and remain damp. We recommend that all floors should have a primary polished floor (vinyl or wood) and carpets should be movable.
• Upholstered sofas and chairs should be avoided.
• Air filters on central air conditioning should be cleaned regularly. Good quality, e.g., electrostatic, filters may be helpful but are no substitute for reducing available mite nests in the house.

In the study by Murray and Ferguson (1983), there was substantial reduction in bronchial hyperreactivity and medication requirements in 10 children who employed dust mite avoidance measures. The regimen was stringent and included removing carpets and sealing heating ducts, as well as removing from the home animals to which the child had positive allergy skin tests. It is important to note, however, that the more complicated the regimen, the less likely it is to be followed (Gaultier et al., 1980). As with programs directed toward other health behaviors, various types of education for allergen avoidance will need to be developed and tested in a variety of different settings and populations in order to appropriately assess their effectiveness.

Virtually all education and self-management programs develop the association between allergic factors and asthma. The aim is not only to identify potential triggers of acute episodes of asthma and allergic symptoms but also to teach the value of avoidance of stimuli. In this regard, many of the initial educational and self-management programs had weaknesses. Most programs, for example, contained only a basic outline of allergic factors and asthma and lacked a major description of the breadth of allergic stimuli and their potential role in precipitating asthma. No program incorporated an intervention component whereby experienced scientists visited the homes of patients and determined what allergic factors in their environments could induce or exacerbate asthma attacks. In addition, the educational materials were uniformly applied to all patients, regardless of the unique characteristics of their asthma. It is anticipated that future educational programs will tailor self-management skills specifically to each patient in order to enhance both the acquisition and the performance of self-management competencies.

Computer-assisted instruction is another approach that has been tested. These methods may not be suitable for all patient populations, but one study of adults with asthma that used a supplemental computer-assisted reinforcement of instructions on dust mite allergen avoidance measures demonstrated better adherence than that achieved by the traditional instruction program (Huss et al., 1991). The study group was well educated and had been evaluated at a tertiary medical center; direct observation and a self-rating scale were used to measure adherence. Among the results was the finding that the observation checklist was a better discriminator of adherence among groups of patients than the self-rating scale, giving credence to the belief that these measures are superior to self-report methods for evaluating adherence.

Central to the hypothesis that education about allergen avoidance reduces morbidity of respiratory diseases such as asthma is the demonstration of threshold levels of exposure to common indoor allergens in individuals at risk. Recent progress in the immunochemical detection of common indoor allergens such as those associated with cats, dust mites, and cockroaches makes it possible to estimate exposure to these allergens and to begin to define threshold levels of exposure that will cause sensitization and increased symptoms. The example given earlier in this report describes the case of dust mites and the fact that researchers have suggested that exposure to greater than 2 µg of *Der p* I mite allergen (or 100 mites)/g of dust increases the risk that children will develop sensitization and asthma (Platts-Mills et al., 1991a). In a prospective study in a cohort of British children at risk of allergic disease because of family history, Sporik and colleagues (1990) reported a trend toward sensitization to dust mite allergen by age 11 with exposure at age 1 to more than 10 µg of *Der p* I/g of dust. The age at which wheezing first occurred was inversely related to the level of exposure at age 1 for those children who became sensitized.

These reports suggest strongly that in addition to genetic factors, exposure to certain allergens in early childhood is an important determinant of subsequent development of respiratory diseases such as asthma. If this is indeed the case, avoidance of indoor allergens and the role of education in promoting avoidance measures in the general public must necessarily assume a high priority. When public education is undertaken, however, it must be based on specific information about which allergens are troublesome in which circumstances, and for which children.

Not all studies have demonstrated beneficial effects on allergic diseases from allergen avoidance (Burr et al., 1980; Gillies et al., 1987; Korsgaard, 1982). In these investigations, however, the avoidance procedures were less stringent than those employed in the "successful" studies cited above. Some argue that it may be difficult to convince parents of children with mild to moderate asthma that more stringent methods are warranted.

Flaws in the First-Generation Programs

The most successful educational and self-management programs have attempted to integrate the expertise of medical and behavioral scientists. Analysis of data from these programs shows significant differences between experimental and control groups and pre-/postintervention within-group changes. This suggests that behavioral techniques are being used by patients to help manage their asthma.

From a behavioral point of view, first-generation asthma education and self-management programs had three major flaws. First, despite the data that have been obtained, knowledge is limited as to exactly what skills patients performed to help control their asthma. As noted by Kanfer and Schefft (1988), "The effectiveness of self-management therapies must be judged on process measures, not just product or outcome measures."

The second flaw was the failure of first-generation programs to incorporate what was then the current state of the art with respect to psychological techniques, into intervention protocols (Thoresen and Kirmil-Gray, 1983). The changes that have occurred in the dimensions of behavioral science over the past two decades (Creer et al., 1992; Sulzer-Azaroff and Mayer, 1991) require the integration of this knowledge into future programs.

The final flaw noted in earlier education and self-management (first generation) programs was their failure to incorporate proven methods and techniques of self-management. A strong experimental foundation underlies self-management methods that have been applied to change a number of health-related behaviors, including substance abuse, smoking, and overeating, and to treat chronic disorders, including diabetes, hypertension, recurrent headache, and chronic pain (Holroyd and Creer, 1986). With few exceptions, asthma education and self-management programs emphasized instruction provided to patients and assessed the output or outcome measures. They failed to analyze, on a point-by-point basis, exactly what skills or changes in behavior were incorporated by patients into their asthma management regimen.

EDUCATING HEALTH CARE PROVIDERS

Allergy curricula in medical schools vary significantly. Often the topic of allergy receives little attention—an unfortunate deficiency considering the relationship of allergy to asthma. The high incidence of asthma, the serious impairment that can result from it, and the rising prevalence and mortality rates clearly show the need for broad education regarding its diagnosis and appropriate management. When one considers that a significant percentage of hospital admissions for asthma can be prevented by educating physicians and patients in the proper control measures, the need for an emphasis on education about asthma becomes obvious.

The education of the patient often comes primarily from the physician at the time of diagnosis and initiation of appropriate therapeutic regimens. Valuable components of this information include the nature of the allergic disease, the allergens responsible, appropriate measures for avoidance of allergens (when avoidance is possible), and optimum use of pharmacotherapy. Patients must also learn that, as in any chronic disease, there are unproven forms of therapy that are useless and sometimes expensive, and that employing such therapies may postpone the use of more appropriate and effective diagnostic and therapeutic regimens.

Studies show that when physicians' management and counseling behavior changes so does the behavior of their patients, and patients' health status improves (Inui et al., 1976; Maiman et al., 1988). Information is limited, however, and further research is needed to clarify (1) how physicians counsel and educate patients regarding allergens, and (2) how effective this is in terms of patient avoidance of allergens.

In general, less is known about the impact of providing education to health care providers than to patients and the general public. The recently developed *Guidelines for the Diagnosis and Management of Asthma* (NHLBI, 1991), however, emphasizes the importance of a partnership between the patient's family and the clinician. This partnership can be instrumental in helping patients understand asthma, as well as in learning and practicing the skills necessary to manage asthma. Evaluation of the effects of this national education program for physicians and health care professionals awaits its widespread dissemination and acceptance in the medical community.

Attempts to evaluate both the knowledge and behavior patterns of clinicians are critical to asthma management; however, developing outcome measures for such evaluations presents a challenge and deserves further research. In one negative study conducted by mailed questionnaire in which the outcomes were morbidity experienced by patients and their reported use of asthma-specific drugs, small-group education of general practitioners in the management of asthma was not effective (P. T. White et al., 1989). There was no difference between the morbidity of patients treated by the group of physicians receiving education and the group of patients treated by the control group. The deficiencies of the study included the use of mailed questionnaires, which are highly subjective; the lack of objective outcome measures; and the lack of direct evaluation of the subjects by the physicians. Studies such as this emphasize the need for validated outcome measures to evaluate the effects of educational interventions for physicians and other health care professionals.

The basis for early recognition of allergic disease and development of a plan for therapeutic management depends on the appropriate education of the physician. The majority of health care of the allergic patient will be delivered by primary care providers who are also pediatricians, internists, or family practitioners. For more serious cases, a subspecialist in allergy-immunology (for allergy and asthma), or pulmonology (for asthma) may be required. Thus, allergic diseases should be emphasized at several levels of medical education. The mechanisms of allergic diseases should be taught during the basic science years of medical school, and the diagnosis and therapy of allergic diseases should be emphasized during the clinical years. Postgraduate training in family practice, pediatrics, and internal medicine should include the diagnosis and management of allergic disease, because of the high incidence and prevalence of these medical problems. Finally, fellowships in allergy and immunology provide the basis for subspecialty practice and of future faculty.

CONCLUSIONS AND RECOMMENDATIONS

A plethora of education and self-management programs exists for asthma and allergy. These programs are readily available to anyone through federal agencies such as NHLBI, or private groups such as the Asthma and Allergy Foundation of America, the American Lung Association, and pharmaceutical companies. Despite methodological weaknesses in the initial application of some of these programs, many have produced a number of positive results.

In the future, there should be closer linkage between the acquisition of knowledge about asthma self-management and the subsequent performance of these skills. This goal can be achieved by combining available educational components with additional elements as required to tailor specific programs to individual patients, who have varying degrees of severity of disease and are sensitized to different allergens. The result of such an approach should be not only more germane educational materials for patients but a reduction in the amount of information they need to learn and remember. Because memory repeatedly has been shown to be a significant factor in decisionmaking, particularly with an intermittent condition such as asthma, emphasizing basic self-management skills leads to better asthma management decisions. Better decisionmaking, in turn, enhances the performance of self-management skills (Creer, 1990).

There are few scientifically tested educational tools and materials designed to teach physicians and other health care providers about allergic diseases and asthma, although numerous meetings, symposia, and other presentations address the role and contribution of indoor allergens to health problems and the importance of reducing exposure to these agents. There have also been few attempts to assess the long-term effectiveness of these materials or programs. As noted, *Guidelines for the Diagnosis and Management of Asthma* (NHLBI, 1991) presents not only detailed recommendations to diagnose and manage asthma but also emphasizes the importance of education and of identifying causative agents, such as indoor allergens, that may initiate and exacerbate asthma symptoms. Evaluation of the impact of these guidelines await their wide dissemination and acceptance. Meanwhile, efforts should continue toward improving what is already known to be effective in the prevention and control of allergic disease and disseminating that knowledge to health care providers, allergy patients, and the public at large.

Education is an important component in the prevention and control of allergen-induced diseases. Considering that a large percentage of hospital admissions for asthma can be prevented by educating physicians and patients in the proper control measures, the need for emphasis on education becomes obvious. By disseminating information to physicians, to health care providers, to patients, and to building design, construction and operations professionals (discussed in Chapter 7), prevention of diseases associated with indoor allergens becomes not only realistic, but may offer a cost-effective means of reducing morbidity.

Patients

In developing and implementing educational interventions, consideration should be given to identifying populations such as those with severe asthma who are more motivated and more likely to benefit from intervention.

Recommendation: Identify population groups most likely to benefit from educational and allergen-avoidance interventions. This effort should be based on an understanding of what allergens serve as risk factors for different individuals.

Socioeconomic, educational, and ethnic characteristics are important variables that should be considered in developing effective educational intervention programs. Programs that focus on these factors in tailoring self-management programs should greatly enhance both the acquisition and the performance of self-management competencies.

Recommendation: Develop focused educational programs for allergic populations with different socioeconomic and educational characteristics. Such programs should help patients:

- understand allergic-disease risk factors;
- predict the occurrence of such risk factors;
- adopt behaviors required to avoid or control these factors; and
- develop self-management skills to translate and use the knowledge they acquire to control allergic risk factors in different contexts.

A relapse prevention component should be included in these programs as well as follow-up studies to assess patient acquisition of allergy-related knowledge and the need for additional educational efforts.

Health Care Providers

Curricula vary in medical schools, often with little focus given to the topic of allergy diagnosis, prevention, and control—an unfortunate situation that should be corrected, especially considering the relationship of allergy to asthma. In addition, improved medical education is important because the majority of health care of the allergic patient is delivered by primary care providers, and the primary care provider is often the patient's main source of information about allergy control.

Recommendation: Incorporate the diagnosis and management of allergic diseases in the curricula and training materials for medical school students, residents in primary care practice, and subspecialists who will subsequently care for patients with allergen-based allergic disease. Nurses, physician assistants, and other non-physician health care providers should receive similar education and training.

Allergic disease should receive additional emphasis at all levels of medical education, across specialties, and in clinical practice. One mechanism to help promote this concept would be to enlist the support and interest of scientific and medical societies.

Recommendation: Encourage scientific societies with expertise in allergy, pulmonary medicine, public health, and occupational and environmental medicine to continue to assess and promote the development of primary prevention strategies for allergic disease.

Engineers, Architects, and Building Maintenance Personnel

As discussed in chapter 7, concerns about the design and operation of heating, ventilation, and air conditioning systems have focussed traditionally on the comfort of the building occupants and the efficiency of the operation of the equipment. It is important that those with responsibility for the design, construction, and maintenance of buildings also have an understanding of the potential health effects associated with indoor environments, and the impact that design and operation of the systems can have on those effects.

Recommendation: Educate those with responsibilities for the design and maintenance of indoor environments about the magnitude and severity of diseases caused by indoor allergens.

Engineers, architects, contractors, and building maintenance personnel receive limited if any education about the health implications of the design, construction, and maintenance of buildings. Improved education in these areas is important to reducing the incidence, prevalence, and severity of adverse health effects associated with indoor environmental exposures.

Recommendation: Develop educational processes and accountability procedures for architects, engineers, contractors, and building maintenance personnel with respect to the health implications of the design, construction, and operation of buildings.

An interdisciplinary approach to the prevention and control on the adverse health effects associated with indoor exposures, including indoor allergens, is important. Such an approach should improve education in all areas of expertise and result in reduced health risks for building occupants.

Recommendation: Develop interdisciplinary educational programs for health care and building design, construction, and operations professionals.

References

ACGIH (American Conference of Governmental Industrial Hygienists). 1986. Documentation of the Threshold Limit Values and Biologic Exposure Indices, 5th ed. Cincinnati, Ohio: ACGIH.

ACGIH. 1989. Air sampling instruments for evaluation of atmospheric contaminants. In: American Conference of Governmental Industrial Hygienists, 7th ed. Cincinnati, Ohio: ACGIH.

Adams, K. F., H. A. Hyde, and D. A. Williams. 1968. Woodlands as a source of allergens, with special reference to basidiospores. Acta Allergol 23(3):265–281.

Adkinson, N. F. 1986. Measurement of total serum immunoglobulin E and allergen specific immunoglobulin E antibody. In: Manual of Clinical Laboratory Immunology, N. R. Rose, H. Friedman, and J. L. Fahey, eds. Washington, D.C.: American Society for Microbiology, pp. 664–674.

Ager, B. P., and J. A. Tickner. 1983. The control of microbiological hazards associated with air-conditioning and ventilation systems. Annals of Occupational Hygiene 27(4):341–358.

Ahman, M. 1992. Nasal peak flow rates in work related nasal blockage. Acta Oto-Laryngologica 112:839–844.

Aisner, J., S. Schimpff, J. E. Bennett, V. M. Young, and P. H. Wiernik. 1976. Aspergillus infections in cancer patients: Association with fireproofing materials in a new hospital. Journal of the American Medical Association 235:411–412.

Akimenko, V. V., I. Andersen, M. D. Lebowitz, and T. Lindvall. 1986. The "sick building" syndrome. In: Proceedings of the Third International Conference on Indoor Air Quality and Climate, vol. 6, B. Berglund, T. Lindvall, and J. Sundell, eds. Stockholm, Sweden: Swedish Council for Building Research, pp. 87–97.

Allen, G. I., L. Breslow, A. Weissman, and H. Nisselson. 1954. Interviewing versus diary keeping in eliciting information in a morbidity survey. American Journal of Public Health 44(7):919.

Altman, L. C. 1984. Clinical Allergy and Immunology. Boston, Mass.: G. K. Hall Medical Publishers.

Andersen, A. A. 1958. New sampler for the collection, sizing, and enumeration of viable airborne particles. Journal of Bacteriology 76:471–484.

Andersen, I., and J. Korsgaard. 1986. Asthma and the indoor environment: Assessment of health implications of high indoor relative humidity. Environment International 12:121–127.

Anderson, R. L., D. C. Mackel, B. S. Stoler, and G. F. Mallison. 1982. Carpeting in hospitals: An epidemiological evaluation. Journal of Clinical Microbiology 15(3):408–415.

Antonicelli, L., M. B. Biol, S. Pucci, C. Schou, and F. Bonifazi. 1991. Efficacy of an air-cleaning device equipped with a high efficiency particulate air filter in house dust mite respiratory allergy. Allergy 46:594–600.

Arbeiter, H. I. 1967. How prevalent is allergy among United States school children? Clinical Pediatrics 6:140–142.

Arkes, H. R. 1981. Impediments to accurate clinical judgment and possible ways to minimize their impact. Journal of Consulting and Clinical Psychology 49:323–330.

Arlian, L. G., I. L. Bernstein, and J. S. Gallagher. 1982. The prevalence of house dust mites, *Dermatophagoides* spp. and associated environmental conditions in homes in Ohio. Journal of Allergy and Clinical Immunology 69:527–532.

Arnow, P. M., R. L. Anderson, D. Maninous, and E. J. Smith. 1978. Pulmonary aspergillosis during hospital renovation. American Review of Respiratory Disease 118:49–53.

Arruda, L. K., T. Platts-Mills, J. W. Fox, and M. D. Chapman. 1990. Aspergillus fumigatus Allergen I, a major IgE-binding protein, is a member of the mitogillin family of cytotoxins. Journal of Experimental Medicine 172:1529–1532.

Arruda, L. K., M. C. Rizzo, M. D. Chapman, E. Fernandez-Caldas, D. Baggio, T. Platts-Mills, and C. K. Naspitz. 1991. Exposure and sensitization to dust mite allergens among asthmatic children in São Paulo, Brazil. Clinical and Experimental Allergy 21:433–439.

ASHRAE (American Society of Heating, Refrigerating, and Air-conditioning Engineers). 1976. Standard 52-1968(RA 76). Method of testing air cleaning devices used in general ventilation for removing particulate matter. Atlanta, Ga.: ASHRAE

ASHRAE. 1981. Public review draft 1981. Thermal environmental conditions for human occupancy. Atlanta, Ga.: ASHRAE

ASHRAE. 1989. Fundamental Handbook. In: Thermal Insulation and Vapor Retarders-Fundamentals, Atlanta, Ga.: ASHRAE, chap. 20.

ASHRAE. 1989a. ASHRAE Handbook: Fundamentals, Robert A. Parsons, ed. New York, N.Y.: ASHRAE.

ASHRAE. 1989b. Standard 62-1989: Ventilation for Acceptable Indoor Air Quality. Atlanta, Ga.: ASHRAE.

ASHRAE. 1993. Standard 55-1992. Thermal environmental conditions for human occupancy. Atlanta, Ga.: ASHRAE

ATS (American Thoracic Society). 1987. Standardization of spirometry: 1987 update. Official statement of the American Thoracic Society. American Review of Respiratory Disease 136:1285–1298.

ATS. 1991. Statement on lung function testing: Selection of reference values and interpretive strategies. American Review of Respiratory Disease 144:1202–1218.

Aukrust, L., and S. Borch. 1979. Partial purification and characterization of two *Cladosporium herbarum* allergens. International Archives of Allergy and Applied Immunology 60:68.

Axelsson, I. G., S. G. Johansson, and K. Wrangsjo. 1987a. IgE-mediated anaphylactoid reactions to rubber. Allergy 42:46–50.

Axelsson, I. G., S. G. Johansson, and O. Zetterstrom. 1987b. A new indoor allergen from a common non-flowering plant. Allergy 42(8):604–611.

Axelsson, I. G., S. G. Johansson, P. H. Larsson, and O. Zetterstrom. 1991. Serum reactivity to other indoor ficus plants in patients with allergy to weeping fig (*Ficus benjamina*). Allergy 46(2):92–98.

Baldo, B. A., and R. S. Baker. 1988. Inhalant allergies to fungi: reactions to bakers' yeast (*Saccharomyces cerevisiae*) and identification of Bakers' yeast enolase as an important allergen. International Archives of Allergy and Applied Immunology 86:201–208.

Banaszak, E. F., W. H. Thiede, and J. N. Fink. 1970. Hypersensitivity pneumonitis due to contamination of an air conditioner. New England Journal of Medicine 283:271.

Bandura, A. 1986. Social Foundation of Thought and Action: A Social Cognitive Theory. Englewood Cliffs, N. J.: Prentice-Hall.

Barbee, R., M. D. Lebowitz, H. Thompson, and B. Burrows. 1976. Immediate skin-test reactivity in a general population sample. Annals of Internal Medicine 84(2):129–133.

Barbee, R. A., R. Dodge, M. D. Lebowitz, and B. Burrows. 1985. The epidemiology of asthma. Chest 87S:21S–25S.

Barbee, R. A., W. Kaltenborn, M. D. Lebowitz, and B. Burrows. 1987. Longitudinal changes in allergen skin test reactivity in a community population sample. Journal of Allergy and Clinical Immunology 79:16–24.

Barneston, R. S., T. G. Merrett, and A. Ferguson. 1981. Studies on hyperimmunoglobulinaemia E in atopic diseases with particular reference to food allergens. Clinical and Experimental Immunology 46:54–60.

Bascom, R., D. Proud, A. G. Togias, S. P. Peters, P. S. Norman, A. Kagey-Sobotka, L. M. Lichtenstein, and R. M. Naclerio. 1986. Nasal provocation: An approach to study the mediators of allergic and non-allergic rhinitis. In: XII International Congress of Allergology and Clinical Immunology, 20–25 October at Washington, D.C. Mt. Prospect, Ill.: American College of Allergists, pp. 113–120.

Bascom, R., R. M. Naclerio, T. K. Fitzgerald, A. Kagey-Sobotka, and D. Proud. 1990. Effect of ozone inhalation on the response to nasal challenge with antigen of allergic subjects. American Review of Respiratory Disease 141:594–601.

Bates, D. Y., and R. Sitzo. 1987. Air pollution and hospital admissions in Southern Ontario: The acid summer haze effect. Environmental Research 43:317–331.

Baur, X., and G. Fruhmann. 1981. Specific IgE antibodies in patients with isocyanate asthma. Chest 80:73–76.

Baur, X., J. Behr, M. Dewair, W. Ehret, G. Fruhmann, C. Vogelmeier, W. Weiss, and V. Zinkernagel. 1988. Humidifier lung and humidifier fever. Lung 166(2):113–124.

Baxter, C. S., H. E. Wey, and W. R. Burg. 1981. A prospective analysis of the potential risk associated with inhalation of aflatoxin-contaminated grain dusts. Food and Cosmetics Toxicology 19:763–769.

Beall, G. N. 1983. Allergy and Clinical Immunology. New York, N.Y.: Wiley.

Beasley, R., R. Roche, J. A. Roberts, and S. T. Holgate. 1989a. Cellular events in the bronchi in mild asthma and after bronchial provocation. American Review of Respiratory Disease 139:806–817.

Beasley, R., M. Cushley, and S. T. Holgate. 1989b. A self-management plan in the treatment of adult asthma. Thorax 44:200–204.

Beaumont, F., H. F. Kauffman, H. J. Sluiter, and K. DeVries. 1985. Sequential sampling of fungal air spores inside and outside the homes of mould-sensitive, asthmatic patients: A search for a relationship to obstructive reactions. Annals of Allergy 55(5):740–746.

Bernstein, D. I., and C. R. Zeiss. 1989. Guidelines for preparation and characterization of chemical-protein conjugate antigens. Journal of Allergy and Clinical Immunology 84:820–822.

Bernstein, D. I., R. Patterson, and C. R. Zeiss. 1982. Clinical and immunologic evaluation of trimellitic anhydride and phthalic anhydride exposed workers using a questionnaire with comparative analysis of enzyme-linked immunosorbent and radioimmunoassay studies. Journal of Allergy and Clinical Immunology 69:311–318.

Bernstein, D. I., J. S. Gallagher, L. D'Souza, and I. L. Bernstein. 1984. Heterogeneity of specific IgE responses in workers sensitized to acid anhydride compounds. Journal of Allergy and Clinical Immunology 74(6):794–801.

Bernstein, I. L. 1972. Enzyme allergy in populations exposed to long-term, low-level concentrations of household laundry products. Journal of Allergy and Clinical Immunology 49(4):219–237.

Bernstein, I. L. 1981. Occupational asthma. Clinics in Chest Medicine 2:255–272.

Bernstein, I. L. 1982. Isocyanate-induced pulmonary diseases: A current perspective. Journal of Allergy and Clinical Immunology 70:24–31.

Bernstein, I. L. 1988. Proceedings of the task force on guidelines for standardizing old and new technologies used for the diagnosis and treatment of allergic diseases. Journal of Allergy and Clinical Immunology 82:487–526.

Bernton, H. S., T. F. McMahon, and H. Brown. 1972. Cockroach asthma. British Journal of Diseases of the Chest 66:61–66.

Biagini, R. E., S. L. Klincewicz, G. M. Henningsen, B. A. MacKenzie, J. S. Gallagher, D. I. Bernstein, and I. L. Bernstein. 1990. Antibodies to morphine in workers exposed to opiates at a narcotics manufacture facility and evidence for similar antibodies in heroin abusers. Life Sciences 47:897–908.

Bielory, L., and M. A. Kaliner. 1985. Anaphylactoid reactions to radiocontrast materials. International Anesthesiology Clinics 23:97–118.

Bierman, C. W., and D. S. Pearlman. 1988. Allergic Diseases from Infancy to Adulthood, 2nd ed. Philadelphia, Pa.: W. B. Saunders Company.

Bischoff, E., A. Fischer, and B. Liebenberg. 1990. Assessment and control of house dust mite infestation. Clinical Therapy 12:216–220.

Bisset, J. 1987. Fungi associated with urea-formaldehyde foam insulation in Canada. Mycopathologia 99(1):47–56.

Blackley, C. H. 1873. Experimental Researches on the Causes and Nature of Catarrhus Aestivus (hay fever, hay asthma). London, England: Balliere, Tindall & Cox.

Block, G. T., and M. Chan-Yeung. 1982. Asthma induced by nickel. Journal of the American Medical Association 247:1600–1603.

Boey, H., R. Rosenbaum, J. Castracane, and L. Borish. 1989. Interleukin-4 is a neutrophil activator. Journal of Allergy and Clinical Immunology 83:978.

Borish, L., J. J. Mascali, and L. J. Rosenwasser. 1991. IgE-dependent monokine production by human peripheral blood mononuclear phagocytes. Journal of Immunology 146:63.

Bornehag, C. G. 1991. Problems associated with the replacement of casein-based self-leveling compound. In: IAQ '91 Healthy Buildings, Atlanta, Ga.: American Society of Heating, Refrigerating and Air-Conditioning Engineers (ASHRAE), pp. 273–275.

Boushey, H., M. J. Holtzman, J. R. Sheller, and J. A. Nadel. 1980. State of the art: Bronchial hyperreactivity. American Review of Respiratory Disease 121:389–413.

Bousquet, J. 1988. In vivo methods for study of allergy: Skin tests, techniques, and interpretation. In: Allergy Principles and Practice, 3rd ed., E. Middleton, Jr., C. E. Reed, E. F. Ellis, N. F. Adkinson, Jr., and J. W. Yunginger, eds. St. Louis, Mo.: C. V. Mosby Company, pp. 419–436.

Boxer, M. B., L. C. Grammer, K. E. Harris, D. E. Roach, and R. Patterson. 1987. Six-year clinical and immunologic follow-up of workers exposed to trimellitic anhydride. Journal of Allergy and Clinical Immunology 80:147–152.

Bresnitz, E. A., and K. M. Rest. 1988. Epidemiologic studies of effects of oxidant exposure on human populations. In: Air Pollution, the Automobile, and Public Health, A. Y. Watson, R. R. Bates, and D. Kennedy, eds. Washington, D.C.: National Academy Press, pp. 389–413.

Britton, W. J., A. J. Woolcock, J. K. Peat, C. J. Sedgewick, D. M. Lloyd, and S. R. Leeder. 1986. Prevalence of bronchial hyperresponsiveness in children: The relationship between asthma and skin reactivity to allergens in two communities. International Journal of Epidemiology 15:202–209.

Broder, I. P., P. Barlow, and R. J. M. Horton. 1962. The epidemiology of asthma and hay fever in a total community, Tecumseh, Michigan. I. Description of study and general findings. Journal of Allergy and Clinical Immunology 33:513–523.

Brown, J., T. Jardetzky, M. Saper, B. Samraoui, P. Njorkman, and D. Wiley. 1988. A hypothetical model of the foreign antigen combining site of class II histocompatibility complex molecule. Nature 332:845–853.

Bruijnzeel-Koomen, C. A. F. M., D. F. van Wichen, C. J. F. Spry, P. Venge, and P. L. B. Bruijnzeel. 1988. Active participation of eosinophils in patch test reactions to inhalant allergens in patients with atopic dermatitis. British Journal of Dermatology 118:229–238.

Brunekreef, B., D. W. Dockery, F. E. Speizer, J. H. Ware, J. D. Spengler, and B. J. Ferris. 1989. Home dampness and respiratory morbidity in children. American Review of Respiratory Disease 140:1363–1367.

Brunekreef, B., L. de Rijk, A. L. Verhoeff, and R. Samson. 1990. Classification of dampness in homes. In: Indoor Air '90: The Fifth International Conference on Indoor Air Quality and Climate, vol. 2. Toronto, Canada: Canada Mortgage and Housing Corp., pp. 15–20.

Buckley, J. M., and D. S. Pearlman. 1988. Controlling the environment for allergic diseases. In: Allergic Diseases from Infancy to Adulthood, 2nd ed., C. W. Bierman and D. S. Pearlman, eds. Philadelphia, Pa.: W. B. Saunders Company, pp. 239–252.

Burge, H. 1990. Bioaerosols: Prevalence and health effects in the indoor environment. Journal of Allergy and Clinical Immunology 86:687–705.

Burge, H., and T. Platts-Mills. 1991. Indoor Biological Aerosols. Research Triangle Park, N.C.: Office of Health and Environmental Assessment. U.S. Environmental Protection Agency.

Burge, H., W. R. Solomon, and J. R. Boise. 1977. Comparative merits of eight popular media in aerometric studies of fungi. Journal of Allergy and Clinical Immunology 60(3)199–203.

Burge, H., W. R. Solomon, and J. R. Boise. 1980. Microbial prevalence in domestic humidifiers. Applied and Environmental Microbiology 39(4):840–844.

Burge, H., M. E. Hoyer, W. R. Solomon, E. G. Simmons, and J. Gallup. 1989. Quality control factors for Alternaria allergens. Mycotaxon 34(1):55–63.

Burge, H., M. L. Muilenberg, and J. Chapman. 1991. Crop plants as a source for medically important fungi. In: Microbial Ecology of Leaves, J. Andrews and S. Hirano, eds. New York, N.Y.: Springer-Verlag, pp. 222–236.

Burr, M. L., B. V. Dean, T. G. Merrett, E. Neale, A. S. St. Leger, and E. R. Verrier-Jones. 1980. Effects of anti-mite measures on children with mite-sensitive asthma: A controlled trial. Thorax 35:506–512.

Burrows, B., and M. D. Lebowitz. 1992. The Beta-agonist dilemma [editorial]. New England Journal of Medicine 326:560–561.

Burrows, B., and F. D. Martinez. 1989. Bronchial responsiveness, atopy, smoking, and chronic obstructive pulmonary disease. American Review of Respiratory Disease 140:1515–1517.

Burrows, B., M. D. Lebowitz, and R. Barbee. 1976. Respiratory disorders and allergy skin-test reaction. Annals of Internal Medicine 84(2):134–139.

Burrows, B., F. M. Hasan, R. A. Barbee, M. Halonen, and M. D. Lebowitz. 1980. Epidemiological observations on eosinophilia and its relationship to respiratory disorders. American Review of Respiratory Disease 122:709–719.

Burrows, B., F. D. Martinez, M. Halonen, R. A. Cline, and M. G. Barbee. 1989. Association of asthma with serum IgE levels and skin-test reactivity to allergens. New England Journal of Medicine 320:271–277.

Burrows, B., M. D. Lebowitz, R. A. Barbee, and M. G. Cline. 1991. Findings prior to diagnoses of asthma among the elderly in a longitudinal study of a general population sample. Journal of Allergy and Clinical Immunology 88(6):870–877.

Bush, R. K., and S. L. Kagen. 1989. Guidelines for the preparation and characterization of high molecular weight allergens used for the diagnosis of occupational lung disease. Journal of Allergy and Clinical Immunology 84:814–819.

Butcher, B. T., R. N. Jones, C. E. O'Neil, H. W. Glindmeyer, J. E. Diem, V. Dharmarajan, H. Weill, and J. E. Salvaggio. 1977. Longitudinal study of workers employed in the manufacture of toluene diisocyanate. American Review of Respiratory Disease 116:411–421.

Butcher, B. T., R. M. Karr, C. E. O'Neil, M. R. Wilson, and V. Dharmarajan. 1979. Inhalation challenge and pharmacologic studies of TDI workers. Journal of Allergy and Clinical Immunology 64:146–152.

Butcher, B. T., C. E. O'Neil, M. A. Reed, and J. E. Salvaggio. 1980. Radioallergosorbent testing of toluene diisocyanate-reactive individuals using p-tolyl isocyanate antigen. Journal of Allergy and Clinical Immunology 66:213–216.

Butcher, B. T., I. L. Bernstein, and H. J. Schwartz. 1989. Guidelines for the clinical evaluation of occupational asthma due to small molecular weight chemicals. Journal of Allergy and Clinical Immunology 84:834–838.

Call, R. S., T. F. Smith, E. Morris, M. D. Chapman, and T. A. E. Platts-Mills. 1992. Risk factors for asthma in inner city children. Journal of Pediatrics 121:862–866.

Calley, A. C. 1979. Some aspects of pollen analysis in relation to archaeology. Kiya 44:95–100.

Campbell, A. R., M. C. Swanson, E. Fernandez-Caldas, C. E. Reed, J. J. May, and D. S. Pratt. 1989. Aeroallergens in dairy barns near Cooperstown, New York, and Rochester, Minnesota. American Review of Respiratory Disease 140(2):317–320.

Carrillo, T., M. Cuevas, T. Muñoz, M. Hinojosa, and I. Moneo. 1986. Contact urticaria and rhinitis from latex surgical gloves. Contact Dermatitis 15:69–72.

CFR (Code of Federal Regulations). 1987. Labor 29: 1910, 1915, 1917, 1918, 1926, and 1928. Hazard Communication, Final Rule, Occupational Safety and Health Administration. Washington, D.C.: U.S. Government Printing Office.

CFR. 1991. Labor 29: 1900–1910. Washington, D.C.: U.S. Government Printing Office.

Chai, H., R. S. Farr, F. Luza, D. A. Mathison, J. A. McLean, R. R. Rosenthal, A. L. Sheffer, S. L. Spector, and R. G. Townley. 1975. Standardization of bronchial inhalation challenge procedures. Journal of Allergy and Clinical Immunology 56: 323–327.

Chan-Yeung, M. 1990. A clinician's approach to determine the diagnosis, prognosis, and therapy of occupational asthma. Medical Clinics of North America 74:811–822.

Chan-Yeung, M., and S. Lam. 1986. Occupational asthma: State of the art. American Review of Respiratory Disease 133:686–703.

Chan-Yeung, M., S. Vedal, S. Lam, et al. 1985. Immediate skin test reactivity and its relationship to age, sex, smoking, and occupational exposure. Archives of Environmental Health 40:53–57.

Chapman, M. D., and T. Platts-Mills. 1980. Purification and characterization of the major allergen from *Dermatophagoides pteronyssinus*-antigen P1. Journal of Immunology 125:587–592.

Chapman, M. D., S. Rowntree, E. B. Mitchell, M. C. Di Prisco de Fuenmajor, and T. Platts-Mills. 1983. Quantitative assessments of IgG and IgE antibodies to inhalant allergens in patients with atopic dermatitis. Journal of Allergy and Clinical Immunology 72:27–33.

Chapman, M. D., W. M. Sutherland, and T. Platts-Mills. 1984. Recognition of two *Dermatophagoides pteronyssinus*-specific epitopes on antigen P1 by using monoclonal antibodies: Binding to each epitope can be inhibited by serum from dust mite-allergic patients. Journal of Allergy and Clinical Immunology 133:2488–2495.

Charpin, C., P. Mata, D. Charpin, M. N. Lavant, C. Allasia, and D. Vervloet. 1991. *Fel d* I allergen distribution in cat fur and skin. Journal of Allergy and Clinical Immunology 88:77–82.

Charpin, D., J. P. Kleisbauer, A. Lanteaume, H. Razzouk, D. Vervloet, M. Toumi, F. Faraj, and J. Charpin. 1988a. Asthma and allergy to house-dust mites in populations living in high altitudes. Chest 93:758–761.

Charpin, D., J. P. Kleisbauer, A. Lanteaume, D. Vervloet et al. 1988b. Is there an urban factor in asthma and allergy? [in French]. Revue Des Maladies Respiratories (Paris) 5:109–114.

Charpin, D., J. Birnbaum, E. Haddi, G. Genard, M. Toumi, and D. Vervloet. 1990a. Altitude and allergy to house dust mites: An epidemiological study in primary school children. Journal of Allergy and Clinical Immunology 85:185.

Charpin, D., J. Birnbaum, E. Haddi, A. N'Guyen, J. Fondarai, and D. Vervloet. 1990b. Evaluation of the acaricide ACARDUST in the treatment of dust mite allergens [in French]. Revue Française Allergologique 30:149.

Charpin, D., J. Birnbaum, E. Haddi, G. Genard, A. Lanteaume, M. Toumi, F. Faraj, K. Van der Brempt, and D. Vervloet. 1991. Altitude and allergy to house-dust mites: A paradigm of the influence of environmental exposure on allergic sensitization. American Review of Respiratory Disease 143:983–986.

Chatham, M. E., E. R. Bleecker, P. L. Smith, R. R. Rosenthal, P. Mason, and P. S. Norman. 1982a. A comparison of histamine, methacholine, and exercise airways reactivity in normal and asthmatic subjects. American Review of Respiratory Disease 126:235–240.

Chatham, M., E. R. Bleecker, P. Mason, P. L. Smith, and P. Norman. 1982b. A screening test for airways reactivity: An abbreviated methacholine inhalation challenge. Chest 82:15–18.

Chester, E. H., and H. J. Schwartz. 1979. Study session on occupational asthma. Journal of Allergy and Clinical Immunology 64:665–666.

Chida, T. 1986. A study on dose-response relationship of occupational allergy in a pharmaceutical plant. Japanese Journal of Industrial Health 28:77–86.

Christensen, L. T., C. D. Schmidt, and L. Robbins. 1975. Pigeon breeder's disease: A prevalence study and review. Clinical Allergy 5:417–430.

Chua, K. Y., G. A. Stewart, W. R. Thomas, R. J. Simpson, R. J. Dilworth, T. M. Plozza, and K. J. Turner. 1988. Sequence analysis of cDNA coding for a major house dust mite allergen, Der p I: Homology with cysteine proteases. Journal of Experimental Medicine 167:175–182.

Chua, K. Y., C. R. Doyle, R. J. Simpson, K. J. Turner, G. A. Stewart, and W. R. Thomas. 1990. Isolation of cDNA coding for the major mite allergen Der p II by IgE plaque immunoassay. International Archives of Allergy and Applied Immunology 91:118–123.

Clark, R. P., D. C. Cordon-Nesbitt, S. Malka, T. D. Preston, and L. Sinclair. 1976. The size of airborne dust particles precipitating bronchospasm in house dust sensitive children. Journal of Hygiene (Cambridge) 77:321–325.

Clarke, C. W., and P. W. Aldons. 1979. The nature of asthma in Brisbane. Clinical Allergy 9:147–152.

Clutterbuck, E. J., E. M. A. Hirst, and C. J. Sanderson. 1989. Human interleukin 5 (IL-5) regulates the production of eosinophils in human bone marrow cultures: Comparison and interaction with IL-1, IL-3, IL-6 and GMCSF. Blood 73:1504.

Cockroft, D. W., and F. G. Horgreave. 1990. Airway hyperresponsiveness. Relevance of random population data to clinical usefulness. American Review of Respiratory Disease 142:497–500.

Cockroft, D. W., K. Y. Murdock, and B. A. Berscheid. 1984. Relationship between atopy and bronchial responsiveness to histamine in a random population. Annals of Allergy 53:26–29.

Cohen, A. A., S. Bromberg, R. W. Buechley, L. T. Heiderscheit, and C. M. Shy. 1972. Asthma and air pollution from a coal-fueled power plant. American Journal of Public Health 62:1181–1188.

Cole, P. 1982. Upper respiratory airflow. In: The Nose, Upper Airway Physiology and the Atmospheric Environment, D. F. Proctor and I. Andersen, eds. New York, N.Y.: Elsevier Biomedical.

Cone, J. E., and M. J. Hodgson, eds. 1989. Problem Buildings: Building-Associated Illness and the Sick Building Syndrome. Occupational Medicine State of the Art Review 4(4):10–12.

Cooke, R. A., J. H. Barnard, S. Hebald, and A. Stull. 1935. Serologic evidence of immunity with coexisting sensitization in a type of human allergy [hayfever]. Journal of Experimental Medicine 62:733–750.

Crapo, R. O., and A. H. Morris. 1981. Standardized single breath normal values for carbon monoxide diffusing capacity. American Review of Respiratory Disease 123:185–190.

Crapo, R. O., A. H. Morris, and R. M. Gardner. 1981. Reference spirometric values using techniques and equipment that meet ATS recommendations. American Review of Respiratory Disease 123:659–664.

Creer, T. L. 1990. Strategies for judgment and decision-making in the management of childhood asthma. Pediatric Asthma, Allergy, and Immunology 4:253–264.

Creer, T. L., J. K. Wigal, H. Kotses, and P. Lewis. 1990. A critique of 19 self-management programs for childhood asthma. II. Comments regarding the scientific merit of the programs. Pediatric Asthma, Allergy, and Immunology 4:41–55.

Creer, T. L., H. Kotses, and J. K. Wigal. 1992. A second-generation model for asthma self-management. Pediatric Asthma, Allergy, and Immunology 6:143–165.

Cromwell, O., J. Pepys, W. E. Parish, and E. G. Hughes. 1979. Specific IgE antibodies to platinum salts in sensitized workers. Clinical Allergy 9:109–117.

Cutten, A. E., S. M. Hasmain, B. P. Segedin, T. R. Bai, and E. J. McKay. 1988. The basidiomycete Canoderma and asthma: Collection, quantitation and immunogenicity of the spores. New Zealand Medical Journal 101(847, pt. 1):361–363.

Dabrowski, A. J., X. Van Der Brempt, M. Soler, N. Seguret, P. Lucciano, D. Charpin, and D. Vervloet. 1990. Cat skin as an important source of Fel d I allergen. Journal of Allergy and Clinical Immunology 86:462–465.

Dales, R., H. Zwanenburg, and R. Burnett. 1990. Canadian air quality health survey: Influence of home dampness and molds on respiratory health. In: Indoor Air '90: The Fifth International Conference on Indoor Air Quality and Climate. Toronto, Canada: Canada Mortgage and Housing Corp., pp. 145–147.

Dandeau, J. P., J. Le Mao, M. Lux, J. Rabillon, and B. David. 1982. Antigens and allergens in Dermatophagoides farinae mite. II. Purification of Ag 11 a major allergen in Dermatophagoides farinae. Immunology 46:679–687.

Davies, R. J., and J. Pepys. 1975. Asthma due to inhaled chemical agents: The macrolide antibiotic spiramycin. Clinical Allergy 1:99–107.

Davies, R. J., D. J. Hendrick, and J. Pepys. 1974. Asthma due to inhaled chemical agents: Ampicillin, benzyl penicillin, 6 amino penicillanic acid and related substances. Clinical Allergy 4:227–247.

Davies, R. J., A. D. Blainey, and J. Pepys. 1983. Occupational asthma. In: Allergy Principles and Practice, 2nd ed., vol 2., E. Middleton, Jr., C. Reed, and E. Ellis. St. Louis, Mo.: C. V. Mosby Company, pp. 1037–1040.

de Blay, F., M. D. Chapman, and T. Platts-Mills. 1991a. Airborne cat allergen (Fel d I): Environmental control with the cat in situ. American Review of Respiratory Disease 143(6):1334–1339.

de Blay, F., P. W. Heymann, M. D. Chapman, and T. Platts-Mills. 1991b. Airborne dust mite allergens: Comparison of Group II allergens with Group I mite allergen and cat allergen Fel d I. Journal of Allergy and Clinical Immunology 88:919–926.

Decker, H., L. Buchanan, L. Hall, and D. Goddard. 1963. Air filtration of microbial particles. American Journal of Public Health 53:1982–1988.

Decker, J. L., and M. A. Kaliner. 1988. Understanding and Managing Asthma. New York, N.Y.: Avon Books.

de Groot, H., K. G. H. Goei, P. van Swieten, and R. C. Aalberse. 1991. Affinity purification of a major and a minor allergen from dog extract: Serologic activity of affinity-purified Can f I and of Can f I-depleted extract. Journal of Allergy and Clinical Immunology 87:1056–1065.

Dekker, C., R. Dales, S. Bartlett, B. Brunekreef, and H. Zwanenburg. 1991. Childhood asthma and the indoor environment. Chest 100(4):922–926.

Demoly, P., J. Bousquet, J. C. Mandersheid, S. Dreborg, H. Dhivert, and F. B. Michel. 1991. Precision of skin prick and puncture tests with nine methods. Journal of Allergy and Clinical Immunology 88:758–762.

de Saint-Georges-Gridelet, D., F. H. Kniest, G. Schober, A. Penaud, and J. E. M.H. van Bronswijk. 1988. Lutte chimique contre les acariens de la poussière de maison. Notes Préliminaire Revue Française Allergologique 28:131–138.

De Zubiria, A., W. E. Horner, and S. B. Lehrer. 1990. Evidence for cross-reactive allergens among basidiomycetes: Immunoprint-inhibition studies. Journal of Allergy and Clinical Immunology 86(1):26–33.

DHHS (Department of Health and Human Services). 1987. Guidelines for Construction and Equipment of Hospital and Medical Facilities. U.S. Department of Health and Human Services and American Institute of Architects Committee on Architecture for Health, Washington, D.C.: American Institute of Architects Press.

Di Pede, C., G. Giegi, J. J. Quackenboss, P. Boyer-Pfersdorf, and M. D. Lebowitz. 1991. Respiratory symptoms and risk factors in an Arizona population sample of Anglo and Mexican-American whites. Chest 99:916–922.

Discher, D. P., G. D. Kleinman, and F. J. Foster. 1975. Pilot Study for Development of an Occupational Disease Surveillance Method. National Institute for Occupational Safety and Health, Office of Health Surveillance and Biometrics. HEW Publication No. (NIOSH) 75-162. Washington, D.C.: U.S. Government Printing Office.

Dolovich, J., and D. C. Little. 1972. Correlates of skin test reactions to *Bacillus subtilis* enzyme preparations. Journal of Allergy and Clinical Immunology 49(1):43–53.

Dolovich, J., S. L. Evans, and E. Nieboer. 1984. Occupational asthma from nickel sensitivity. I. Human serum albumin in the antigenic determinant. British Journal of Industrial Medicine 41:51–55.

Dorward, A. J., M. J. Colloff, N. S. MacKay, C. McSharry, and N. C. Thomson. 1988. Effect of house dust mite avoidance measures on adult atopic asthma. Thorax 43:98–105.

Dowse, G. K., K. J. Turner, G. A. Stewart, M. P. Alpers, and A. J. Woolcock. 1985. The association between *Dermatophagoides* mites and the increasing prevalence of asthma in village communities in Papau New Guinea highlands. Journal of Allergy and Clinical Immunology 75:75–83.

DuBois, A. B., B. Y. Botelho, and J. H. Comroe, Jr. 1956. A new method for measuring airway resistance in man using a body plethysmograph: Values in normal subjects and in patients with respiratory disease. Journal of Clinical Investigation 35:327–335.

Duffort, O. A., J. Carreira, G. Nitti, F. Polo, and M. Lombardero. 1991. Studies on the biochemical structure of the major cat allergen. *Felis domesticus* I. Molecular Immunology 28:301–309.

Dybendal, T., W. C. Wedberg, and S. Elsayed. 1991. Dust from carpeted and smooth floors. IV. Solid material, proteins, and allergens collected in the different filter stages of vacuum cleaners after 10 days of use in school. Allergy 46(6):427–435.

Dykewicz, M. S., P. Laufer, R. Patterson, M. Roberts, and H. M. Sommers. 1988. Woodsman's disease: Hypersensitivity pneumonitis from cutting live trees. Journal of Allergy and Clinical Immunology 81(2):455–460.

Dykewicz, M. S., R. Patterson, D. W. Cugell, K. E. Harris, and W. Af. 1991. Serum IgE and IgG to formaldehyde-human serum albumin: Lack of relation to gaseous formaldehyde exposure and symptoms. Journal of Allergy and Clinical Immunology 87:48–57.

Eggleston, P. A., C. A. Newill, A. A. Ansari, A. Pustelnik, S. R. Lou, R. I. Evans III, D. G. Marsh, J. L. Longbottom, and M. Corn. 1989. Task related variation in airborne concentrations of laboratory animal allergens: Studies with *Rat n* I. Journal of Allergy and Clinical Immunology 84:347–352.

Eggleston, P. A., A. A. Ansari, B. Zeimann, N. F. Adkinson, Jr., and M. Corn. 1990. Occupational challenge studies with laboratory workers allergic to rats. Journal of Allergy and Clinical Immunology 86:63–72.

Ehnert, B., S. Lau, A. Weber, and U. Wahn. 1991. Reduction of mite allergen exposure and bronchial hyper-reactivity [abstract]. Journal of Allergy and Clinical Immunology 87:320.

Eisen, E. A. 1987. Standardizing Spirometry: Problems and Prospects. In: Occupational Pulmonary Disease, L. A. Rosenstock, ed. Occupational Medicine-State of the Art Reviews 2(2):213–225.

Eller, P. M. 1984. NIOSH Manual of Analytic Methods. Cincinnati, Ohio: U.S. Department of Health and Human Services.

Engelberg, A. L. 1988. Guide to the Evaluation of Permanent Impairment. Chicago, Ill.: American Medical Association.

Enright, P. 1992. Surveillance for lung disease: Quality assurance using computers and a team approach. W. S. Beckett, and R. Bascom eds. Occupational Lung Disease. Occupational Medicine-State of the Art Reviews. Philadelphia, Pa.: Hanley and Belfus.

Enright, P. L., L. R. Johnson, J. E. Connett, H. Voelker, and A. S. Buist. 1991. Spirometry in the Lung Health Study. I. Methods and Quality Control. American Review of Respiratory Disease 143(6):1215–1223.

EPA (U.S. Environmental Protection Agency). 1991a. Building Air Quality: A Guide for Building Owners and Facility Managers. ANR-445W. Washington D.C.: U.S. Environmental Protection Agency, Indoor Air Division.

EPA. 1991b. Introduction to Indoor Air Quality. EPA/400/3-91/003. Washington, D.C.: U.S. Environmental Protection Agency

ERS (European Respiratory Society). In press. Statement of the ERS on the use of peak expiratory flow measurements. European Respiratory Journal. Supplement.

Evans, R. III. 1992. Asthma among minority children. Chest 101(6):368S–371S.

Executive Committee, American Academy of Allergy and Immunology. 1989. Position statement: Beta-adrenergic blockers, immunotherapy, and skin testing. Journal of Allergy and Clinical Immunology 84:129–130.

Fernandez de Corres, L., J. L. Corrales, D. Muñoz, and I. Leanizbarrutia. 1984. Dermatitis alergicas de contacto por plantas. Allergologia et Immunopathologia (Madrid) 12(4):313–319.

Fernandez de Corres, L., I. Leanizbarrutia, D. Muñoz, and J. L. Corrales. 1985. Dermatitis alergica de contacto por ajo, primula, frullania y compuestas. Allergologia et Immunopathologia (Madrid) 13(4):291–299.

Findlay, S. R., E. Stotsky, K. Leitermann, Z. Hemady, and J. L. Ohman. 1983. Allergens detected in association with airborne particles capable of penetrating into the peripheral lung. American Review of Respiratory Disease 128:1008–1012.

Fink, J. N. 1982. Evaluation of the patient for occupational immunologic lung disease. Journal of Allergy and Clinical Immunology 70:11–14.

Fink, J. N. 1988. Hypersensitivity pneumonitis. In: Allergy Principles and Practice, 3rd ed., E. Middleton, Jr., C. Reed, E. Ellis, N. F. Adkinson, Jr., and J. W. Yunginger, eds. St. Louis, Mo.: C. V. Mosby Company.

Fink, J. N., E. F. Banaszak, W. H. Thiede, and J. J. Barboriak. 1971. Interstitial pneumonitis due to hypersensitivity to an organism contaminating a heating system. Annals of Internal Medicine 74:80–83.

Fink, J. N., V. L. Moore, and J. J. Barboriak. 1975. Cell-mediated hypersensitivity in pigeon breeders. International Archives of Allergy and Applied Immunology 49:831–836.

Finklea, J. F., J. H. Farmer, G. J. Love, et al. 1974a. Aggravation of asthma by air pollutants: 1970–1971 New York studies. In: Health Consequences of Sulfur Dioxides. U. S. Environmental Protection Agency, Office of Research and Development. EPA-650/1-74-004. Washington, D.C.: U.S. Government Printing Office.

Finklea, J. F., J. G. French, G. R. Lowrimore, et al. 1974b. Prospective surveys of acute respiratory disease in volunteer families. In: Health Consequences of Sulfur Dioxides,

1970–1971. U.S. Environmental Protection Agency, Office of Research and Development. EPA-650/1-74-004. Washington, D.C.: U.S. Government Printing Office.

Finnegan, M. J., C. A. C Pickering, and P. S. Burge. 1984. The sick building syndrome: Prevalence studies. British Medical Journal 289:1573–1575.

Finnegan, M. J., C. A. Pickering, P. S. Davies, P. K. Austwick, and D. C. Warhurst. 1987. Amoebae and humidifier fever. Clinical Allergy 17(3):235–242.

Fireman, P., and R. G. Slavin, eds. 1991. Atlas of Allergies. Philadelphia, Pa.: J. B. Lippincott Company.

Fisher, A. A. 1987. Contact urticaria and anaphylactoid reaction due to corn starch surgical glove powder. Contact Dermatitis 16:224.

Ford, S. A., B. A. Baldo, D. Geraci, and D. Bass. 1986. Identification of *Parietaria judaica* pollen allergens. International Archives of Allergy and Applied Immunology 79(2):120–126.

Forstrom, L. 1980. Contact urticaria from latex surgical gloves. Contact Dermatitis 6:33–34.

Fraser, D. W., T. R. Tsai, W. Orenstein, W. E. Parkin, H. J. Beecham, R. G. Sharrar, J. Harris, G. F. Mallison, S. M. Martin, J. E. McDade, C. C. Shepard, and P. S. Brachman. 1977. Legionnaires' disease: Description of an epidemic of pneumonia. New England Journal of Medicine 297(22):1189–1197.

Freedman, S. O., and P. Gold. 1976. Clinical Immunology, 2nd ed. Hagerstown, Md.: Harper & Row.

Freeman, G. L., and S. Johnson. 1964. Allergic diseases in adolescents. I. Description of survey; prevalence of allergy. American Review of Respiratory Disease 107:549–559.

Freidhoff, L. R., D. A. Meyers, W. B. Bias, G. A. Chase, R. Hussain, and D. G. Marsh. 1981. A genetic-epidemiologic study of human immune responsiveness to allergens in an industrial population. I. Epidemiology of reported allergy and skin-test positivity. American Journal of Medical Genetics 9:323–340.

Gabriel, M., M. K. Ng, W. G. L. Allan, L. E. Hill, and A. J. Nunn. 1977. Study of prolonged hyposensitization with *D. pteronyssinus* extract in allergic rhinitis. Clinical Allergy 7:325–339.

Galen, R. S. 1986. Use of predictive value theory in clinical immunology. In: Manual of Clinical and Laboratory Immunology, N. R. Rose, H. Friedman, and J. L. Fahey, eds. Washington, D.C.: American Society for Microbiology, pp. 966–970.

Gaultier, M., P. Gervais, S. Dally, and O. Diamant-Berger. 1980. Environmental etiological factors in asthma in adults. Annales de Medicine Interne (Paris) 131:91–94.

Gelber, L. E., L. H. Seltzert, J. K. Bouzoukis, S. M. Pollart, M. D. Chapman, and T. Platts-Mills. In press. Sensitization and exposure to indoor allergens (dust mite, cat and cockroach) as risk factors for asthma. American Review of Respiratory Disease

Gelfand, H. H. 1983. Respiratory allergy due to chemical compounds encountered in the rubber, lacquer, shellac, and beauty culture industries. Journal of Allergy 34:374.

Gell, P. G. H., and R. R. A. Coombs. 1968. Clinical Aspects of Immunology, 2nd ed. Philadelphia, Pa.: Davis.

Gell, P. G. H., R. R. A. Coombs, and P. J. Lachmann, eds. 1975. Clinical Aspects of Immunology. Oxford, U.K.: Blackwell.

Geller-Bernstein, C., N. Keynan, A. Bejerano, A. Shomer-Ilan, and Y. Waisel. 1987. Positive skin tests to fern spore extracts in atopic patients. Annals of Allergy 58(2):125–127.

Gergen, P, J., and P. C. Turkeltaub. 1991. The association of allergen skin test reactivity and respiratory disease among whites in the U.S. population: Data from the second National Health and Nutrition Examination Survey, 1976–1980 (NHANES II). Archives of Internal Medicine 151:487–492.

Gergen, P. J., and P. C. Turkeltaub. 1992. The association of individual allergen reactivity with respiratory disease in a national sample: Data from the second National Health and

Nutrition Examination Survey, 1976–1980 (NHANES II). Journal of Allergy and Clinical Immunology 90:579–588.

Gergen, P. J., and K. B. Weiss. 1992. The increasing problem of asthma in the United States [editorial]. American Review of Respiratory Disease 146:823–824.

Gergen, P. J., P. C. Turkeltaub, and M. G. Kovar. 1987. The prevalence of allergic skin test reactivity to eight common aeroallergens in the U.S. population: Results from the Second National Health and Nutrition Examination Survey. Journal of Allergy and Clinical Immunology 80:669–679.

Giannini, E. H., W. T. Northey, and C. R. Leathers. 1975. The allergenic significance of certain fungi rarely reported as allergens. Annals of Allergy 35(6):372–376.

Gillespie, D. N., M. J. Dahlberg, and J. W. Yunginger. 1985. Inhalant allergy to wild animal (deer and elk). Annals of Allergy 55:122–125.

Gillies, D. R. N., J. M. Littlewood, and J. K. Sarsfield. 1987. Controlled trial of house dust mite avoidance in children with mild to moderate asthma. Clinical Allergy 17:105–111.

Glindmeyer, H., J. Diem, J. Hughes, R. N. Jones, and H. Weill. 1981. Factors influencing the interpretation of FEV_1 declines across the working shift. Chest 79(suppl.):71–72.

Godfrey, R. C., and M. Griffiths. 1976. The prevalence of immediate positive skin tests to *Dermatophagoides pteronyssinus* and grass pollen in school children. Clinical Allergy 6:79–82.

Gold, W. M., and H. A. Boushey. 1988. Pulmonary function testing. In: Textbook of Respiratory Medicine, J. F. Murray, and J. Nadel, eds. Philadelphia, Pa.: W. B. Saunders Company.

Grammer, L. C., K. E. Harris, M. J. Chandler, D. Flaherty, and R. Patterson. 1987. Establishing clinical and immunologic criteria for diagnosis of occupational lung disease with phthalic anhydride and tetrachlorophthalic anhydride exposures as a model. Journal of Occupational Medicine 29:806–811.

Grammer, L. C., R. Patterson, and C. R. Zeiss. 1989. Guidelines for the immunologic evaluation of occupational lung disease. Journal of Allergy and Clinical Immunology 84:805–814.

Grammer, L. C., M. A. Shaughnessy, R. A. Davis, and R. Patterson. 1991a. Clinical and immunologic evaluation of 95 workers exposed to TMXDI and TMI [abstract]. Journal of Allergy and Clinical Immunology 87:200.

Grammer, L. C., K. E. Harris, K. R. Sonenthal, R. E. Roach, and C. Ley. 1991b. Identification and control of trimellitic anhydride immunologic lung disease: An 11 year experience [abstract]. Clinical Research 39:703a.

Grammer, L. C., K. E. Harris, D. W. Cugell, and R. Patterson. 1992. Evaluation of a worker with possible formaldehyde asthma [abstract]. Journal of Allergy and Clinical Immunology 89:203.

Gravesen, S. 1978. Identification and prevalence of culturable mesophilic microfungi in house dust from 100 Danish homes. Comparison between airborne and dust-bound fungi. Allergy 33(5):268–272.

Gravesen, S., L. Larsen, and P. Skov. 1983. Aerobiology of schools and public institutions: Part of a study. Ecological Disorders 2:411–413.

Graziano, F. M., and R. F. Lemanske. 1989. Clinical Immunology. Baltimore, Md.: Williams & Wilkins.

Green, W. F., N. R. Nicholas, C. M. Salome, and A. J. Woolcock. 1989. Reduction of house dust mites and mite allergens: Effects of spraying carpets and blankets with Allersearch DMA, an acaricide combined with an allergen reducing agent. Clinical and Experimental Allergy 19:203–207.

Greenberger, P. A. 1988. Allergic bronchopulmonary aspergillosis. In: Allergy Principles and Practice, G. Middleton, C. E. Reed, E. F. Ellis, N. F. Adkinson, Jr., and J. W. Yunginger, eds. St. Louis, Mo.: C. V. Mosby Company, pp. 1219–1236.

Greenberger, P. A., R. Patterson, and C. M. Tapio. 1985. Prophylaxis against repeated radiocontrast media reactions in 857 cases. Archives of Internal Medicine 145:2197–2200.

Gregg, I. 1983. Epidemiological aspects and the role of infection. In: Asthma, T. J. H. Clark and S. Godfrey, eds. Philadelphia, Pa.: W. B. Saunders Company, pp. 160–183, 242–284.

Gregg, I. 1989. Some historical aspects of asthma. Lecture given to Southampton History of Medicine Society, February 1989.

Guerin, B., and R. D. Watson. 1988. Skin tests. Clinical Review of Allergy 6:211–227.

Haahtela, T., F. Bjorksten, M. Helskala, and I. I. Suoniemi. 1980. Skin prick test reactivity to common allergens in Finnish adolescents. Allergy 35:425–431.

Hagy, G. W., and G. A. Settipane. 1969. Bronchial asthma, allergic rhinitis, and allergy skin tests among college students. Journal of Allergy and Clinical Immunology 44:323–332.

Haida, M., H. Okudaira, T. Ogita, K. Ito, T. Miyamoto, T. Nakajima, and O. Hongo. 1985. Allergens of the house dust mite Dermatophagoides farinae: Immunochemical studies of four allergenic fractions. Journal of Allergy and Clinical Immunology 75(6):686–692.

Halonen, M., D. Stern, C. G. Ray, A. L Wright, L. M. Taussing, and F. M. Martinez. 1992. The predictive relationship between umbilical cord serum IgE levels and subsequent incidence of lower respiratory ilnesses and eczema in infants. American Review of Respiratory Disease 146:866–870.

Hanson, P. J., and R. Penny. 1974. Pigeon breeder's disease: Study of the cell-mediated immune response to pigeons by the lymphocyte culture technique. International Archives of Allergy and Applied Immunology 47:498–507.

Hausen, B. M., and K. H. Schulz. 1988. Immediate inhalative allergy due to the nectar of Abutilon striatum thompsonii. Allergologie 11(2):47–51.

Helm, R. M., D. L. Squillace, and J. W. Yunginger. 1988. Production of a proposed International Reference Standard Alternaria extract. II. Results of a collaborative trial. Journal of Allergy and Clinical Immunology 81:651–663.

Henderson, F.W. 1993. Personal Communication, January 26, 1993.

Hendrick, D. J., and D. J. Lane. 1977. Occupational formalin asthma. British Journal of Industrial Medicine 34:11–18.

Heymann, P. W., M. D. Chapman, and T. Platts-Mills. 1986. Antigen Der f I from the dust mite Dermatophagoides farinae: Structural comparison with Der p I from D. pteronyssinus and epitope specificity of murine IgG and human IgE antibody responses. Journal of Immunology 137:2841–2847.

Heymann, P. W., M. D. Chapman, R. C. Aalberse, J. W. Fox, and T. Platts-Mills. 1989. Antigenic and structural analysis of Group II allergens (Der p II and Der p II) from house dust mites (Dermatophagoides spp). Journal of Allergy and Clinical Immunology 83:1055–1067.

Hilberg, O., A. C. Jackson, D. L. Swift, and O. F. Pedersen. 1989. Acoustic rhinometry: Evaluation of nasal cavity geometry by acoustic reflection. Journal of Applied Physiology 66:295–303.

Hirsch, D. J., S. R. Hirsch, and J. H. Kalbfleisch. 1978. Effect of central air conditioning and meteorologic factors on indoor spore counts. Journal of Allergy and Clinical Immunology 62(1):22–26.

Hodgson, M. J., P. Morey, J. S. Simon, T. D. Waters, and H. N. Fink. 1987. An outbreak of recurrent acute and chronic hypersensitivity pneumonitis in office workers. American Journal of Epidemiology 125:631–638.

Holroyd, K. A., and T. L. Creer. 1986. Self-management of Chronic Disease: Handbook of Clinical Interventions and Research. Orlando, Fla: Academic Press.

Hood, M. A. 1989. Gram negative bacteria as bioaerosols. In: Biological Contaminants in Indoor Environments, ASTM STP 1071, P. Morey, J. Feeley, and J. Otten J, eds. Philadelphia, Pa.: American Society for Testing Materials, pp. 60–70.

Horn, N., and P. Lind. 1987. Selection and characterization of monoclonal antibodies against a major allergen in *D. pteronyssinus*: Species-specific and common epitopes in three *Dermatophagoides* species. International Archives of Allergy and Applied Immunology 83:404–409.

Horner, W. E., M. D. Ibanez, V. Liengswangwong, J. Sastre, and S. B. Lehrer. 1988. Identification and analysis of basidiospore allergens from puffballs. Journal of Allergy and Clinical Immunology 82:787–795.

Horner, W. E., M. D. Ibanez, and S. B. Lehrer. 1989. Immunoprint analysis of Calvatia cyathiformis allergens. I. Reactivity with individual sera. Journal of Allergy and Clinical Immunology 83(4):784–792.

Horwood, L. J., D. M. Fergusson, and F. T. Shannon. 1985. Social and familial factors in the development of early childhood asthma. Journal of Pediatrics 75: 859–868.

Hughes, A. M. 1976. The Mites of Stored Food and Houses. London, England: Her Majesty's Stationery Office.

Hulett, A. C., and R. J. Dockhorn. 1979. House dust mite (*D. farinae*) and cockroach allergy in a midwestern population. Annals of Allergy 42:160–165.

Hunninghake, G. W., J. E. Gadek, O. Kawanami, V. J. Ferrans, and R. G. Crystal. 1979. Inflammatory and immune processes in the human lung in health and disease: evaluation by bronchoalveolar lavage. American Journal of Pathology 97:149.

Hunter, C., C. Grant, B. Flannigan, and A. Bravery. 1988. Mould in buildings: The air spora of domestic buildings. International Biodeterioration 24:81–101.

Huss, K., M. Salerno, Sr., and R. W. Huss. 1991. Computer-assisted reinforcement of instruction: Effects on adherence in adult atopic asthmatics. Research in Nursing and Health 14:259–267.

IAQU. 1991. IAQ professionals debate duct cleaning. Indoor Air Quality Update 4(10):1–7.

Inui, T. S., E. L. Yourtee, and J. W. Williamson. 1976. Improved outcomes in hypertension after physician tutorials: A controlled trial. Ann Intern Med 84:646–651.

Ishii, A., M. Takaoka, M. Ichinoe, Y. Kabasawa, and T. Ouchi. 1979. Mite fauna and fungal flora in house dust from homes of asthmatic children. Allergy 34(6):379–387.

Ito, K., T. Miyamoto, T. Shibuya, K. Kamei, K. Mano, T. Taniai, and M. Sasa. 1986. Skin test and radioallergosorbent test with extracts of larval and adult midges of *Tokunagayusurika akamusi Tokunaga* (Diptera: Chironomidae) in asthmatic patients of the metropolitan area of Tokyo. Annals of Allergy 57:199–204.

Jacobs, R. L. M., C. P. M. Andrews, and F. O. Jacobs. 1989. Hypersensitivity pneumonitis treated with an electrostatic dust filter. Annals of Internal Medicine 110:115–118.

Johansson, S. G., H. Bennich, and T. Berg. 1971. In vitro diagnosis of atopic allergy. 3. Quantitative estimation of circulating IgE antibodies by the radioallerosorbent test. International Archives of Allergy and Applied Immunology 41:443–451.

Johansson, S. G., and T. Foucard. 1978. IgE in immunity and disease. In: Allergy Principles and Practice, E. Middleton, C. E. Reed, and E. F. Ellis, eds. St. Louis, Mo.: C. V. Mosby Company, p. 551.

Johnson, A. M. 1986. Immunoprecipitation in gels. In: Manual of Clinical and Laboratory Immunology, N. R. Rose, H. Friedman, and J. L. Fahey, eds. Washington, D.C.: American Society for Microbiology.

Johnson, C. L., I. L. Bernstein, J. S. Gallagher, P. F. Bonventre, and S. M. Brooks. 1980. Familial hypersensitivity pneumonitis induced by *Bacillus subtilis*. American Review of Respiratory Disease 122:339–348.

Jones, R. T., M. E. Bubak, V. A. Gosselin, and J. W. Yunginger. 1992. Relative latex allergen contents of several commercial latex gloves [abstract]. Journal of Allergy and Clinical Immunology 89(1):225.

Kabat, E. A., and M. M. Mayer. 1967. Experimental Immunochemistry. Springfield, Ill.: Charles C Thomas, pp. 559–565.

Kanfer, F. H., and B. K. Schefft. 1988. Guiding the Process of Therapeutic Change. Champaign, Ill.: Research Press.

Kang, B., D. Vellody, H. Homburger, and J. W. Yunginger. 1979. Cockroach cause of allergic asthma: Its specificity and immunologic profile. Journal of Allergy and Clinical Immunology 63:80–86.

Kang, B., M. Wilson, K. H. Price, and T. Kambara. 1991. Cockroach-allergen study: Allergen patterns of three common cockroach species probed by allergic sera collected in two cities. Journal of Allergy and Clinical Immunology 87:1073–1080.

Kaplan, A. P. 1985. Allergy. New York, N.Y.: Churchill Livingstone.

Kapyla, M. 1985. Frame fungi on insulated windows. Allergy 40(8):558–564.

Karlsson, S., F. Banhidi, Z. Banhidi, and A. C. Albertsson. 1984. Accumulation of malodorous amines and polyamines due to clostridial putrefaction indoors. In: Indoor Air: Proceedings of the 3rd International Conference on Indoor Air Quality and Climate. Stockholm, Sweden: Swedish Council for Building Research, pp. 287–293.

Karol, M. H. 1983. Concentration-dependent immunologic response to toluene diisocyanate (TDI) following exposure. Toxicology and Applied Pharmacology 68:229–241.

Kauffman, H. F., S. van der Heide, F. Beaumont, J. G. de Monchy, and K. de Vries. 1984. The allergenic and antigenic properties of spore extracts of *Aspergillus fumigatus*: A comparative study of spore extracts with mycelium and culture filtrate extracts. Journal of Allergy and Clinical Immunology 73(5, pt. 1):567–573.

Kawai, T., M. Tamura, and M. Murao. 1984. Summer-type hypersensitivity pneumonitis. Chest 85:311–317.

Kay, A. B., M. O. Gad El Rab, J. Steward, and H. H. Erwa. 1978. Widespread IgE-mediated hypersensitivity in northern Sudan of the chironomid *Cladotanytarsus lewisi* ("green nimitti"). Clinical and Experimental Immunology 34:106–110.

Kelly, K., M. Sitlock, and J. P. Davis. 1991. Anaphylactic reactions during general anesthesia among pediatric patients—United States, January 1990–1991. Morbidity and Mortality Weekly Report 40:437–443.

Kendrick, B. 1985. The Fifth Kingdom. Waterloo, Ontario: Mycologue Publications.

Kern, R. A. 1921. Dust sensitization in bronchial asthma. Medical Clinics of North America 5:751–758.

Kino, T., J. Chihara, K. Fukuda, Y. Sasaki, Y. Shogaki, and S. Oshima. 1987. Allergy to insects in Japan. III. High frequency of IgE antibody responses to insects (moth, butterfly, caddis fly, and chironomid) in patients with bronchial asthma and immunochemical quantitation of insect-related airborne particles smaller than 10 microns in diameter. Journal of Allergy and Clinical Immunology 79:857–866.

Klaustermeyer, W. B. 1983. Practical Allergy and Immunology. New York, N.Y.: Wiley.

Klink, M., M. G. Cline, M. Halonen, and B. Burrows. 1990. Problems in defining normal limits for serum IgE. Journal of Allergy and Clinical Immunology 85:440–444.

Knudson, R. J., M. D. Lebowitz, C. J. Holberg, and B. Burrows. 1983. Changes in the normal maximal expiratory flow-volume curve with growth and aging. American Review of Respiratory Disease 127:725–734.

Knysak, D. 1989. Animal allergens. In: Immunology and Allergy Clinics of North America, W. R. Solomon, ed. Philadelphia, Pa.: W. B. Saunders Company, pp. 357–364.

Koivikko, A. and J. Savolainen. 1988. Mushroom allergy. Allergy 43(1):1–10.

Korenblat, P. E., and H. J. Wedner. 1984. Allergy: Theory and Practice. Orlando, Fla.: Grune and Stratton.

Korsgaard, J. 1982. Preventive measures in house-dust allergy. American Review of Respiratory Disease 125:80–84.

Korsgaard, J. 1983a. House dust mites and absolute indoor humidity. Allergy 38:85–92.

Korsgaard, J. 1983b. Mite asthma and residency: A case-control study on the impact of exposure to house-dust mites in dwellings. American Review of Respiratory Disease 128:231–235.

Koshte, V. L., S. L. Kagen, and R. C. Aalberse. 1989. Cross-reactivity of IgE antibodies to caddis fly with arthropoda and mollusca. Journal of Allergy and Clinical Immunology 84:174–183.

Kozak, P., J. Gallup, L. H. Cummins, and S. A. Gillman. 1980a. Currently available methods for home mold surveys. I. Description of techniques. Annals of Allergy 45:85–89.

Kozak, P., J. Gallup, L. H. Cummings, and S. A. Gillman. 1980b. Currently available methods for home mold surveys. II. Examples of problem homes surveyed. Annals of Allergy 45:167–176.

Kreiss, K. 1988. The epidemiology of building-related complaints and illness. In: Textbook of Respiratory Medicine, J. F. Murray and J. A. Nadel, eds. Philadelphia, Pa.: W. B. Saunders Company, pp. 1069–1106.

Kreiss, K. 1990. The sick building syndrome: Where is the epidemiologic basis? American Journal of Public Health 80:1172–1173.

Krzyzanowski, M., A. E. Camilli, and M. D. Lebowitz. 1990. Relationships between pulmonary function and changes in chronic respiratory symptoms: Comparison of Tucson and Cracow longitudinal studies. Chest 98:62–70.

Kuehn, T., D. Pru, D. Vesley, et al. 1991. Matching Filtration to Health Requirements. ASHRAE publication 3505.

Kumar, P., M. Lopez, W. Fan, K. Cambre, and R. C. Elston. 1990. Mold contamination of automobile air conditioner systems. Annals of Allergy 64(2pt1):174–177.

Laatikainen, T. P., P. L. Korhonen, et al. 1991. Methods for evaluating dust accumulation in ventilation duct. In: IAQ '91 Healthy Buildings, Atlanta, Ga.: American Society of Heating, Refrigerating and Air-Conditioning Engineers.

Lalonde, M. 1974. A New Perspective on the Health of Canadians: A Working Document. Ottawa: Information Canada.

Lam, S., and M. Chan-Yeung. 1980. Ethylenediamine induced asthma. American Review of Respiratory Disease 121:151–155.

Lasser, E., C. Berry, L. Talner, L. C. Santini, E. K. Lang, F. H. Gerber, and H. O. Stolberg. 1987. Pretreatment with corticosteroids to alleviate reactions to intravenous contrast material. New England Journal of Medicine 317:845–849.

Last, J. M. 1986. Epidemiology and health information. In: Public Health and Preventive Medicine, East Norwalk, Conn.: Appleton Century Crofts, chap. 2.

Lau, S., G. Falkenhorst, A. Weber, I. Werthman, P. Lind, P. Buettner-Goetz, and U. Wahn. 1989. High mite-allergen exposure increases the risk of sensitization in atopic children and young adults. Journal of Allergy and Clinical Immunology 84:718–725.

Laurent, A., C. F. Cannell, and K. H. Marquis. 1972. Reporting health events in household interviews. U.S. National Center for Health Statistics. Washington, D.C.: U.S. Government Printing Office.

Lau-Schadendorf, S., A. F. Rusche, A. K. Waber, P. Buettner-Goetz, and U. Wahn. 1991. Short term effect of solidified benzyl benzoate on mite-allergen concentration in house dust. Journal of Allergy and Clinical Immunology 87:41–47.

Lawlor, G., and T. J. Fischer. 1981. Manual of Allergy and Immunology: Diagnosis and Therapy. Boston, Mass.: Little, Brown.

Lawther, P. J., R. E. Waller, and M. Henderson. 1970. Air pollution and exacerbations of bronchitis. Thorax 25:525.

Lebowitz, M. D. 1977. The relationship of socio-environmental factors on the prevalence of obstructive lung problems and other chronic conditions. Journal of Chronic Diseases 30:599–611.

Lebowitz, M. D. 1989. The trends in airway obstructive disease morbidity in the Tucson epidemiological study. American Review of Respiratory Disease 140:S35–S41.

Lebowitz, M. D. 1991. The use of peak expiratory flow rate measurements in respiratory disease. Pedatric Pulmonology 11(2):166–174.

Lebowitz, M. D., R. Barbee, and B. Burrows. 1975. Allergic skin test: Sensitivity, specificity and reliability. American Journal of Epidemiology 102:452.

Lebowitz, M. D., and B. Burrows. 1977. Quantitative relationships between cigarette smoking and chronic productive cough. International Journal of Epidemiology 6:107–113.

Lebowitz, M. D., E. J. Cassell, and J. R. McCarroll. 1972a. Health and the urban environment. XI. The incidence and burden of minor illness in a healthy population: Methods, symptoms and incidence. American Review of Respiratory Disease 106:824–834.

Lebowitz, M. D., E. J. Cassell, and J. R. McCarroll. 1972b. Health and the urban environment. XII. The incidence and burden of minor illness in a healthy population: Duration, severity and burden. American Review of Respiratory Disease 106:835–841.

Lebowitz, M. D., R. J. Knudson, and B. Burrows. 1975. Tucson epidemiologic study of obstructive lung diseases. American Journal of Epidemiology 102:137–163.

Lebowitz, M. D., R. J. Knudson, G. Robertson, and B. Burrows. 1982. Significance of intraindividual changes in maximum expiratory flow volume and peak expiratory flow measurements. Chest 81:566–570.

Lebowitz, M. D., R. Barbee, and B. Burrows. 1984. Family concordance of IgE, atopy and disease. Journal of Allergy and Clinical Immunology 73:259–266.

Lebowitz, M. D., C. J. Holberg, B. Boyer, and C. Hayes. 1985. Respiratory symptoms and peak flow associated with indoor and outdoor air pollutants in the Southwest. Journal of the Air Pollution Control Association 35:1154.

Lebowitz, M. D., J. Quackenboss, A. E. Camilli, D. Bronnimann, C. J. Holberg, and B. Boyer. 1987. The epidemiological importance of intraindividual changes in objective pulmonary responses. European Journal of Epidemiology 3(4):390–398.

Lebowitz, M. D., J. J. Quackenboss, M. L. Soczek, S. D. Colome, and P. J. Lioy. 1989. Workshop: Development of Questionnaires and Survey Instruments. ASTM Special Technical Publication 1002. Philadelphia, Pa.: American Society for Testing Materials.

Lebowitz, M. D., C. J. Holberg, and F. Martinez. 1990. A longitudinal study of risk factors in asthma and chronic bronchitis in children. European Journal of Epidemiology 6(4):341–347.

Lebowitz, M. D., M. K. O'Rourke, J. J. Quackenboss, and C. Di Pede. 1991. Bronchial responsiveness: Its use in epidemiological studies. Journal of Allergy and Clinical Immunology 87(1):340.

Lehrer, S. R., M. Lopez, B. T. Butcher, J. Olson, M. Reed, and J. E. Salvaggio. 1986. Basidiomycete mycelia and spore-allergen extracts: Skin test reactivity in adults with symptoms of respiratory allergy. Journal of Allergy and Clinical Immunology 78:478–485.

Lessof, M. H. 1984. Allergy: Immunological and Clinical Aspects. New York, N.Y.: Wiley.

Levitt, R. C., W. Mitzner, and S. R. Kleeberger. 1990. A genetic approach to the study of lung physiology: Understanding biological variability in airway responsiveness. American Journal of Physiology (Lung Cellular and Molecular Physiology) 258:L157–L164.

Licorish, K., H. S. Novey, P. Kozak, R. D. Fairshter, and A. F. Wilson. 1985. Role of Alternaria and Penicillium spores in the pathogenesis of asthma. Journal of Allergy and Clinical Immunology 76(6):819–825.

Lidd, D., and R. S. Farr. 1962. Primary interaction between ^{131}I-labeled ragweed pollen and antibodies in the sera of humans and rabbits. Journal of Allergy and Clinical Immunology 33:45–58.

Liengswangwong, V., J. E. Salvaggio, F. L. Lyon, and S. B. Lehrer. 1987. Basidiospore allergens: Determination of optimal extraction methods. Clinical Allergy 17(3):191–198.

Lind, P. 1985. Purification and partial characterization of two major allergens from the house dust mite Dermatophagoides pteronyssinus. Journal of Allergy and Clinical Immunology 76:753–761.

Lind, P., P. S. Norman, M. Newton, H. Lowenstein, and B. Schwartz. 1987. The prevalence of indoor allergens in the Baltimore area: House dust-mite and animal-dander antigens measured by immunochemical techniques. Journal of Allergy and Clinical Immunology 80:541–547.

Lindgren, S., L. Belin, S. Dreborg, R. Einnarsson, and I. Pahlman. 1988. Breed-specific dog-dandruff allergens. Journal of Allergy and Clinical Immunology 82:196–204.

Lockey, R. F. 1979. Allergy and Clinical Immunology. Garden City, N.Y.: Medical Examination Publishing Co.

Lockey, R. F., and S. C. Bukantz. 1987. Principles of Immunology and Allergy. Philadelphia, Pa.: W. B. Saunders Company.

Lockey, R. F., L. M. Benedict, P. C. Turkeltaub, and S. C. Bukantz. 1987. Fatalities from immunotherapy (IT) and skin testing (ST). Journal of Allergy and Clinical Immunology 79:660–677.

Lofdahl, C. G., and N. Svedmyr. 1991. Beta-agonists: Friends or foes [editorial]. European Respiratory Journal 4:1161–1165.

Lopez, M., and J. Salvaggio. 1988. Hypersensitivity pneumonitis. In: Textbook of Respiratory Medicine, J. F. Murray and J. Nadel, eds. Philadelphia, Pa.: W. B. Saunders Company.

Lopez, M., J. Salvaggio, and B. Butcher. 1976. Allergenicity and immunogenicity of Basidiomycetes. Journal of Allergy and Clinical Immunology 57(5):480–488.

Lopez, M., J. R. Voigtlander, S. B. Lehrer, and J. E. Salvaggio. 1989. Bronchoprovocation studies in basidiospore-sensitive allergic subjects with asthma. Journal of Allergy and Clinical Immunology 84(2):242–246.

Lotz, W. A. 1989. Moisture problems in buildings in hot humid climates. ASHRAE Journal April:26–27.

Lowenstein, H., B. Markussen, and B. Weeks. 1976. Isolation and partial characterization of three major allergens of horse hair and dandruff. International Archives of Allergy and Applied Immunology 51:48–67.

Lstiburek, J. W. 1989. Historical perspective on North American wood frame construction. Presented at CPSC/ALA Workshop on Biological Pollutants in the Home, July 10–11, Alexandria, Va.

Luczynska, C. M., L. K. Arruda, T. Platts-Mills, J. D. Miller, M. Lopez, and M. D. Chapman. 1989. A two-site monoclonal antibody ELISA for the quantitation of the major *Dermatophagoides* spp. allergens *Der p* I and *Der f* I. Journal of Immunological Methods 118:227–235.

Luczynska, C. M., Y. Li, M. D. Chapman, and T. Platts-Mills. 1990. Airborne concentrations and particle size distribution of allergen derived from domestic cats (*Felis domesticus*). Measurements using cascade impactor, liquid impinger, and a two-site monoclonal antibody assay for *Fel d* I. American Review of Respiratory Disease 141(2):361–367.

Lundblad, F. P. 1991. House dust mite allergy in an office building. Applied Occupational and Environmental Hygiene 6:94–96.

Luoma, R. 1984. Environmental allergens and morbidity in atopic and non-atopic families. Acta Paediatrica Scandinavica 73:448–453.

Lushniak, B. D., C. M. Reh, J. S. Gallagher, and D. I. Bernstein. 1990. Indirect assessment of exposure to MDI by evaluation of specific immune response to MDI-HSA in foam workers [abstract]. Journal of Allergy and Clinical Immunology 85:251.

MacDonald, M. M., R. H. Dunstan, and F. M. Dewey. 1989. Detection of low-MR glycoproteins in surface washes of some fungal cultures by gel-filtration HPLC and by monoclonal antibodies. Journal of General Microbiology 135(pt. 2):375-383.

Macher, J., and J. Girman. 1989. Multiplication of microorganisms in an evaporative air cooler and possible indoor air contamination. Paper 89-83.6. Air and Waste Management Association.

Macher, J. M., and M. W. First. 1984. Personal air samplers for measuring occupational exposures to biological hazards. American Industrial Hygiene Association Journal 45(2):76–83.

Maiman, L. A., M. H. Becker, G. S. Liptak, L. F. Nazarian, and K. A. Rounds. 1988. Improving pediatricians' compliance-enhancing practices. American Journal of Diseases of Children, 142:773–779.

Mak, H., P. Johnston, H. Abbey, and R. Talamo. 1982. Prevalence of asthma and health service utilization of asthmatic children in an inner city. Journal of Allergy and Clinical Immunology 70:367–372.

Malo, J. L., A. Cartier, M. Doepner, E. Nieboer, S. Evans, and J. Dolovich. 1982. Occupational asthma caused by nickel sulfate. Journal of Allergy and Clinical Immunology 69:55–59.

Marcus, D., and C. A. Alper. 1986. Methods for allotyping complement proteins. In: Manual of Clinical and Laboratory Immunology, N. R. Rose, H. Friedman, and J. L. Fahey, eds. Washington, D.C.: American Society for Microbiology, pp. 185–190.

Marinkovich, V., and A. Hill. 1975. Hypersensitivity alveolitis. Journal of the American Medical Association 231:944–947.

Marquis, K. H. 1978. Evaluation of health diary data in the health insurance study. In: NCHSR Research Proceedings Series Health Survey Research Methods. Washington, D.C.: U.S. Government Printing Office.

Martin, T. R., and R. G. Goodman. 1990. The role of lung mononuclear cells in asthma. Immunology and Allergy Clinics of North America 10:295.

Mathison, D. A., D. D. Stevenson, and R. A. Simon. 1982. Asthma and the home environment [editorial]. Annals of Internal Medicine 97:128–130.

Matson, S. C., M. C. Swanson, C. E. Reed, and J. W. Yunginger. 1983. IgE and IgG-immune mechanisms do not mediate occupation related respiratory or systemic symptoms in hog farms. Journal of Allergy and Clinical Immunology 72:299–304.

Matsumura, Y. 1970a. The effects of ozone, nitrogen dioxide, and sulfur dioxide on the experimentally induced allergic respiratory disorder in guinea pigs. I. The effect on sensitization with albumin through the airway. American Review of Respiratory Disease 102:430–437.

Matsumura, Y. 1970b. The effects of ozone, nitrogen dioxide, and sulfur dioxide on the experimentally induced allergic respiratory disorder in guinea pigs. II. The effects of ozone on the absorption and the retention of antigen in the lung. American Review of Respiratory Disease 102:438–443.

Matsumura, Y. 1970c. The effects of ozone, nitrogen dioxide, and sulfur dioxide on the experimentally induced allergic respiratory disorder in guinea pigs. III. The effect of the occurrence of dyspneic attacks. American Review of Respiratory Disease 102:444–447.

Mazur, G., W. M. Becker, and X. Baur. 1987. Epitope mapping of major insect allergens (chironomid hemoglobins) with monoclonal antibodies. Journal of Allergy and Clinical Immunology 80:876–883.

McCarroll, J. R., et al. 1966. Health profiles versus environmental pollutants. American Journal of Public Health 56:266.

McConnell, L. H., J. N. Fink, D. P. Schleuter, and M. G. Schmidt. 1973. Asthma caused by nickel sensitivity. Annals of Internal Medicine 78:888–890.

Meding, B., and S. Fregert. 1984. Contact urticaria from natural latex gloves. Contact Dermatitis 10:52–53.

Melan, H. L. 1972. Principles of immunologic management of allergic diseases due to extrinsic allergens. In: Allergic Diseases: Diagnosis and Management, R. Patterson, ed. Philadelphia, Pa.: J. B. Lippincott Company, pp. 293–304.

Michel, O., R. Ginanni, J. Duchateau, F. Vertongen, B. Le Bon, and R. Sergysels. 1991. Domestic endotoxin exposure and clinical severity of asthma. Clinical and Experimental Allergy 21:441–448.

Middleton, E., C. E. Reed, E. F. Ellis, N. R. Adkinson, Jr., and J. W. Yunginger, eds. 1988. Allergy: Principles and Practice, 3rd ed. St. Louis, Mo.: C. V. Mosby Company.

Miller, A., K. Kaminsky, J. D. Miller, and R. Hamilton. 1992. Dust mite allergen in blankets, and the effects of hot water washing [abstract]. Journal of Allergy and Clinical Immunology 89:257.

Milton, D., R. Gere, H. Feldman, and I. Greaves. 1989. A precise sensitive Limulus test for airborne endotoxin. American Review of Respiratory Disease 139:A387.

Mitchell, E. B., S. Wilkins, J. Deighton, and T. Platts-Mills. 1985. Reduction of house dust mite allergen levels in the home: Use of the acaracide pirimiphos-methyl. Clinical Allergy 15:235–240.

Mittal, A., M. K. Agarwal, and D. N. Shivpuri. 1979. Studies on allergenic algae of Delhi area: Botanical aspects. Annals of Allergy 42(4):248–252.

Miyamoto, T., S. Oshima, T. Ishizaka, and S. Sato. 1968. Allergic identity between the common floor mite (*Dermatophagoides farinae*, Hughes, 1961) and house dust as a causative agent in bronchial asthma. Journal of Allergy 42:14–28.

Molfino, N. A., S. C. Wright, I. Katz, S. Tarlo, F. Silverman, P. A. McClaan, J. P. Szalai, M. Raisenne, A. S. Slulsky, and N. Zamel. 1991. Effect of low concentrations of ozone on inhaled allergen responses in asthmatic subjects. Lancet 338:199–203.

Molhave, L. 1987. The sick buildings: A subpopulation among the problem buildigs. In: Proceedings of the Fourth International Conference on Indoor Air Quality and Climate, vol. 2, B. Siefert, H. Esdorn, M. Fischer, H. Ruden, and J. Wegner, eds. West Berlin, Germany: Institute for Water, Soil, and Hygiene, pp. 469–473.

Mooney, H. W. 1962. Methodology in Two California Health Surveys. U.S. Public Health Service. Washington, D.C.: U.S. Government Printing Office.

Morey, P. 1984. Case presentations: Problems caused by moisture in occupied spaces of office buildings. Annals of American Conference of Governmental Industrial Hygienists (ASHRAE) 10:121–127.

Morey, P. 1988. Microorganisms in buildings and HVAC systems: A summary of 21 environmental studies. In: Proceedings, IAQ '88, Altanta, Ga.: American Society of Heating, Refrigerating and Air-Conditioning Engineers, pp. 10–24.

Morey, P. 1990a. Bioaerosols in the indoor environment: Current practices and approaches. In: Indoor Air Quality International Symposium, R. Gammage and D. Weeks, eds. Akron, Ohio: American Industrial Hygiene Association, pp. 51–72.

Morey, P. 1990b. Practical aspects of sampling for organic dusts and microorganisms. American Journal of Industrial Medicine 18(3):273–278.

Morey, P. 1992. Microbiological contamination in buildings: Precautions during remediation activities. In: IAQ 92 Environments for people, Atlanta, GA.: American Society of Heating, Refrigerating and Air Conditioning Engineers, pp. 94–100.

Morey, P., and D. Shattuck. 1989. Role of ventilation in the causation of building-associated illness. Occupational Medicine-State of the Art Reviews 4:625–642.

Morey, P., and J. Singh. 1991. Indoor air quality in nonindustrial occupational environments. In: Patty's Industrial Hygiene and Toxicology, 4th ed. New York, N.Y.: Wiley, pp. 531–594.

Morey, P., and C. Williams. 1991. Is porous insulation inside a HVAC system compatible with a healthy building? In: IAQ '90 Healthy Buildings, Atlanta, Ga.: American Society of Heating, Refrigerating and Air-Conditioning Engineers, pp. 128–135.

Morey, P., M. Hodgson, W. Sorenson, G. Kullman, W. Rhodes, and G. Visvesvara. 1986. Environmental studies in moldy office buildings. ASHRAE Transactions 92(1):399–419.

Morgenstern, J., I. J. Griffith, A. W. Brauer, B. L. Rogers, J. F. Bond, M. D. Chapman, and M. C. Kuo. 1991. Amino acid sequence of *Fel d* I, the major allergen of the domestic cat: Protein sequence and cDNA cloning. Proceedings of the National Academy of Sciences 88:9690–9694.

Morrill, C. G., D. W. Dickey, P. C. Weiser, R. A. Kinsman, H. Chai, and S. L. Spector. 1981. Calibration and stability of standard and mini-Wright peak flow meters. Annals of Allergy 46:70–73.

Morris, J. F., A. Koski, and L. C. Johnson. 1971. Spirometric standards for healthy nonsmoking adults. American Review of Respiratory Disease 103:57–76.

Mossman, T. R., and R. L. Coffman. 1989. TH2 and TH2 cells: Different patterns of lymphokine secretion lead to different functional properties. Annual Reviews of Immunology 7:145.

Muller, C. F., A. Waybur, and E. R. Weinerman. 1952. Methodology of a family health study. Public Health Reports 67:1149.

Muranaka, M., S. Suzuki, K. Koizumi, S. Takafuji, T. Miyamota, R. Ikemori, and H. Tokiwa. 1986. Adjuvant activity of diesel-exhaust particulates for the production of IgE antibody in mice. Journal of Allergy and Clinical Immunology 77:616–623.

Murray, A. B., and A. C. Ferguson. 1983. Dust-free bedrooms in the treatment of asthmatic children with house dust or house dust mite allergy: A controlled trial. Pediatrics 71:418–422.

Murray, A. B., A. C. Ferguson, and B. J. Morrison. 1985. Sensitization to house dust mites in different climatic areas. Journal of Allergy and Clinical Immunology 76:108–112.

NADCA. National Air Duct Cleaners Association. 1991. Cleaning Standards for Air Conveyance Systems. Revised Draft Standards. Washington, D.C.: NADCA.

Nagano, T., K. Ohara, S. Yamamoto, and T. Sugai. 1982. Contact dermatitis from plants these 7 years and presentation of some representative cases. Skin Research 24(2):218–224.

Nathan, S. P., M. D. Lebowitz, and R. J. Knudson. 1979. Spirometric testing: Number of tests required and data selection. Chest 76:384–388.

NCHS (National Center for Health Statistics). 1986. Current Estimates from the National HIS, U.S. NCHS Series 10, No. 160; DHHS Publ. No. (PHS) 86-1588. Hyattsville, Md.: NCHS.

NCHS. 1992. 1990 Summary: National Hospital Discharge Survey, Advance data, No. 210, 18 Feb. 1992. DHHS Pub. No. (PHS) 92-1250. Hyattsville, Md.: NCHS.

Nelson, H. S. 1983. Diagnostic procedures in allergy. I. Allergy skin testing. Annals of Allergy 51:411–417.

Nelson, H. S., and R. M. Skufca. 1991. Double-blind study of suppression of indoor fungi and bacteria by the PuriDyne biogenic air purifier. Annals of Allergy 66:263–266.

Nelson, H. S., S. R. Hirsch, J. L. Ohman, T. Platts-Mills, C. E. Reed, and W. R. Solomon. 1988. Recommendations for the use of residential air-cleaning devices in the treatment of allergic respiratory diseases. Journal of Allergy and Clinical Immunology 82:661–669.

Nelson, S. B., R. M. Gardner, and R. O. Crapo. 1990. Performance evaluation of contemporary spirometers. Chest 97:288–297.

Nevalainen, A., H. Heinonen-Tanski, and R. Savolainen. 1990. Indoor and outdoor air occurrence of *Pseudomonas* bacteria. In: Proceedings of the Fifth International Conference on Indoor Air Quality and Climate, Toronto, Canada 2:51–53.

Newacheck, P. W., P. P. Budetti, and N. Halfon. 1986. Trends in activity-limiting chronic conditions among children. American Journal of Public Health 76:178–194.

Newball, H. H. 1975. The unreliability of the maximal midexpiratory flow as an index of acute airway changes. Chest 67(3):311–314.

NHLBI (National Heart, Lung and Blood Institute). 1982. Epidemiology of Respiratory Diseases. NIH Pub. No. 82-2019. Bethesda, Md.: NHLBI.

NHLBI. 1991. Guidelines for the Diagnosis and Management of Asthma. Expert Panel Report of the National Asthma Education Program. NIH Pub. No. 91-3042. Bethedsa, Md.: NHLBI.

NHLBI. 1992. International Consensus Report on the Diagnosis and Treatment of Asthma. European Respiratory Journal 5:601–641.

NIAID (National Institute of Allergy and Infectious Diseases). 1979. Asthma and Other Allergic Diseases. NIH Pub. No. 79-387. Bethesda, Md.: NIAID.

Nicholas, W. M. 1983. Occupational asthma. Current Concepts in Allergy and Clinical Immunology 14:1.

NIOSH (National Institute for Occupational Safety and Health). 1978. Criteria for a Recommended Standard of Occupational Exposure to Diisocyanates. HEW Pub. No. 78-215. Washington, D.C.: U.S. Government Printing Office.

NKB (Nordic Committee on Building Regulations). 1990. Indoor Climate Air Quality. NKB.

Nordman, H., H. Keshinen, and M. Tuppurainen. 1985. Formaldehyde asthma: Rare or overlooked? Journal of Allergy and Clinical Immunology 75:91–99.

Norman, P. S. 1986. Skin testing. In: Manual of Clinical and Laboratory Immunology, N. R. Rose, H. Friedman, and J. L. Fahey, eds. Washington, D.C.: American Society for Microbiology, pp. 660–663.

Noster, U., B. M. Hausen, G. Felten, and K. H. Schulz. 1976. Mushroom worker's lung caused by inhalation of spores of the edible fungus *Pleurotus Florida* ("oyster mushroom"). Deutsche Medizinische Wochenschrift 101(34):1241–1245.

Novey, H. S., M. Habib, and I. D. Wells. 1983. Asthma and IgE antibodies induced by chromium and nickel salts. Journal of Allergy and Clinical Immunology 72:407–412.

NRC (National Research Council). 1981. Indoor Pollutants. Washington, D.C.: National Academy Press.

NRC. 1983a. An Assessment of the Health Risks of Morpholine and Diethylaminoethanol. Washington, D.C.: National Academy Press.

NRC. 1983b. Risk Assessment in the Federal Government: Managing the Process. Washington, D.C.: National Academy Press.

NRC. 1985. Epidemiology and Air Pollution. Washington, D.C.: National Academy Press.

NRC. 1986. Environmental Tobacco Smoke. Washington, D.C.: National Academy Press.

NRC. 1987a. Counting Injuries and Illnesses in the Workplace: Proposals for a Better System. Washington, D.C.: National Academy Press.

NRC. 1987b. Policies and Procedures for Control of Indoor Air Quality in Existing Buildings. Washington, D.C.: National Academy Press.

NRC. 1991. Human Exposure Assessment for Airborne Pollutants, Advances and Opportunities. Washington, D.C.: National Academy Press.

NRC. 1992a. Biologic Markers in Immunotoxicology. Washington, D.C.: National Academy Press.

NRC. 1992b. Multiple Chemical Sensitivities: Addendum to Biologic Markers in Immunotoxicology. Washington, D.C.: National Academy Press.

Nutter, A. F. 1979. Contact urticaria to rubber. British Journal of Dermatology 101:597–598.

Ohman, J. L., and J. R. Lorusso. 1987. Cat allergen content of commercial house dust extracts: Comparison with dust extracts from cat-containing environment. Journal of Allergy and Clinical Immunology 79:955–959.

Ohman, J. L., and B. Sundin. 1987. Standardized allergenic extracts derived from mammals. Clinical Reviews in Allergy 5:37–47.

Ohman, J. L., F. C. Lowell, and K. J. Bloch. 1973. Allergens of mammalian origins: Characterization of allergen extracted from cat pelts. Journal of Allergy and Clinical Immunology 52:231–241.

Ohman, J. L., S. Kendall., and F. C. Lowell. 1977. IgE antibody to cat allergens in an allergic population. Journal of Allergy and Clinical Immunology 60:317–323.

O'Hallaren, M. T., J. W. Yunginger, K. P. Offord, M. J. Somers, E. J. O'Connell, D. J. Ballard, and M. I. Sachs. 1991. Exposure to an aeroallergen as a possible precipitating factor in respiratory arrest in young patients with asthma. New England Journal of Medicine 324(6):359–363.

O'Neil, C. E., J. M. Hughes, B. T. Butcher, J. E. Salvaggio, and S. B. Lehrer. 1988. Basidiospore extracts: Evidence of common antigenic/allergenic epitopes. International Archives of Allergy and Applied Immunology 85:161–166.

O'Rourke, M. K. 1989. Comparative pollen calendars from Tucson, Arizona: Durham vs. Burkard samples. Aerobiologia 6:136–140.

O'Rourke, M. K. 1992. Exposure to Aeroallergens. Presentation at Second Meeting of the Institute of Medicine's Committee on Health Effects of Indoor Allergens, 20–22 February, at Arnold and Mabel Beckman Center, Irvine, Calif.

O'Rourke, M. K., and S. L. Buchmann. 1986. Pollen yield from olive trees cvs. Manzanillo and Swan Hill in closed urban environments. Journal for the American Society of Horticultural Science 111(2):980–984.

O'Rourke, M. K., and M. D. Lebowitz. 1984. A comparison of regional atmospheric pollen with pollen collected at and near homes. Grana 23:55–64.

O'Rourke, M. K., and M. D. Lebowitz. In press. The importance of environmental allergens in the development of allergic and chronic obstructive diseases. In: Environmental Respiratory Diseases, S.L. Demeter, E. Corlisco, and C. Zenz, eds. New York, N.Y.: Van Norstrand Reinholt.

O'Rourke, M. K., J. J. Quackenboss, and M. D. Lebowitz. 1989. An epidemiological approach investigating respiratory disease response in sensitive individuals to indoor and outdoor pollen exposure in Tucson, Arizona. Aerobiologia 5:104–110.

O'Rourke, M. K., J. Quackenboss, and M. D. Lebowitz. 1990. Indoor pollen and mold characterization from homes in Tucson, Arizona, USA. In: Indoor Air '90: The Fifth International Conference on Indoor Air Quality and Climate, vol. 2. Toronto, Canada: Canada Mortgage and Housing Corp., pp. 9–14.

Owen, S., M. Morganstern, J. Hepworth, and A. Woodcock. 1990. Control of house dust mite antigen in bedding. Lancet 335:396–397.

Palmgren, U., G. Strom, B. Blomquist, and P. Malmberg. 1986. Collection of airborne microorganisms on Nuclepore filters, estimation and analysis—CAMNEA method. Journal of Applied Bacteriology 61(5):401–406.

Pan, P. M., H. Burge, H. J. Su, and J. D. Spengler. 1992. Central versus room air conditioning for reducing exposure to airborne fungus spores. Journal of Allergy and Clinical Immunology 89(1, pt. 2):283.

Park, H. S., M. K. Lee, B. O. Kim, K. J. Lee, J. H. Roh, Y. H. Moon, and C. S. Hong. 1991. Clinical and immunologic evaluations of reactive dye-exposed workers. Journal of Allergy and Clinical Immunology 87:639–649.

Parker, C. W. 1980. Clinical Immunology. Philadelphia, Pa.: W. B. Saunders Company.

Pasanen, A. L., P. Kalliokoski, P. Pasanen, T. Salmi, and A. Tossavainen. 1989. Fungi carried from farmers' work into farm homes. American Industrial Hygiene Association Journal 50(12):631–633.

Pasanen, P., M. Hujanen, P. Kalliokoski, et al. 1991. Criteria for changing ventilation filters. In: IAQ '91 Healthy Buildings, Atlanta, Ga.: American Society for Heating, Refrigerating and Air-Conditioning Engineers, pp. 383–385.

Pattemore, P. K., M. I. Asher, A. C. Harrison, E. A. Mitchell, H. H. Rea, and A. W. Stewart. 1990. The interrelationship among bronchial hyperresponsiveness, the diagnosis of asthma, and asthma symptoms. American Review of Respiratory Disease 142:549–554.

Patterson, R. 1985. Allergic Diseases, 3rd ed. Philadelphia, Pa.: J. B. Lippincott Company.

Patterson, R., C. R. Zeiss, M. Roberts, J. J. Pruzansky, D. Wolkonsky, and R. Chacon. 1978. Human anti-hapten antibodies in trimellitic anhydride inhalation reactions: Immunoglobulin classes of anti-trimellitic anhydride antibodies and hapten inhibition studies. Journal of Clinical Investigation 62:971–978.

Patterson, R., W. Addington, A. S. Banner, G. E. Byron, M. Franco, F. A. Herbert, M. Brooke, N. Jacob, J. Pruzansky, M. Rivera, M. Roberts, D. Yawn, and C. R. Zeiss. 1979. Antihapten antibodies in workers exposed to trimellitic anhydride fumes: A potential immunopathogenetic mechanism for the trimellitic anhydride pulmonary disease-anemia syndrome. American Review of Respiratory Disease 120:1259–1267.

Patterson, R., K. M. Nugent, K. E. Harris, and M. E. Eberle. 1990. Immunologic hemorrhagic pneumonia caused by isocyanates. American Review of Respiratory Disease 141:226–230.

Pazur, J. H., B. L. Liu, and F. J. Miskiel. 1990. Comparison of the properties of glucoamylases from *Rhizopus niveus* and *Aspergillus niger. biotechnol.* Applied Biochemistry 12(1):63–78.

Peart, A. F. W. 1952. Canada's sickness survey: Review of methods. Canada Journal of Public Health 43(10):401.

Peat, J. K., W. J. Britton, C. M. Salome, and A. J. Woolcock. 1987. Bronchial hyperresponsiveness in two populations of Australian schoolchildren. III. Effect of exposure to environmental allergens. Clinical Allergy 17:291–300.

Pecegueiro, M., and F. M. Brandao. 1985. Airborne contact dermatitis to plants. Contact Dermatitis 13(4):277–279.

Pennock, B. E., R. M. Rogers, and D. R. McCaffree. 1981. Changes in measured spirometric indices: What is significant? Chest 80:97–99.

Pepys, J., and B. J. Hutchcroft. 1975. Bronchial provocation test in etiology, diagnosis and analysis of asthma. American Review of Respiratory Disease 112:829–859.

Pepys, J., and P. A. Jenkins. 1965. Precipitin (FLH) test in farmer's lung. Thorax 20:21–35.

Pepys, J., C. A. C Pickering, and E. G. Hughes. 1972. Asthma due to inhaled chemical agents: Complex salts of platinum. Clinical Allergy 2:391–396.

Pepys, J., W. E. Parish, O. Cromwell, and E. G. Hughes. 1979. Passive transfer in man and the monkey of type I allergy due to heat labile and heat stable antibody to complex salts of platinum. Clinical Allergy 9:99–108.

Perelmutter, L., A. Lavallee, and G. S. Wiberg. 1972. The effect of laundry detergent containing *B. subtilis* enzyme in producing sensitization reactions in guinea pigs. International Archives of Allergy and Applied Immunology 42(2):214–224.

Perks, W. H., P. S. Burge, M. Rehahn, and M. Green. 1979. Work-related respiratory disease in employees leaving an electronic factory. Thorax 34:19.

Perrin, B., A. Cartier, H. Ghezzo, L. Grammer, K. Harris, H. Chan, M. Chan-Yeung, and J. Luc-Malo. 1991. Reassessment of the temporal patterns of bronchial obstruction after exposure to occupational sensitizing agents. Journal of Allergy and Clinical Immunology 87(3):630–639.

Perrin, B., F. Lagier, J. L'Archeveque, A. Cartier et al. 1992. Occupational asthma: Validity of monitoring of peak expiratory flow rates and non-allergic bronchial responsiveness as compared to specific inhalation challenge. European Respiratory Journal 5:40–48.

Petry, R. W., M. J. Voss, L. A. Kroutel, W. Crowley, R. K. Bush, and W. W. Busse. 1985. Monkey dander asthma. Journal of Allergy and Clinical Immunology 75:268–271.

Pickering, C. A. C. 1972. Inhalation tests with chemical allergens: Complex salts of platinum. Proceedings of the Royal Society of Medicine 65:272–274.

Pierson, T. K., R. G. Hetes, and D. F. Naugle. 1991. Risk characterization framework for noncancer end points. Environmental Health Perspectives 95:121–129.

Platt, S. D., C. J. Martin, S. M. Hunt, and C. W. Lewis. 1989. Damp housing mould growth and symptomatic health state. British Medical Journal 298:1673–1678.

Platts-Mills, T., and M. Chapman. 1987. Dust mites: Immunology, allergic disease, and environmental control [review]. Journal of Allergy and Clinical Immunology 80:755–775. Published erratum appears in Journal of Allergy and Clinical Immunology 82(5, pt. 1):841, 1988.

Platts-Mills, T., and M. Chapman. 1991. Allergen standardization. Journal of Allergy and Clinical Immunology 87:621–625.

Platts-Mills, T., and A. L. de Weck. 1989. Dust mite allergens and asthma: A world wide problem. Journal of Allergy and Clinical Immunology 83:416–427.

Platts-Mills, T., E. R. Tovey, E. B. Mitchell, H. Moszoro, P. Nock, and S. R. Wilkins. 1982. Reduction of bronchial hyperreactivity during prolonged allergen avoidance. Lancet 2:675–678.

Platts-Mills, T., P. W. Heymann, J. L. Longbottom, and S. R. Wilkins. 1986. Airborne allergens associated with asthma: Particle sizes carrying dust mite and rat allergens measured with a cascade impactor. Journal of Allergy and Clinical Immunology 77:850–857.

Platts-Mills, T., M. L. Hayden, M. D. Chapman, and S. R. Wilkins. 1987. Seasonal variation in dust mite and grass-pollen allergens in dust from the houses of patients with asthma. Journal of Allergy and Clinical Immunology 79:781–791.

Platts-Mills, T., M. Chapman, S. Pollart, P. Heymann, and C. Luczynska. 1988. Establishing health standards for indoor levels of foreign proteins. Paper 88-110.1. Dallas, Tex.: Air Pollution Control Association.

Platts-Mills, T., G. W. Ward, R. Sporik, L. Gelber, M. D. Chapman, and P. Heymann. 1991a. Epidemiology of the relationship between exposure to indoor allergens and asthma. International Archives of Allergy and Applied Immunology 94:339–345.

Platts-Mills, T., M. D. Chapman, B. Mitchell, P. W. Heymann, and B. Duell. 1991b. Role of inhalant allergens in atopic eczema. In: Handbook of Atopic Eczema, T. Ruzieka, J. Ring and B. Prrzyb, eds. New York, N.Y.: Springer-Verlag.

Platts-Mills, T., W. R. Thomas, R. C. Aalberse, D. Vervloet, and M. D. Chapman. 1992. Dust mite allergens and asthma: Report of a second international workshop. Journal of Allergy and Clinical Immunology 89(5):1046–1060.

Plaut, M., J. H. Pierce, C. J. Watson, et al. 1989. Mast cell lines produce lymphokines in response to cross-linkage of $F_{C_E}RI$ or to calcium ionophores. Nature 339:64.

Polla, B. S., R. de Haller, G. Nerbollier, and R. Rylander. 1988. Maladie des humidificateurs: Role des endotoxines et des anticorps précipitants. Schweizerische Medizinische Wochenschrift 118(37):1311–1313.

Pollart, S. M., M. J. Reid, J. A. Fling, M. D. Chapman, and T. Platts-Mills. 1988. Epidemiology of emergency room asthma in northern California: Association with IgE antibody to ryegrass pollen. Journal of Allergy and Clinical Immunology 82:224–230.

Pollart, S. M., M. D. Chapman, G. P. Fiocco, G. Rose, and T. Platts-Mills. 1989. Epidemiology of acute asthma: IgE antibodies to common inhalant allergens as a risk factor for emergency room visits. Journal of Allergy and Clinical Immunology 83:875–882.

Pollart, S. M., T. F. Smith, E. C. Morris, L. E. Gelber, T. Platts-Mills, and M. D. Chapman. 1991. Environmental exposure to cockroach allergens: Analysis with monclonal antibody-based enzyme immunoassays. Journal of Allergy and Clinical Immunology 87:505–510.

Popa, V., and J. Singleton. 1988. Provocative dose and discriminate analysis in histamine bronchoprovocation: Are the current predictive data satisfactory? Chest 94:466–475.

Price, J. A., and J. L. Longbottom. 1986. Allergy to rabbits. I. Specificity and non-specificity of RAST and crossed-radioelectrophoresis due to presence of light chains in rabbit allergenic extracts. Allergy 41:603–612.

Price, J. A., I. Pollock, S. A. Little, J. L. Longbottom, and J. O. Warner. 1990. Measurements of airborne mite antigen in homes of asthmatic children. Lancet 336:895–897; see also Letter, Lancet 337:1038, 1991.

Quackenboss, J., M. D. Lebowitz, and C. Hayes. 1989a. Epidemiological study of respiratory responses to indoor/outdoor air quality. Environment International 15:493.

Quackenboss, J., M. D. Lebowitz, C. Hayes, and C. L. Young. 1989b. Respiratory responses to indoor/outdoor air pollutants: Combustion products, formaldehyde, and particulate matter. In: Combustion Processes and the Quality of the Indoor Air Environment, J. Harper, et al., eds. Pittsburgh, Pa.: Air Pollution Control Association, pp. 280–293.

Quackenboss, J., M. Krzyzanowski, and M. D. Lebowitz. 1991a. Exposure assessment approaches to evaluate respiratory health effects of particulate matter and nitrogen dioxide. Journal of Exposure Analysis and Environmental Epidemiology 1:83–107.

Quackenboss, J., M. D. Lebowitz, and M. Krzyzanowski. 1991b. The normal range of diurnal changes in peak expiratory flow rates: Relationship to symptoms and respiratory disease. American Review of Respiratory Disease 143:323–330.

Quinlan, P., J. M. Macher, L. E. Alevantis, and J. E. Cone. 1989. Protocol for the comprehensive evaluation of building-associated illness. Occupational Medicine—State of the Art Reviews 4:771–797.

Rankin, J. A., M. Hitchcock, W. Merrill, et al. 1982. IgE-dependent release of leukotriene C4 from alveolar macrophages. Nature 297:329.

Reed, C. E., and M. C. Swanson. 1987. Antigens and allergic asthma. Chest 91(6, suppl.):161S–165S.

Reed, C. E., M. Swanson, M. Lopez, A. M. Ford, J. Major, W. B. Witmer, and T. B. Valdes. 1983. Measurement of IgG antibody and airborne antigen to control an industrial outbreak of hypersensitivity pneumonitis. Journal of Occupational Medicine 25:207–210.

Regli, S., J. B. Rose, C. N. Haas, and C. P. Gerba. 1991. Modeling the risk from Giardia and viruses in drinking water. Journal of American Water Works Association 83(11):76–84.

Reisman, R. E., P. M. Mauriello, G. B. Davis, J. W. Georgitis, and J. M. DeMasi. 1990. A double-blind study of the effectiveness of a high-efficiency particulate air (HEPA) filter in the treatment of patients with perennial allergic rhinitis and asthma. Journal of Allergy and Clinical Immunology 85(6):1050–1057.

Renzetti, A. D., Jr., E. R. Bleecker, G. R. Epler, R. N. Jones, R. E. Kanner, and L. H. Repsher. 1986. American Thoracic Society Ad Hoc Committee on Impairment/Disability Criteria: Evaluation of impairment/disability secondary to respiratory disorders. American Review of Respiratory Disease 133:1205–1209.

Reynolds, H. Y. 1988. Bronchoalveolar lavage. In: Textbook of Respiratory Medicine, J. F. Murray, and J. Nadel, eds. Philadelphia, Pa.: W. B. Saunders Company.

Richman, P. G., and D. S. Cissell. 1984. Total protein in the ninhydrin method. In: Methods of the Laboratory of Allergenic Products, Bethesda, Md.: U.S. Food and Drug Administration, p. 186.

Riedel, F., M. Kramer, C. Scheibenbogen, and C. H. L. Rieger. 1988. Effects of SO_2 exposure on allergic sensitization in the guinea pig. Journal of Allergy and Clinical Immunology 82:527–534.

Roberts, A. E. 1951. Platinosis: A five year study of the effects of soluable platinum salts on employees in a platinum laboratory and refinery. Archives of Industrial Hygiene and Occupational Medicine 4:549–559.

Robertson, G. 1988. Sources, nature, and symptomology of indoor air pollutants. In: Proceedings of Healthy Buildings 1988, vol. 3, B. Berglund, and T. Lindvall, eds. Stockholm, Sweden: Swedish Council for Building Research, pp. 507–516.

Robinson, D. S., Q. Hamid, S. Ying, et al. 1992. Predominant TH2-like bronchoalveolar T-lymphocyte population in atopic asthma. New England Journal of Medicine 326:298–304.

Roitt, I. 1988. Essential Immunology. London, England: Blackwell Scientific.

Rose, C., and T. E. King, Jr. 1992. Controversies in hypersensitivity pneumonitis. American Review of Respiratory Disease 145:1–2.

Rose, C. S. 1992. Water-related lung diseases. In: Occupational Lung Disease. Occupational Medicine—State of the Art Reviews, W. S. Beckett and R. Bascom, eds. Philadelphia, Pa.: Hanley and Belfus, pp. 271–286.

Rose, G., J. A. Woodfolk, M. L. Hayden, and T. Platts-Mills. 1992. Testing of methods to control mite allergen in carpets fitted to a concrete slab [abstract]. Journal of Allergy and Clinical Immunology 89:315.

Rose, J. B., and C. P. Gerba. 1991. Assessing potential health risks from viruses and parasites in reclaimed water in Arizona and Florida. Water Science and Technology 23(10–12):2091–2098.

Rosenwasser, L. J. 1986. Monocyte and macrophage function. In: Manual of Clinical Laboratory Immunology. N. R. Rose, H. Friedman, and J. L. Fahey, eds. American Society for Microbiology, Washington, D.C., p. 321.

Rost, G. A. 1932. Ueber Erfagrungen mit der allergenfreien Kammer nach Storm van Leeuwen: Insbesonderen der Spatperiode der exsudativen Diathese. Archives of Dermatology and Syphilis 155:297.

Saad, R., and A. A. el-Gindy. 1990. Fungi of the house dust in Riyadh, Saudi Arabia. Zentralblatt fuer Mikrobiologie 145(1):65–68.

Sakaguchi, M., S. Inouye, H. Yasueda, T. Irie, S. Yoshizawa, and T. Shida. 1989. Measurement of allergens associated with dust mite allergy. II. Concentrations of airborne mite allergens (Der I and Der II) in the house. International Archives of Allergy and Applied Immunology 90(2):190–193.

Salvaggio, J. 1979. NIAID Task Force Report: Asthma and Other Allergic Diseases. U.S. Department of HEW. NIH Publication No. 79-387. Washington, D.C.: U.S. Government Printing Office, p. 330.

Samet, J. 1990. Environmental Controls and Lung Disease: Report of the ATS (American Thoracic Society) Workshop on Environmental Controls and Lung Disease, Santa Fe, New Mexico, March 24–26, 1988. American Review of Respiratory Disease 142:915–939.

Samson, R. A. 1985. Occurrence of moulds in modern living and working environments. European Journal of Epidemiology 1(1):54–61.

Samter, M. 1988. Immunologic Diseases, 4th ed. Boston, Mass.: Little, Brown.

Santilli, J., W. J. Rockwell, and R. P. Collins. 1990. Individual patterns of immediate skin reactivity to mold extracts. Annals of Allergy 65(6):454–458.

Savolainen, J., M. Viander, R. Einarsson, and A. Koivikko. 1989. Allergenic variability of different strains of Candida albicans. International Archives of Allergy and Applied Immunology 90(1):61–66.

Savolainen, J., M. Viander, R. Einarsson, and A. Koivikko. 1990. Immunoblotting analysis of concanavalin A-isolated allergens of Candida albicans. Allergy 45(1):40–46.

SBCCI (Southern Building Code Congress International). 1990. 1988 Standard Mechanical Code, 1989/1990 Revisions. Birmingham, Ala.: SBCCI.

Schicht, H. 1972. The diffusion of microorganisms by air conditioning installations. Steam and Heating Engineer October:6–13.

Schlesselman, J. J. 1982. Case Control Studies. Oxford: Oxford University Press 32:4.

Schlicting, H. E. 1969. The importance of airborne algae and protozoa. Journal of the Air Pollution Control Association 19:946.

Schmidt, E. A., B. M. Cannan, R. C. Mulhall, and D. L. Coleman. 1986. Brief report: Effects of ultra high speed floor burnishing on air quality in health care facilities. Infection Control 7:501–505.

Schoettlin, C. E., and E. Landau. 1961. Air pollution and asthmatic attacks in the Los Angeles area. Public Health Reports 76:545.

Schou, C., G. N. Hansen, T. Lintner, and H. Lowenstein. 1991a. Assay for major dog allergen Can f I: Investigation of house dust samples and commercial dog extracts. Journal of Allergy and Clinical Immunology 88:847–853.

Schou, C., U. G. Svendsen, and H. Lowenstein. 1991b. Purification and characterization of the major dog allergen, Can f I. Clinical and Experimental Allergy 21:321–328.

Schumacher, M. J., R. D. Griffith, and M. K. O'Rourke. 1988. Recognition of pollen and other particulate aeroantigens by immunoblot microscopy. Journal of Allergy and Clinical Immunology 82(4):608–616.

Schwartz, H. J., K. M. Citron, E. H. Chester, J. Kaimal, P. B. Barlow, G. L. Baum, and M. R. Schuyler. 1978. A comparison of sensitization to *Aspergillus* antigens among asthmatics in Cleveland and London. Journal of Allergy and Clinical Immunology 62:9–14.

Schwartz, J., D. Gold, D. W. Dockery, S. T. Weiss, and F. E. Speizer. 1990. Predictors of asthma and persistent wheeze in a national sample of children in the United States. American Review of Respiratory Disease 142:555–562.

Sears, M. R., G. P. Hervison, M. D. Holdaway, C. J. Hewitt, E. M. Flannery, and P. A. Silva. 1989. The relative risks of sensitivity to grass pollen, house dust mite, and cat dander in the development of childhood asthma. Clinical and Experimental Allergy 19:419–424.

Seaton, A., B. Cherrie, and J. Turnbull. 1988. Rubber glove asthma. British Medical Journal 296:531–532.

Sharma, O. P., and O. J. Balchum. 1983. Key facts in pulmonary disease. New York, N.Y.: Churchill Livingstone.

Shen, H. D., W. L. Lin, R. J. Chen, and S. H. Han. 1990. Cross-reactivity among antigens of different air-borne fungi detected by ELISA using five monoclonal antibodies against *Penicillium notatum*. Chung Hua I Hsueh Tsa Chih 46(4):196–201.

Sheppard, D. 1988a. Occupational asthma and byssinosis. In: Textbook of Respiratory Medicine, J. F. Murray and J. Nadel, eds. Philadelphia, Pa.: W. B. Saunders Company.

Sheppard, D. 1988b. Sulfur dioxide and asthma—A double-edged sword? Journal of Allergy and Clinical Immunology 82:961–964.

Sher, A., R. L. Coffman, S. Hieny, P. Scott, and A. W. Cheever. 1990. Interleukin 5 (IL-5) is required for the blood tissue eosinophilia but not granuloma formation induced by infection with Schistosoma mansoni. Proceedings of the National Academy of Sciences USA 87:61.

Shield, S., M. S. Blaiss, and S. Gross. 1992. Prevalence of latex sensitivity in children evaluated for inhalant allergy [abstract]. Journal of Allergy and Clinical Immunology 89(1):224.

Shields, M. C., K. W. Griffin, and W. L. McNabb. 1990. The effect of a patient education program on emergency room use for inner-city children with asthma. American Journal of Public Health 80:36–38.

Shoaf, C. R. 1991. Current assessment practices for noncancer end points. Environmental Health Perspectives 95:111–119.

Shulan, D. J., J. M. Weiler, F. Koontz, and H. B. Richerson. 1985. Contamination of intradermal skin test syringes. Journal of Allergy and Clinical Immunology 76:226–227.

Sibbald, B. 1980. Genetic basis of sex differences in the prevalence of asthma. British Journal of Diseases of the Chest 74:93–94.

Sibbald, B., M. E. C Horn, E. A. Brain, and I. Gregg. 1980. Genetic factors in childhood asthma. Thorax 35:(9)671–674.

Silverman, H. I. 1989. Rubber anaphylaxis [letter]. New England Journal of Medicine 321:837.

Siraganian, R. P., and W. A. Houk. 1986. Histamine release assay. In: Manual of Clinical Laboratory Immunology, N. R. Rose, H. Friedman, and J. L. Fahey, eds. Washington, D.C.: American Society for Microbiology, pp. 675–684.

Sjostedt, L., and S. Willers. 1989. Predisposing factors in laboratory animal allergy: A study of atopy and environmental factors. American Journal of Industrial Medicine 16:199–208.

Slater, J. E. 1989. Rubber anaphylaxis. New England Journal of Medicine 320:1126–1130.

Slater, J. E., and S. K. Chhabra. 1992. Latex antigens. Journal of Allergy and Clinical Immunology 89(3):673–678.

Slater, J. E., C. Shaer, and L. A. Mostello. 1990a. Rubber-specific IgE in children with spina bifida [abstract]. Journal of Allergy and Clinical Immunology 85(pt. 2):293.

Slater, J. E., L. A. Mostello, C. Shaer, and R. W. Honsinger, Jr. 1990b. Type I hypersensitivity to rubber. Annals of Allergy 65:411–414.

Slater, J. E., L. A. Mostello, and C. Shaer. 1991. Rubber-specific IgE in children with spina bifida. Journal of Urology 146:578–579.

Slavin, R. 1989. Nasal polyps and sinusitis. In: Allergy Principles and Practice, 3rd ed., E. Middleton, C. Reed, E. F. Ellis, N. F. Adkinson, Jr., and J. W. Yunginger, eds. St. Louis, Mo.: C. V. Mosby Company, pp. 1299–1303.

Slovak, A. J. M. 1981. Occupational asthma caused by a plastics blowing agent, azodicarbonamide. Thorax 36:906–909.

Slovak, A. J. M. 1987. Achieved objectives in laboratory animal allergy research: Their significance for policy and practice. New England Regional Allergy Proceedings 8:189–194.

Sly, R. M. 1988. Mortality from asthma 1979–1984. Journal of Allergy and Clinical Immunology 82:705–717.

Smith, A. B., D. I. Bernstein, M. A. London, J. Gallagher, G. A. Ornella, S. K. Gelletly, K. Wallingford, and M. A. Newman. 1990. Evaluation of occupational asthma from airborne egg protein exposure in multiple settings. Chest 98:398–404.

Smith, J. M. 1988. Epidemiology and natural history of asthma, allergic rhinitis, and atopic dermatitis (eczema). In: Allergy Principles and Practice, 3rd ed., E. Middleton, C. E. Reed, E. F. Ellis, N. F. Adkinson, Jr., and J. W. Yunginger, eds. St. Louis, Mo.: C. V. Mosby Company, pp. 891–929.

Smith, J. M., and L. A. Knowles. 1965. Epidemiology of asthma and allergic rhinitis. I. In a rural area. II. In a university-centered community. American Review of Respiratory Disease 92:15–18.

Smith, J. M., M. E. Disney, J. D. Williams, and Z. A. Goels. 1969. Clinical significance of skin reactions to mite extracts in children with asthma. British Medical Journal 1:723–726.

Snider, G. L. 1988. Obstructive diseases: Chronic bronchitis and emphysema. In: Textbook of Respiratory Medicine, J. F. Murray and J. A. Nadel, eds. Philadelphia, Pa.: W. B. Saunders Company, pp. 1069–1106.

Snyder, S. L., and P. Z. Sobocinski. 1975. An improved 2,4,6-trinitrobenzenesulfonic acid method for the determination of amines. Annals of Biochemistry 64:284–288.

Solomon, W. R. 1974. Fungus aerosols arising from cold-mist vaporizers. Journal of Allergy and Clinical Immunology 54:222–228.

Solomon, W. R. 1990. Airborne microbial allergens: Impact and risk assessment. Toxicology and Industrial Health 6(2):309–324.

Solomon, W. R., and K. P. Matthews. 1988. Aerobiology and inhalant allergens. In: Allergy Principles and Practice, 3rd ed., E. Middleton, Jr., C. E. Reed, E. F. Ellis, N. F. Adkinson, Jr., and J. W. Yunginger, eds. St. Louis, Mo.: C. V. Mosby Company, pp. 312–372.

Solomon, W. R., H. A. Burge, and J. R. Boise. 1980a. Exclusion of particulate allergens by window air conditioners. Journal of Allergy and Clinical Immunology 65:304–308.

Solomon, W. R., H. A. Burge, and J. R. Boise. 1980b. Comparative particle recoveries by the retracting Rotorod, Rotoslide, and Burkard Spore Trap sampling in a compact array. International Journal of Biometeorology 24:107–116.

Spanner, D., J. Dolovich, S. Tarlo, G. Sussman, and K. Buttoo. 1989. Hypersensitivity to natural latex. Journal of Allergy and Clinical Immunology 83:1135–1137.

Sparrow, D., G. O'Connor, and S. T. Weiss. 1988. The relation of airways responsiveness and atopy to the development of chronic obstructive lung disease. Epidemiologic Reviews 10:29–47.

Spengler, J. D., and K. Sexton. 1983. Indoor air pollution: A public health perspective. Science 221:9–17.

Spiegelman, J., and H. Friedman. 1968. The effect of central air filtration and air conditioning on pollen and microbial contamination. Journal of Allergy and Clinical Immunology 42:193–203.

Spiegelman, J., F. Friedman, and G. I. Blumstien. 1963. The effect of central air conditioning on pollen, mold and mocrobial contamination. Journal of Allergy and Clinical Immunology 34:426–286.

Spivacke, C. A., and E. F. Grove. 1925. Studies in hypersensitiveness. XIV. A study of the atopen in house dust. Journal of Immunology 10:465.

Sporik, R., S. T. Holgate, T. Platts-Mills, and J. Cogswell. 1990. Exposure to house dust mite allergen (Der p I) and the development of asthma in childhood: A prospective study. New England Journal of Medicine 323:502–507.

Sporik, R. B., T. Platts-Mills, and J. J. Cogswell. In press. Exposure and sensitization to house dust mite allergen of children admitted to hospital with asthma. Clinical and Experimental Allergy.

Stankus, R. P., J. E. Morgan, and J. E. Salvaggio. 1982. Immunology of hypersensitivity pneumonitis. CRC Critical Reviews in Toxicology 11:15–32.

Stites, D. P., and A. I. Terr. 1991. Basic and Clinical Immunology, 7th ed. East Norwalk, Conn.: Appleton & Lange.

Stolwijk, J. J. 1984. The "sick building" syndrome. In: Proceedings of the Third International Conference on Indoor Air Quality, vol. 1, B. Berglund, T. Lindvall, and J. Sundell, eds. Stockholm, Sweden: Swedish Council for Building Research, pp. 23–29.

Stone, C. A., G. C. Johnson, J. D. Thornton, B. J. Macauley, P. W. Holmes, and E. H. Tai. 1989. Leucogyrophana pinastri, a wood decay fungus as a probable cause of an extrinsic allergic alveolitis syndrome. Australian and New Zealand Journal of Medicine 19(6):727–729.

Storm van Leeuwen, W., W. Einthoven, and W. Kremer. 1927. The allergen proof chamber in the treatment of bronchial asthma and other respiratory diseases. Lancet 1:1287–1289.

Strachan, D. P., B. Flannigan, E. M. McCabe, and F. McGarry. 1990. Quantification of airborne moulds in the homes of children with and without wheeze. Thorax 45(5):382–387.

Streifel, A., D. Stevens, and F. Rhame. 1987. In-hospital source of airborne penicillum spores. Journal of Allergy and Clinical Immunology 25:1–4.

Strom, G., B. Hellstrom, and A. Kumlin. 1990. The sick building syndrome: An effect of microbial growth in building constructions? In: Indoor Air '90: The Fifth International Conference on Indoor Air Quality and Climate, vol. 4. Toronto, Canada: Canada Mortgage and Housing Corp., pp. 173–178.

Su, H. J., H. Burge, and J. D. Spengler. 1989. Microbiological contamination in the residential environment. Joint Canadian and PanAmerican Symposium on Aerobiology and Health. Ottawa, Ontario, Canada, June 7–9, 1989.

Su, H. J., H. Burge, and J. D. Spengler. 1990. Indoor saprophytic aerosols and respiratory health. Journal of Allergy and Clinical Immunology 85(1, pt. 2):248.

Su, H. J., H. Burge, and J. D. Spengler. 1992. Association of airborne fungi and wheeze/asthma symptoms in school-age children [abstract]. Journal of Allergy and Clinical Immunology 89(part 2):251.

Sulzberger, M. B. 1971. Atopic dermatitis, Part III. In: Dermatology in General Medicine, T. B. Fitzpatrick, et al., eds. New York: McGraw-Hill Book Company, pp. 687–697.

Sulzer-Azaroff, B., and G. R. Mayer. 1991. Behavior Analysis for Lasting Change. Fort Worth, Tex.: Holt, Rinehard and Winston, Inc.

Svartengren, M., R. Falk, L. Linnman, K. Philipson, and P. Camner. 1987. Deposition of large particles in human lung. Experimental Lung Research 12:75–88.

Swanson, M. C., M. K. Agarwal, J. W. Yunginger, and C. E. Reed. 1984. Guinea-pig-derived allergens: Clinicoimmunologic studies, characterization, airborne quantitation, and size distribution. American Review of Respiratory Disease 129:844–849.

Swanson, M. C., M. K. Agarwal, and C. E. Reed. 1985. An immunochemical approach to indoor aeroallergen quantitation with a new volumetric air sampler: Studies with mite, roach, cat, mouse, and guinea pig antigens. Journal of Allergy and Clinical Immunology 76(5):724–729.

Swanson, M. C., A. R. Campbell, M. J. Klauck, and C. E. Reed. 1989. Correlations between levels of mite and cat allergens in settled and airborne dust. Journal of Allergy and Clinical Immunology 83(4):776–783.

Swanson, M. C., A. R. Campbell, M. T. O'Hollaren, and C. E. Reed. 1990. Role of ventilation, air filtration, and allergen production rate in determining concentrations of rat allergens in the air of animal quarters. American Review of Respiratory Disease 141(6):1578–1581.

Swanson, M. C., M. E. Bubak, L. W. Hunt, and C. E. Reed. 1992. Occupational respiratory allergic disease from latex [abstract]. Journal of Allergy and Clinical Immunology 89(1):227.

Sweet, L., J. Anderson, Q. Callies, and E. Coates. 1971. Hypersensitivity pneumonitis related to a home furnace humidifier. Journal of Allergy and Clinical Immunology 48:171–178.

Tarlo, S., A. Radkin, and R. S. Tobin. 1988. Skin testing with extracts of fungal species derived from the homes of allergy clinic patients in Toronto, Canada. Clinical Allergy 18:45–52.

Tarlo, S. M., L. Wong, J. Roos, and N. Booth. 1990. Occupational asthma caused by latex in a surgical glove manufacturing plant. Journal of Allergy and Clinical Immunology 85(3):626–631.

Task Group on Lung Dynamics. 1966. Deposition and retention models for internal dosimetry of the human respiratory tract. Health Physics 12:173–207.

Taylor, R. N. 1986. Quality control in immunoserology. In: Manual of Clinical and Laboratory Immunology, N. R. Rose, H. Friedman, and J. L. Fahey, eds. Washington, D.C.: American Society for Microbiology, pp. 951–956.

Thoresen, C. E., and K. Kirmil-Gray. 1983. Self-management psychology and the treatment of childhood asthma. Journal of Allergy and Clinical Immunology 72:596–606.

Thorpe, S. C., D. M. Kemeny, R. C. Panzani, B. McGurl, and M. Lord. 1988. Allergy to castor bean. II. Identification of the major allergen in castor bean seeds. Journal of Allergy and Clinical Immunology 82:67–72.

Topping, M. P., K. M. Venables, C. M. Luczynska, W. Howe, and A. J. Newman Taylor. 1986. Specificity of the human IgE response to inhaled acid anhydrides. Journal of Allergy and Clinical Immunology 77:834–842.

Tovey, E. R. 1982. The quantitation and distribution of house dust mite allergens. Ph.D. Thesis. Sydney University, Sydney, Australia.

Tovey, E. R., M. D. Chapman, C. W. Wells, and T. Platts-Mills. 1981a. The distribution of dust mite allergen in the houses of patients with asthma. American Review of Respiratory Disease 124:630–635.

Tovey, E. R., M. D. Chapman, and T. Platts-Mills. 1981b. Mite faeces are a major source of house dust allergens. Nature 289:592–593.

Tovey, E. R., G. R. Marks, M. Matthews, W. F. Green, and A. Woolcock. 1992. Changes in mite allergen Der p I in house dust following spraying with a tannic acid/acaricide solution. Clinical and Experimental Allergy 22:67–74.

Turjanmaa, K. 1987. Incidence of immediate allergy to latex gloves in hospital personnel. Contact Dermatitis 17:270–275.

Turjanmaa, K., T. Reunala, R. Tuimala, and T. Karkkainen. 1984. Severe IgE mediated allergy to surgical gloves [abstract]. Allergy (Suppl. 2).

Turjanmaa, K., K. Laurila, S. Makinen-Kiljunen, and T. Reunala. 1988. Rubber contact urticaria: Allergenic properties of 19 brands of latex gloves. Contact Dermatitis 19:362–367.

Twarog, F. J., F. J. Picone, R. S. Strunk, J. So, and H. R. Colten. 1976. Immediate hypersensitivity to cockroach: Isolation and purification of the major antigens. Journal of Allergy and Clinical Immunology 59:154.

Twiggs, J. T., M. K. Agarawal, M. J. Dahlberg, and J. W. Yunginger. 1982. Immunochemical measurement of airborne mouse allergens in laboratory animal facility. Journal of Allergy and Clinical Immunology 69:522–526.

Tyndall, R., C. Dudney, D. Katz, et al. 1989. Characterization of microbial content and dispersion from home humidifiers. Report to Consumer Product Safety Commission on Project No. 86-1283, January 25, 1989.

U.S. Department of Energy. 1985. Nonresidential building energy consumption survey: Characteristics of commercial buildings, 1983. DOE/EIA-0246(83). Washington, D.C.: Energy Information Administration.

U.S. Department of Energy. 1986. Residential building energy consumption survey: Housing characteristics of 1984. DOE/EIA-0314(84). Washington, D.C.: Energy Information Administration.

Unanue, E. R. and J. C. Cerottini. 1989. Antigen presentation. FASEB Journal 3:2496.

Van Arsdel, P. P., Jr., and E. B. Larson. 1989. Diagnostic tests for patients with suspected allergic disease: Utility and limitations. Annals of Internal Medicine 110:304–312.

Van As, A. 1982. The accuracy of peak expiratory flow meters. Chest 82(3):263.

Van Bronswijk, J. E. M. H. 1981. House Dust Biology. Zoelmand, The Netherlands: N.I.B. Publishers.

Van Hecke, E., P. Hindryckx, J. M. C. Geuns, and E. Devriese. 1991. Airborne contact dermatitis from coleus in a housewife. Contact Dermatitis 25(2):128–129.

Van Metre, T. E., Jr., D. G. Marsh, N. F. Adkinson, Jr., J. E. Fish, A. Kagey-Sobotka, D. S. Norman, E. B. Radden, Jr., and G. L. Rosenberg. 1986. Dose of cat (*Felis domesticus*) allergen (*Fel d* I) that induces asthma. Journal of Allergy and Clinical Immunology 78:62–75.

Vannier, W. E., and D. H. Campbell. 1961. A starch block electrophoresis study of aqueous house dust extracts. Journal of Allergy 32:36.

Verbrugge, L. M., and C. E. Depner. 1981. Methodological analyses of Detroit health diaries. In: NCHSR Research Proceedings Series Health Survey Research Methods, DHHS Publication No. (PHS) 81-3268. Hyattsville, Md.: U.S. Department of Health and Human Services.

Verhoeff, A. P., J. H. van Wijnen, J. S. Boleij, B. Brunekreef, E. S. van Reenen-Hoekstra, and R. A. Samson. 1990a. Enumeration and identification of airborne viable mould propagules in houses: A field comparison of selected techniques. Allergy 45(4):275–284.

Verhoeff, A. P., J. H. van Wijnen, P. Fischer, B. Brunekreef, J. S. Boleij, E. S. van Reenen, and R. A. Samson. 1990b. Presence of viable mould propagules in the indoor air of houses. Toxicology and Industrial Health 6(5):133–145.

Vervloet, D., P. Bongrand, A. Arnaud, C. Boutin, and J. Charpin. 1979. Objective immunological and clinical data observed during an altitude cure at Briançon in asthmatic children allergic to house dust and Dermatophagoides. Revue Française des Maladies Respiratoires 7:19–27.

Virtanen, T., K. Louhelainen, and R. Montyjarvi. 1986. Enzyme-linked immunosorbent assay (ELISA) inhibition method to estimate the level of airborne bovine epidermal antigen in cowsheds. International Archives of Allergy and Applied Immunology 81:253–257.

Voller, A., and D. Bidwell. 1986. Enzyme-linked immunosorbent assays. In: Manual of Clinical and Laboratory Immunology, N. R. Rose, H. Friedman, and J. L. Fahey, eds. Washington, D.C.: American Society for Microbiology, pp. 99–109.

Voorhorst, R., F. T. M. Spieksma, H. Varekamp, M. J. Leupen, and A. W. Lyklema. 1967. The house dust mite (*Dermatophagoides pteronyssinus*) and the allergens it produces: Identity with the house dust allergen. Journal of Allergy 39:325–339.

Walshaw, M. J., and C. C. Evans. 1986. Allergen avoidance in house dust mite sensitive adult asthma. Quarterly Journal of Medicine 58:199–215.

Warner, J. O., and A. L. Boner. 1988. Allergy and childhood asthma. In: Clinical Immunology and Allergy, A. B. Kay, ed. London, England: Bailliere Tindall, pp. 217–224.

Warner, J. A., and J. L. Longbottom. 1991. Allergy to rabbits. III. Further identification and characterization of rabbit allergens. Allergy 46:481–491.

Wassenaar, D. P. J. 1988a. Effectiveness of vacuum cleaning and wet cleaning in reducing house dust mites, fungi and mite allergen in a cotton carpet: A case study. Experimental and Applied Acarology 4:53–62.

Wassenaar, D. P. J. 1988b. Reducing house dust mites by vacuuming. Experimental and Applied Acarology 4:167–171.

Webb, J. L. 1978. Pollen analysis in Southwestern archaeology. In: Discovering Past Behavior, P. Grebenger, ed., New York, N.Y.: Gordon & Breach, Inc., pp. 13–28.

Weiss, K. B., P. J. Gergen, and E. F. Crain. 1992a. Inner-city asthma: The epidemiology of an emerging U.S. public health concern. Chest 101(6):362–367.

Weiss, K. B., P. J. Gergen, and T. A. Hodgson. 1992b. An economic evaluation of asthma in the United States. New England Journal of Medicine 326:862–866.

Weiss, N. S., and J. M. Rubin. 1980. Practical Points in Allergy, 2nd ed. Garden City, N.Y.: Medical Examination Publishing Company.

Weiss, S., I. Tager, M. Schenker, and F. Speizer. 1983. State of the art: The health effects of involuntary smoking. American Review of Respiratory Disease 128:933–942.

Weiss, S., T. Tosteson, M. Segal, I. Tager, S. Redline, and F. Speizer. 1992. Effects of asthma on pulmonary function in children. American Review of Respiratory Disease 145:58–64.

Weissman, D. N., L. Halmepuro, J. E. Salvaggio, and S. B. Lehrer. 1987. Antigenic/allergenic analysis of basidiomycete cap, mycelia, and spore extracts. International Archives of Allergy and Applied Immunology 84(1):56–61.

Weitzman, M., S. L. Gortmaker, A. M. Sobol, and J. M. Perrin. 1992. Recent trends in the prevalence and severity of childhood asthma. Journal of the American Medical Association 268:2673–2677.

West, M., and E. Hansen. 1989. Determination of material hygroscopic properties that affect indoor air quality. In: IAQ '89, Atlanta, Ga.: American Society of Heating, Refrigerating and Air-Conditioning Engineers, pp. 60–63.

Wharton, G. W. 1976. House dust mites [review]. Journal of Medical Entomology 12:577–621.

Whipp, B. J., and K. Wasserman. 1988. Exercise. In: Textbook of Respiratory Medicine, J. F. Murray and J. Nadel, eds. Philadelphia, Pa.: W. B. Saunders Company.

White, J. 1990. Solving moisture and mould problems. In: Indoor Air '90: The Fifth International Conference on Indoor Air Quality and Climate, vol. 4. Toronto, Canada: Canada Mortgage and Housing Corp., pp. 589–594.

White, P. T., C. A. Pharoah, H. R. Anderson, and P. Freeling. 1989. Randomized controlled trial of small group education on the outcome of chronic asthma in general practice. Journal of the Royal College of General Practitioners 39:182–186.

WHO (World Health Organization). 1982. Estimating Human Exposure to Air Pollutants. Copenhagen/Geneva: WHO.

WHO. 1990. Biological contaminants in indoor air. EURO Reports and Studies 113. Copenhagen: WHO.

Wiberg, G. S., A. Lavallee, and L. Perelmutter. 1972. Allergenicity of enzyme detergents in guinea pigs exposed to simulated home-use conditions. Journal of Allergy and Clinical Immunology 49(3):194–196.

Wide, L., and H. Bennich. 1967. Diagnosis of allergy by an in vitro test for allergen antibodies. Lancet 2:1105–1107.

Wierenga, E. A., M. Snoek, C. deGroot, L. Chretien, J. D. Bos, H. M. Jansen, and M. I. Kapsenberg. 1990. Evidence for compartmentalization of functional subsets of CD4+ T lymphokines in atopic patients. Journal of Immunology 144:4651.

Wigal, J. K., T. L. Creer, H. Kotses, and P. Lewis. 1990. A critique of 19 self-management programs for childhood asthma. I. Development and evaluation of the programs. Pediatric Asthma and Allergy Immunology 4:17–39.

Wilson, M. R., R. M. Karr, and J. E. Salvaggio. 1981. Contribution of immunologic techniques to current understanding of occupational lung disease. In: Occupational Lung Diseases:

Research Approaches and Methods, H. Weill and M. Turner-Warwick, eds. New York, N.Y.: Marcel Dekker, Inc., pp. 125–142.

Windholz, M., S. Budavari, R. F. Blumetti, and E. S. Otterbein, eds. 1983. The Merck Index: An Encyclopedia of Chemicals, Drugs, and Biologicals, 10th ed. Rahway, N.J.: Merck, p. 8148.

Wood, R. A., P. A. Eggleston, P. Lind, L. Ingemann, B. Schwartz, S. Graveson, D. Terry, B. Wheeler, and N. F. Adkinson, Jr. 1988. Antigenic analysis of household dust samples. American Review of Respiratory Disease 137:358–363.

Wood, R. A., M. D. Chapman, N. F. Adkinson, Jr., and P. A. Eggleston. 1989. The effect of cat removal on allergen content in household dust samples. Journal of Allergy and Clinical Immunology 83:730–734.

Wood, R. A., K. E. Mudd, and P. A. Eggleston. 1992. The distribution of cat and dust mite allergens on wall surfaces. Journal of Allergy and Clinical Immunology 89:126–130.

Woods, J. E. 1982. Do buildings make you sick? In: Proceedings of the Third Canadian Building Congress: Achievements and Challenges in Building Sciences and Technologies. Victoria, B.C., 18–20 October.

Woods, J. E. 1983. Sources of Indoor Contaminants. ASHRAE Transactions 89(1A):462–497.

Woods, J. E. 1988. Recent developments for heating, cooling, and ventilating buildings: Trends for assuring healthy buildings. In: Proceedings of Healthy Buildings 1988, vol. 1, B. Berglund and T. Lindvall, eds. Stockholm, Sweden: Swedish Council for Building Research, pp. 99–107.

Woods, J. E. 1989a. Cost avoidance in owning and operating buildings. Occupational Medicine State of the Art Reviews 4:753–770.

Woods, J. E. 1989b. HVAC systems as sources or vectors of microbiological contaminants. Presented at CPSC/ALA Workshop on Biological Pollutants in the Home, July 10–11, Alexandria, Va.

Woods, J. E. 1990. Continuous accountability: A means to assure acceptable indoor environmental quality. In: Proceedings of the Fifth International Conference on Indoor Air Quality and Climate, vol. 5, D. Walkinshaw, ed. Ottawa, Canada: International Conference on Indoor Air Quality, Inc., pp. 85–94.

Woods, J. E. 1991. An engineering approach to controlling indoor air quality. Environmental Health Perspectives 95:14–21.

Woods, J. E., and B. C. Krafthefer. 1986. Filtration as a method for air quality control for occupied spaces. In: Fluid, Filtration: Gas, vol. 1, ASTM STP 975, R. R. Raber, ed. Philadelphia, Pa.: American Society for Testing and Materials.

Woods, J. E., and D. R. Rask. 1988. Heating, ventilation, air-conditioning systems: The engineering approach to methods of control. In: Architectural Design and Indoor Microbial Pollution, R. B. Kundsin, ed. New York, N.Y.: Oxford University Press, pp. 174–197.

Woods, J. E., J. E. Janssen, and B. C. Krafthefer. 1986. Rationalization equivalence between the "ventilation rate" and "air quality". Procedures in ASHRAE Standard 62, Indoor Air Quality 1986. Atlanta, Ga.: American Society for Heating, Refrigerating and Air-Conditioning Engineers, pp. 181–191.

Woods, J. E., G. M. Drewry, and P. R. Morey. 1987. Office worker perceptions of indoor air quality effects on discomfort and performance. In: Proceedings of the Fourth International Conference on Indoor Air Quality and Climate, vol. 2, B. Siefert, H. Esdorn, M. Fischer, H. Ruden, and J. Wegner, eds. West Berlin, Germany: Institute for Water, Soil, and Hygiene, pp. 464–468.

Woods, J. E., P. R. Morey, and D. R. Rask. 1989. Indoor air quality diagnostics: Qualitative and quantitative procedures to improve environmental conditions. In: Design and Protocol for Monitoring Indoor Air Quality, ASTM STP 1002, N. L. Nagda and J. P. Harper, eds. Philadelphia, Pa.: American Society for Testing and Materials. Philadelphia, Pa., pp. 80–98.

Woolcock, A. J. 1988. Asthma. In: Textbook of Respiratory Medicine, J. F. Murray, and J. Nadel, eds. Philadelphia, Pa.: W. B. Saunders Company.

Woolcock, A. J., M. H. Colman, and M. W. Jones. 1978. Atopy and bronchial reactivity in Australian and Melanesian populations. Clinical Allergy 8:155–164.

Woolcock, A. J., K. Yan, and C. M. Salome. 1988. Effect of therapy on bronchial hyperresponsiveness in the long-term management of asthma. Clinical Allergy 18:165–176.

Yaglou, C., and U. Wilson. 1942. Disinfection of Air by Air Conditioning Processes. In: Aerobiology, No. 17. Washington, D.C.: American Association for the Advancement of Science, pp. 129–132. xx

Yasueda, H., H. Mita, Y. Yui, and T. Shida. 1989. Comparative analysis of physicochemical and immunochemical properties of the two major allergens from *Dermatophagoides pteronyssinus* and the corresponding allergens from *Dermatophagoides farinae*. International Archives of Allergy and Applied Immunology 88:402–407.

Ylönen, J., R. Mäntyjärvi, A. Taivainen, and T. Virtanen. 1992. IgG and IgE antibody responses to cow dander and urine in farmers with cow-induced asthma. Clinical and Experimental Allergy 22:83–90.

Yoshizawa, S., F. Sugawara, H. Yasueda, T. Shida, T. Irie, M. Sakaguchi, and S. Inouye. 1991. Kinetics of the falling of airborne mite allergens (*Der* I and *Der* II). Arerugi 40(4):435–438.

Yuuki, T., Y. Okumura, and T. Ando. 1990. Cloning and sequencing of cDNAs corresponding to mite major allergen *Der f* II. Japanese Journal of Allergology 39:557–561.

Zagraninski, R. T., B. R. Leaderer, and J. A. J Stolwijk. 1979. Ambient sulfates, photochemical oxidants, and acute health effects: An epidemiological study. Environmental Research 19:306.

Zeiger, R. S. 1988. Development and prevention of allergic disease in childhood. In: Allergy Principles and Practice, E. Middleton, C. E. Reed, E. F. Ellis, N. F. Adkinson, Jr., and J. W. Yunginger, eds. St. Louis, Mo.: C. V. Mosby Company, pp. 930–968.

Zeiss, C. R., R. Patterson, J. J. Pruzansky, M. M. Miller, M. Rosenberg, and D. Levitz. 1977. Trimellitic anhydride-induced airway syndromes: Clinical and immunologic studies. Journal of Allergy and Clinical Immunology 60:96–103.

Zeiss, C. R., D. Levitz, R. Chacon, P. Wolkonsky, R. Patterson, and J. J. Pruzansky. 1980. Quantitation and new antigenic determinant (NAD) specificity of antibodies induced by inhalation of trimellitic anhydride in man. International Archives of Allergy and Applied Immunology 61:380–388.

Zeiss, C. R., P. Wolkonsky, J. J. Pruzansky, and R. Patterson. 1982. Clinical and immunologic evaluation of trimellitic anhydride workers in multiple industrial settings. Journal of Allergy and Clinical Immunology 70:15–18.

Zeiss, C. R., P. Wolkonsky, R. Chacon, and R. Patterson. 1983. Syndromes in workers exposed to trimellitic anhydride: A longitudinal clinical and immunologic study. Annals of Internal Medicine 98:8–12.

Zeiss, C. R., C. L. Leach, D. Levitz, N. S. Hatoum, P. J. Garvin, and R. Patterson. 1989. Lung injury induced by short-term intermittent trimellitic anhydride (TMA) inhalation. Journal of Allergy and Clinical Immunology 84:219–223.

Zeiss, C. R., J. H. Mitchell, P. F. D Van Peenen, D. Kavich, M. J. Collins, L. Grammer, M. Shaughnessy, D. Levitz, J. Henderson, and R. Patterson. 1992. A clinical and immunologic study of employees in a facility manufacturing trimellitic anhydride (TMA). Allergy Proceedings 13:193–198.

Acronyms

ABPA	allergic bronchopulmonary aspergillosis
AAAI	American Academy of Allergy and Immunology
ACAI	American College of Allergy and Immunology
ACGIH	American Conference of Governmental Industrial Hygienists
AHU	air-handling unit
APC	antigen-presenting cell
ASHRAE	American Society of Heating, Refrigerating and Air Conditioning Engineers
ATSDR	Agency for Toxic Substances and Disease Registry
BAL	bronchoalveolar lavage
BRI	building-related illness
CBER	Center for Biologics Evaluation and Research
CTL	cytolytic T lymphocytes
ELISA	enzyme-linked immunosorbent assay
EPA	U.S. Environmental Protection Agency
FDA	U.S. Food and Drug Administration
FEF	forced expiratory flow
FEF_{25-75}	forced expiratory flow between 25 and 75 percent of forced vital capacity

FEV$_1$ forced expiratory volume exhaled in 1 second
FVC   forced vital capacity

GM-CSF   granulocyte macrophage colony stimulating factor

HEPA filter   high-efficiency particulate arresting filter
HETES   hydroxyeicosatetraenoic acids
HIV   human immunodeficiency virus
HMW   high molecular weight
HP   hypersensitivity pneumonitis
HVAC system   heating, ventilation, and air conditioning system

IFN   interferon
Ig   immunoglobulin
IgE   immunoglobulin E
IL-1   interleukin-1
IUIS   International Union of Immunologic Societies

kDa   kilodalton (molecular mass measure)

LMW   low molecular weight
LOAEL   lowest observed adverse effect level
LRSS   late respiratory systemic syndrome

MCS   multiple chemical sensitivity
MHC   major histocompatibility complex
NAAQS   National Ambient Air Quality Standards

NHLBI   National Heart, Lung, and Blood Institute
NIAID   National Institute of Allergy and Infectious Disease
NIEHS   National Institute of Environmental Health Sciences
NIH   National Institutes of Health
NIOSH   National Institute for Occupational Safety and Health
NOAEL   no observed adverse effect level

OSHA   Occupational Safety and Health Administration
OTA   Office of Technology Assessment
OTS   Office of Toxic Substances

PAF   platelet-activating factor
PDA   pulmonary disease anemia
PEL   permissible exposure limit
PG   prostaglandin

PMN	polymorphonuclear leukocyte
ppb	parts per billion
ppm	parts per million
RAST	radioallergosorbent test
SBS	sick (or tight) building syndrome
TLV	threshold limit value
TMA	trimellitic anhydride
TM-HSA	trimellitic anhydride conjugated with human serum albumin
UF	uncertainty factor
WHO	World Health Organization

Glossary

Acquired immunity. Immunity gained during one's lifetime, not inherited. This can be either:

Active—the result when an antibody is produced by the individual's immune system in response to a naturally acquired infection or vaccination, or

Passive—the result when an antibody is transferred to the individual from another, immune human or animal host.

Agranulocytosis. Absence of granulocytes from the circulating blood, resulting in high fever, great weakness, and ulceration of the mucous membranes.

Allergen. A chemical or biological substance (e.g., pollen, animal dander, or house dust mite proteins) that causes an allergic reaction, characterized by hypersensitivity.

Allergen challenge. Administration of an antigen to a previously sensitized individual to induce a dose-response for clinical or research evaluation.

Allergic broncho-pulmonary aspergillosis (ABPA). An immunologic hypersensitivity reaction in the bronchi caused by colonization of sputum of patients with *Aspergillus fumigatus*. The inflammatory reaction may recur and progress over years to result in destruction of the bronchi and fibrosis of pulmonary tissue.

Allergic broncho-pulmonary fungosis (ABPF). A syndrome similar to ABPA in which the fungal organism is not *Aspergillus fumigatus*. ABPF is far less common than ABPA.

Allergy. A state of immunologically mediated hypersensitivity to a foreign material.

Anaphylaxis. A systemic, immunologically mediated hypersensitivity reaction to a foreign substance. Clinical manifestations are cutaneous, respiratory, and cardiovascular with shock and laryngeal edema as important causes of death when a fatality occurs.

Antibody. A protein molecule formed by the immune system in response to the body's contact with an antigen, having the specific capacity of neutralizing the antigen and creating immunity against certain microorganisms and toxicants. Certain antibodies can cause adverse hypersensitivity (allergic) reactions.

Antigen. A substance that stimulates the production of an antibody when introduced into the body. Antigens are usually high-molecular-weight compounds, such as proteins. However, low-molecular-weight compounds (e.g., drugs or industrial chemicals) can bind to serum proteins and become antigenic.

Asthma. A usually chronic condition characterized by intermittent episodes of wheezing, coughing, and difficulty in breathing, sometimes caused by an allergy to inhaled substances.

Atopen. The exciting cause of any form of atopy.

Atopy. The state of having one or more of a defined group of diseases—allergic rhinoconjunctivitis, allergic asthma, and atopic dermatitis—that are caused by a genetic propensity to produce IgE antibodies to environmental allergens encountered through inhalation, ingestion, and, possibly, skin contact. A broader definition, sometimes used for epidemiologic studies, requires only the presence of IgE antibody, regardless of allergic disease.

Autoantibody. An antibody that reacts with a component of the tissues of the animal making the antibody.

Autoimmunity. A condition resulting from the production of autoantibodies, characterized by cell-mediated or humoral immunologic responses to antigens of one's own body, sometimes with damage to normal components of the body.

B cell. A type of lymphocyte that produces antibodies and originates in bone marrow.

Bioaerosol. An aerosol containing living organisms or particles derived from living organisms.

Bronchoalveolar lavage fluid. The fluid obtained from the lungs by lavage. Lavage is a technique in which an organ is flushed with water to allow analysis of material in the drainage fluid (in this case, cells from the bronchioles and alveoli).

Bronchoconstriction. Narrowing of a bronchus caused by contraction of bronchial smooth muscle.

Bronchoprovocation test. An aerosolized solution of the test allergen is delivered by inhalation through a dosimeter in graded increasing dosages. The response is measured by spirometry.

Bronchospastic. Constriction of the bronchi due to smooth muscle contraction and edema and resulting in asthma.

Bronchus. One of the large conducting air passages of the lung.

Byssinosis. An occupational respiratory disease associated with the inhalation of cotton, flax, or hemp dust. It is characterized initially by chest tightness, shortness of breath, and cough but may lead to permanent lung damage.

Carcinogen. A cancer-causing substance or agent.

Cell-mediated immunity (CMI). Those manifestations of the immune reaction that are characterized by a cellular response, in particular by T lymphocytes and macrophages.

Colony-forming units. The number of colonies on an agar plate (petri dish) after culturing.

Contact sensitization. To stimulate an immune response upon initial skin contact with an antigen, with the consequence of preparing the body for a repeat response upon reexposure to the antigen.

Coprophagia. Feeding on dung.

Cross-reactivity of antigens. The interaction of an antigen with an antibody formed against a different antigen with which the first antigen shares closely related or common antigenic determinants. The effect is to reduce the specificity and sensitivity of the test method.

Cytokine. A substance produced by cells, including T cells, that transmits messages between cells to control and modulate immune responses.

Cytotoxic. Lethal to cells.

Delayed-type hypersensitivity. An inflammatory reaction that occurs 24 to 48 hours after challenge with an antigen and that is a result of lymphocyte-mediated immunity.

Dermatitis. An inflammatory skin condition.

Dermatographism. Urticaria due to physical allergy, in which moderately firm stroking or scratching of the skin with a dull instrument produces a pale, raised welt or wheal, with a red flare on each side.

Dermatophagoides pteronyssinus. A common cosmopolitan sarcoptiform mite commonly found in house dust accumulations and thought to be a contributory cause of atopic house dust asthma.

Dew point temperature. The temperature at which moist air becomes saturated (100 percent relative humidity) with water vapor when cooled at constant pressure.

Dose-response. The quantitative relationship between exposure to a substance, usually expressed as a dose, and the extent of the biologic effect or response.

Droplet nuclei. Aerosols that contain an organism or particle.

Endotoxin. The lipopolysaccharide of the outer membrane of Gram-negative bacteria, extractable from cells with trichloroacetic acid but not naturally released in quantity until cell lysis. The lipid A portion of the lipopolysaccharide is responsible for its toxic effects, which include leukopenia, thrombocytopenia, fever, and shock. Unlike the specific exotoxins, endotoxins from various organisms have similar pathogenic effects. Also known as *bacterial pyrogen.*

Epidemiology. The scientific study of the distribution and occurrence of human diseases, health conditions, and their determinants.

Epitope. Antigenic determinant.

Erythema. Redness of the skin.

Extravasate. To exude from or pass out of a vessel into the tissues, for example, from blood, lymph, or urine.

Fomites. Inanimate objects or substances that function to transfer infectious organisms from one individual to another.

Granulocyte. Any of several types of white blood cells with a granular cytoplasm.

Hematology. The science of blood and its nature, function, and diseases.

Histamine. A substance released by basophils in mast cells during allergic reactions. The pharmacologic effects of histamine include dilating blood vessels and stimulating gastric secretion.

Host. An organism that harbors or provides nourishment, habitat, or transport to another organism, whether symbiont, commensal, or parasite.

Host resistance. The ability of an organism to mount a successful immune response against disease-causing antigens.

Humoral immunity. Immunity associated with and characterized by antibodies that circulate in the blood.

Hygroscopic. Capable of readily taking up and retaining moisture.

Hypersensitivity. Excessive or abnormal reactivity to a stimulus.

Hypersensitivity diseases. Diseases for which a subsequent exposure to an antigen produces a greater effect than that produced on initial exposure. (See discussion in Chapter 5.)

Hypersensitivity pneumonitis. An inflammatory disease of the lung caused by inhalation of foreign substances such as microbial organisms (farmer's lung), indoor antigens (bird breeders' lung), and industrial allergenic chemicals (e.g., acid anhydrides and isocyanates). The diseases are due to immunologic reactions in pulmonary tissue. Also known as *extrinsic allergic alveolitis.*

Immediate-type hypersensitivity. An immune response mediated by immunoglobulin E antibodies, characterized by hives, wheezing, and/or abrupt changes in blood pressure, and occurring within a few minutes or hours after exposure to an antigen.

Immune system. A specialized group of body cells, cell products, tissues, and organs that respond to foreign organisms and substances in the body.

Immunize. To deliberately introduce an antigenic substance (vaccination or active immunization) or antibodies (passive immunization) into an individual, with the aim of decreasing susceptibility to infectious diseases or protecting against toxicants.

Immunocompetence. The capacity to respond immunologically to antigens.

Immunoglobulin(s). A protein (or family of proteins) that participates in the immune reaction as the antibody for a specific antigen. There are five categories of immunoglobulin (Ig) based on structural differences: IgG, IgM, IgA, IgD, and IgE.

Immunology. The study of the immune system concerned with the phenomena that allow an organism to respond to a subsequent exposure to a foreign substance in a way that is distinct from the way it responds to the initial exposure to that same substance.

Immuno-suppression. Suppressing the natural immune response of an organism, thus permitting an individual to accept a foreign substance, such as a transplant, but also increasing the likelihood of infection.

Immunotherapy. (1) The treatment of disease by the administration to the patient of antibody raised in another individual or another species (passive immunotherapy) or by immunizing the patient with antigens appropriate to the disease (active immunotherapy). (2) Therapy, used especially in the treatment of cancer, intended to stimulate the effector mechanisms of the immune response nonspecifically.

Immunotoxic. Having the potential to adversely affect the immune response or damage components of the immune system.

Incidence rate. The number of cases of a disease, abnormality, condition, etc., arising in a defined population during a stated period, expressed as a ratio, such as x cases per 1,000 population per year.

Inhalant. A substance that may be taken into the body through the respiratory system.

Innate immunity. See *natural immunity, nonspecific immunity.*

In vitro. Literally, in glass; pertains to a biological process taking place in an artificial environment, usually a laboratory.

In vivo. Literally, in the living; pertains to a biological process or reaction taking place in a living organism.

Insolation. Solar radiation that has been received.

Isotype. An immunoglobulin heavy or light chain class or subclass characterized by antigenic determinants (isotypic markers) in the constant region. Every normal individual expresses all of the isotypes of its species.

Leukocyte. Any of the small, colorless (white) cells in the blood, lymph, and tissues, important in the body's defenses against infection.

Ligand. Any one of several molecules or ions, identical or different, that bind to the same central entity. For example, the nitrogen atoms and the oxygen molecule bind to the iron of hemoglobin. The hydrogen ions that bind to the same protein molecule are another example.

Lymphocyte. A specialized leukocyte (white cell) formed in the lymphatic tissues important in the synthesis of antibodies.

Lymphoid organs. The principal organs of the immune system, including bone marrow, thymus, spleen, and lymph nodes. They produce, store, and distribute the immune system cells.

Lymphokine. A protein that mediates interactions among lymphocytes and is vital to proper immune function.

Lyse. To break up or rupture a cell membrane.

Macrophage. A type of large, amoeba-like cell found in the blood and lymph that has an important role in host defense mechanisms by ingesting dead tissue, tumor cells, and foreign particles such as bacteria and parasites. The macrophage also plays an important role in antigen processing and presentation.

Mast cells. Connective tissue cells commonly found adjacent to blood vessels and in the lymphatic system, skin, lung, and other tissues. Mast cells are approximately 20 μ in diameter and have cytoplasm filled by numerous (30–100) prominent metachromatic granules that stain black or purplish black with Romanowsky dyes. The granules contain heparin and histamine, and are involved in urticarial reactions. Mast cells are so named because their cytoplasmic granules, originally thought to contain stored nutrients, were likened to mast, i.e., the acorns, nuts, etc. used to fatten livestock.

Methacholine.   Chemical substance that is used as a parasympathomimetic agent and vasodilator.

Metric.   Pertaining to measures based on the meter.

Microbial agents.   Microbiological organisms or parts thereof.

Mitogenesis.   The initiation of cell division, or mitosis.

Monoclonal antibody.   An antibody secreted by a single clone of antibody-producing cells.   Such antibodies have the same combining site, the same light chain, and the same immunoglobulin class, subclass, and allotype, and their production in tissue culture by the hybridoma technique has enormously expanded the availability of highly characterized and specific antibodies of far-reaching practical and experimental importance.   Monoclonal antibodies are occasionally found in human disease, as in cold hemolytic antibody disease, or after immunization of experimental animals, as with bacterial polysaccharides, but most antibody formation in vivo is highly polyclonal.   Monoclonal antibodies are also used widely for the isolation and purification of proteins on a preparative and industrial scale.

Monocyte.   A large, phagocytic, nongranular leukocyte (white cell) containing one nucleus.

Morbidity.   Illness or disease.

Mucous membranes.   The membranes with mucous coverings, as in the nose, mouth, and upper airways.

Mycelial fragments.   The growing, nonspore phase of fungi.

Nasal lavage fluid.   The fluid obtained from the cleaning and washing out of nasal passages.

Natural immunity.   See *nonspecific immunity*.

Natural killer cell.   A type of lymphocyte that attacks cancerous or virus-infected cells without previous exposure to the antigen. Also known as *NK cell*.

Nonspecific immunity.   Immunity that exists from birth and occurs without prior exposure to an antigen; also known as *innate immunity*.

Pathogenic.   Capable of causing disease.

Pathology.   The study of the causes, development, and effects of disease or injury and the associated structural and functional changes.

Peak flow.   A measure of lung function. (See discussion in Chapter 5.)

Peripheral blood.   Circulating blood that is remote from the heart.

Peritrophic.   Designating a tubular chitinous sheath (the *peritrophic membrane*) inside the intestine of many insects. The sheath is continuously secreted at the anterior end of the stomach.

Pesticide.   Any substance or mixture of substances intended to kill or control pests such as insects, fungi, rodents, weeds, or roundworms.

Phagocytosis. Consumption of foreign particles (e.g., bacteria) by cells that surround the particle and then ingest it.

Plenum. A space filled with matter (not a vacuum); an enclosed volume of gas under greater pressure than that surrounding the container.

Pneumoconiosis. A condition characterized by mineral dust deposits in the lungs as a result of occupational or environmental exposure.

Pneumonitis. Inflammation of the lungs.

Prevalence rate. The ratio of the number of morbid cases to the number of people at risk at a specific time and place.

Reagent. Any substance capable of reacting with another, particularly if the reaction produces a change of physical properties whereby the second substance may be detected or measured. For example, dimethylglyoxime is a reagent for nickel, iron, bismuth, etc. The reagent is called selective if it indicates the presence of a small number of compounds or ions, and specific (or characteristic) if it only gives an indication with a single substance.

Reservoir. Any source of infection; a space, container, or depot in which something accumulates or is kept in reserve.

Resistance. The native or acquired ability of an organism to maintain its immunity to ward off disease.

Rhinitis. Inflammation of the lining of the nose.

Rhinoconjunctivitis. A condition that consists of a combination of rhinitis and conjunctivitis.

Rhinomanometry. The measurement of airflow and variations in air pressure within the nose. The resistance to respiration offered by the nasal soft tissues may be calculated from the results of the measurement.

Saprophytic. Obtaining food by absorbing dissolved organic material.

Sensitization. The process by which an immune response is stimulated on first being exposed to an antigen, with the consequence of preparing the body's immune system for a stronger response upon reexposure to the same antigen, as in a hypersensitivity reaction.

Serology. The study of serum, the watery liquid that separates from coagulated blood.

Serum. The clear yellowish fluid obtained when whole blood is separated into its solid and liquid components.

Silicosis. A condition of lung fibrosis that is brought about by prolonged inhalation of silica dust.

Skin tests. Tests for an allergy or infectious disease, performed by a patch test, scratch test, or an intracutaneous injection of an allergen or extract of the disease-causing organism.

Specificity. The quality or state of being specific, as of an antigen to its corresponding antibody.

Susceptibility. The extent to which an individual is liable to infection or to the effects of substances, such as toxicants, allergens, or other influences. The antithesis of resistance.

T cell. A lymphocyte produced in the bone marrow that matures in the thymus and is integral to cell-mediated immunity. T cells regulate the growth and differentiation of other lymphocytes and are involved in antibody production.

Vaccination. The administration of vaccine orally or by injection to protect against a given disease.

Vasodilation. Dilation of the blood vessels.

Wheal and flare. *Wheal*—A smooth, slightly elevated area on the body surface, which is redder or paler than the surrounding skin; it is often attended with severe itching, and is usually evanescent, changing its size or shape, or disappearing within a few hours. It is the typical lesion of urticaria, the dermal evidence of allergy, and in sensitive persons may be provoked by mechanical irritation of the skin. Known as also *hive* and *welt*. *Flare*—(1) The red outermost zone of the "triple response" (Sir Thomas Lewis) urticarial wheal reaction, a manifestation of immediate, as opposed to delayed, allergy or hypersensitivity; (2) a spreading flush or area of redness on the skin, spreading out around an infective lesion or extending beyond the main point of reaction to an irritant; and (3) sudden exacerbation of a disease.

White cell. A colorless cell in the blood, lymph, or tissues that is an important component of the immune system. See *leukocyte*.

Committee and Staff Biographies

COMMITTEE

ROY PATTERSON (Chair) is the Ernest S. Bazley Professor of Medicine and Chief of the Allergy-Immunology Division of the Department of Medicine of Northwestern University Medical School. He is a physician board certified in internal medicine and allergy-immunology. He served as Chairman of the Department of Medicine at Northwestern for 17 years and has been a member of various National Institute of Health and other national committees and societies. His research interests include immunoglobulin E (IgE) mediated allergies, occupational immunologic lung disease, animal models of allergy and asthma, allergic drug reactions, allergic aspergillosis, and investigation and control of asthma. With his research colleagues, he has developed a markedly improved form of allergen immunotherapy and discovered a process that may reduce IgE antibody levels, the cause of common allergies.

HARRIET BURGE (Vice-Chair) is Associate Research Scientist in the Allergy Division at the University of Michigan Medical School and Associate Professor of Environmental Microbiology at the Harvard School of Public Health. Dr. Burge is a mycologist specializing in indoor air quality. She chairs the Bioaerosol Committee of the American Conference of Governmental Industrial Hygienists and the Michigan Occupational Health Standards Commission, is a fellow of the American Academy of Allergy and Immunology and the American College of Allergy and Immunology, and is vice-chair of the Pan American Aerobiology Association. Her research on indoor allergens is supported by government, industry, and private foundations.

REBECCA BASCOM is Director of the Environmental Research Facility and Associate Professor in the Division of Pulmonary and Critical Care Medicine, Department of Medicine, University of Maryland School of Medicine. She is a physician, board certified in internal medicine (pulmonary medicine) and preventive medicine (occupational medicine). Dr. Bascom is a member of the American Thoracic Society (Long Range Planning, Environmental and Occupational Health Assembly), American Academy of Allergy and Immunology (chairing the Committee on Environmental Control and Air Pollution), American College of Occupational and Environmental Medicine, and the American Public Health Association. Her research interests center on differential responsiveness to irritants and the effects of irritants on allergic diseases. Her professional practice centers on occupational and environmental respiratory diseases and on the development of practical work site surveillance programs to assist in the control of respiratory hazards.

EULA BINGHAM is Professor of Environmental Health in the School of Medicine at the University of Cincinnati. Dr. Bingham also serves as a trustee for the Natural Resources Defense Council, the Greater Cincinnati Occupational Health Clinic, and Director of the Ohio Hazardous Substance Institute and is a member of the Institute of Medicine. Her research interests are in environmental health, occupational safety and health, and chemical carcinogenesis. Dr. Bingham has served in numerous public-sector positions including: Department of Labor (Assistant Secretary for Occupational Safety and Health, National Institute for Safety and Health study section), U.S. Food and Drug Administration (Food and Drug Advisory Commission, Environmental Health Advisory Commission), and the U.S. Environmental Protection Agency (Science Advisory Board). Dr. Bingham has also served as a member of several committees and boards for the National Academy of Sciences, National Research Council, and Institute of Medicine. Dr. Bingham is a recipient of the Rockefeller Foundation Public Service Award and the American Public Health Association Alice Hamilton Award.

ROBERT K. BUSH is a Professor of Medicine (CHS) at the University of Wisconsin, Madison, and Chief of Allergy at the William S. Middleton Veterans Affairs Hospital, Madison, Wisconsin. He is an active member of the American Academy of Allergy and Immunology and the American Thoracic Society. He serves on the Allergenic Products Advisory Panel to the U.S Food and Drug Administration and is a member of the editorial board for the *Journal of Allergy and Clinical Immunology*. His research interests include the characterization of fungal and occupational allergens.

LESLIE C. GRAMMER is a Professor of Medicine in the Division of Allergy-Immunology at Northwestern University Medical School. Dr. Grammer's principal research is in allergic responses to low-molecular-weight reactive chemicals. She is board certified in internal medicine, allergy-immunology, diagnostic laboratory immunology, and occupational medicine. She is a fellow of the American Academy of Allergy and Immunology and of the American College of Allergy and Immunology. She serves on the editorial board of the *Journal of Allergy and Immunology*, and she was a member of the Illinois Task Force on Asbestos.

MICHAEL D. LEBOWITZ is a Professor of Medicine (pulmonary, epidemiology, environmental medicine) and Associate Director of the Respiratory Sciences Center at the University of Arizona College of Medicine. He is a fellow of the American College of Chest Physicians and of the American College of Epidemiology, and an elected member of the American Epidemiological Society and other societies. He is on the executive committees of the International Societies of Environmental Epidemiology and Exposure Analysis. He is an associate editor of the *Journal of Exposure Analysis and Environmental Epidemiology* and a section editor of the *Journal of Toxicology and Industrial Health*, and is on the editorial board of *Pediatric Pulmonology*. His research interests are in pulmonary diseases and environmental health. He cochaired the National Research Council/ National Academy of Sciences Committee on Indoor Pollutants and was senior editor of the World Health Organization EHC monograph *Guidelines on Studies in Environmental Epidemiology* and of the World Health Organization/EURO monograph *IAQ: Biological Contaminants*.

FLOYD MALVEAUX is Professor of Microbiology and Medicine and Chairman of the Department of Microbiology at Howard University College of Medicine. He is a physician (allergy) and basic scientist (immunology). His clinical and research interests relate to the epidemiology of asthma morbidity in high-risk populations, design of interventions to decrease this level of morbidity, and the mechanisms of allergens in eliciting bronchial inflammation in asthma. He is a member of the Advisory Council of the National Institute of Allergy and Infectious Diseases, the Board of Trustees of the National Medical Association, the Board of Directors of the Asthma and Allergy Foundation of America, and the editorial boards of *Annals of Allergy* and the *Journal of the National Medical Association*. He also was a member of the expert panel that drafted the *Guidelines for the Diagnosis and Management of Asthma* recently published by the National Heart, Lung, and Blood Institute.

JOHN L. MASON is a consulting engineer and was Vice President of Engineering and Technology for Allied Signal Aerospace Company until his retirement in 1989. For Allied Signal's predecessor, the Garrett Corporation, Dr. Mason led the engineering group that designed the heating, ventilating, air-conditioning and pressurization systems for a majority of the world's aircraft and for the Mercury, Gemini, Apollo, and Skylab spacecraft. He is a Fellow and was 1990 President of the Society of Automotive Engineers and is a member of the National Academy of Engineering.

PHILIP R. MOREY is Manager of Indoor Air Quality Services for Clayton Environmental Consultants, a Marsh & McLellan company. He is chair of the American Society for Testing and Materials D22.05.06 Technical Committee on Bioaerosols in Indoor Air, editor of that committee's conference proceedings on biological contaminants in indoor environments, and was previously Professor of Biological Sciences at Texas Technical University. At Clayton, he directs indoor air quality evaluations in office buildings and hospitals which emphasize ventilation system analysis and sampling for microorganisms.

THOMAS A. W. PLATTS-MILLS is Professor of Medicine and Head of the Division of Allergy and Clinical Immunology at the University of Virginia. Dr. Platts-Mills' clinical and research interests concern the role of indoor allergens in asthma and atopic dermatitis. In particular, studies have focused on exposure to allergens in domestic houses and the methods for reducing exposure. He is a fellow of the American Academy of Allergy and Clinical Immunology and is Chairman of the Environmental and Occupational Interest Section. He is a member of the editorial board of the *Journal of Allergy and Clinical Immunology, Clinical and Experimental Allergy*, and the *Journal of Immunological Methods*.

CAROL RICE is an Associate Professor of Environmental Health at the University of Cincinnati. A certified industrial hygienist, Dr. Rice specializes in evaluating occupational exposures over the working lifetimes of employees. The exposure metrics resulting from the measurement of current exposure and those constructed from historical data are used in occupational epidemiological studies. Most recently, she has contributed to the evaluation of exposure-response relationships for cohorts exposed to refractory ceramic fibers and silica.

LANNY J. ROSENWASSER is Head of the Allergy and Immunology Division at the National Jewish Center for Immunology and Respiratory Medicine and Professor of Medicine at the University of Colorado Health

Sciences Center. Dr. Rosenwasser's clinical and research interests relate to the cellular immune response to allergens in airways and basic issues of cytokine biology in inflammation and immunity. Dr. Rosenwasser has served as an associate editor of the *Journal of Immunology* and *The Journal of Allergy and Clinical Immunology* and was a charter member of the National Institute of Allergy and Infectious Diseases' AIDS Review Committee. He has also served as chairman of the Veterans Affairs Merit Review Board for Immunology.

ABBA TERR is Clinical Professor of Medicine and Director of the Allergy Clinic of Stanford University Medical Center. His principal research interests are allergy informatics, occupational allergy, and the regulation of atopic immunoglobulin E antibody production. He is a physician board certified in internal medicine and allergy and maintains a private consulting practice of allergy and immunology. He is coauthor of the textbook *Basic and Clinical Immunology*, and he serves on a number of national societies and committees.

M. DONALD WHORTON is Vice President of ENSR Consulting and Engineering, the environmental consulting division of American NuKEM, an integrated environmental services company. Dr. Whorton is a physician epidemiologist specializing in occupational and environmental health issues. He is a member of the Institute of Medicine and has served or is serving on other Institute of Medicine or National Research Council committees. He is a fellow of the American College of Environmental and Occupational Medicine, chairman of its Occupational Medical Practices Committee, and a fellow of The American Public Health Association.

JAMES E. WOODS, JR., is the William E. Jamerson Professor of Building Construction in the College of Architecture and Urban Studies at Virginia Polytechnic Institute and State University. Dr. Woods is a registered professional engineer and specializes in research on control of the indoor environment. He is a fellow of the American Society of Heating, Refrigerating and Air Conditioning Engineers. He has served as a member of the Science Advisory Board of the U.S. Environmental Protection Agency on the Committee of Indoor Air Quality and Total Human Response.

STAFF

ANDREW M. POPE is a Senior Staff Officer and Study Director in the Institute of Medicine's Division of Health Promotion and Disease Prevention. His primary interests focus on the occupational and environmental influences on human health, with expertise in physiology, toxicology, and epidemiology. As a Research Fellow in the Division of Pharmacology and Toxicology at the U.S. Food And Drug Administration, Dr. Pope's research focused on the neuroendocrine and reproductive effects of various environmental substances in food-producing animals. During his tenure at the National Academy of Sciences, and since 1989 at the Institute of Medicine, Dr. Pope has directed and edited numerous reports on occupational and environmental issues; topics include injury control, disability prevention, biologic markers, neurotoxicology, and indoor allergens.

POLLY A. BUECHEL served as the Project Assistant from October 1991 to October 1992. Previously, Ms. Buechel was a staff assistant for the Brookings Institution on a study of the World Bank. She received a B.A. degree in international relations from the University of California, Davis.

Index